Reticulate Evolution and Humans

Reticulate Evolution and Humans

Origins and Ecology

Michael L. Arnold
Department of Genetics, University of Georgia, Athens, Georgia, USA

OXFORD
UNIVERSITY PRESS

Great Clarendon Street, Oxford OX2 6DP

Oxford University Press is a department of the University of Oxford.
It furthers the University's objective of excellence in research, scholarship,
and education by publishing worldwide in

Oxford New York

Auckland Cape Town Dar es Salaam Hong Kong Karachi
Kuala Lumpur Madrid Melbourne Mexico City Nairobi
New Delhi Shanghai Taipei Toronto

With offices in

Argentina Austria Brazil Chile Czech Republic France Greece
Guatemala Hungary Italy Japan Poland Portugal Singapore
South Korea Switzerland Thailand Turkey Ukraine Vietnam

Oxford is a registered trade mark of Oxford University Press
in the UK and in certain other countries

Published in the United States
by Oxford University Press Inc., New York

First published 2009

British Library Cataloguing in Publication Data
Data available

Library of Congress Cataloging in Publication Data
Data available

Typeset by Newgen Imaging Systems (P) Ltd., Chennai, India
Printed in Great Britain
on acid-free paper by
CPI Antony Rowe

ISBN 978–0–19–953958–1 (Hbk.)

10 9 8 7 6 5 4 3 2 1

Picasso liked to add, combine, and stick things together just as much as he liked to strip, undo, prune, and purify: "The supreme art is to summarize," André Masson would write, referring to the long evolution of image into symbol from naturalism...and then into quasi-abstraction...whose graphic quality adopts the simple rhythm of a prehistoric figure.

(Léal, Piot, and Bernadac, *The Ultimate Picasso*, 2003, p. 360)

To Frances, Brian, and Jenny. You make my life rich.

Preface

This book is an exploration of how the transfer of genes between divergent lineages—through a diverse array of mechanisms—has affected, and continues to affect, humans. In particular, it is a journey into the data that support the hypothesis that *Homo sapiens* as well as those organisms on which it depends for survival and battles against for existence are marked by mosaic genomes. This mosaicism reflects the rampant (as reflected by the proportion of organisms that illustrate this process) exchange of genetic material during evolutionary diversification. This is the underlying hypothesis for this book. I hope to show in the following chapters that it also reflects the consistent observation made when the genomes of organisms are mined for genetic variation.

Chapter 1 provides a basis for much of the terminology used throughout. It also illustrates many of the concepts, processes, and mechanisms that characterize reticulate evolution. In Chapters 2–4, I will illustrate how genetic exchange has impacted greatly the genetic variation and evolutionary trajectories of primates in general, and the clade containing our own genus and those genera with which we share the closest ancestry in particular. Chapters 5–7 contain a description of the reticulate evolutionary history of organisms that benefit worldwide populations of *H. sapiens* through the provision of shelter, clothing, and sustenance. In Chapter 8, I turn to the question of how genetic exchange may have led to the origin and evolution of those viruses, bacteria, and so on, which breach physical and physiological defenses to cause epidemics and pandemics among humans. Finally, in Chapter 9, I will briefly direct the reader to consider the evidence presented in the previous chapters to draw general conclusions and to suggest ways in which the findings presented might be applied.

To paraphrase one of my Mom's favorite maxims, I believe that any author worth the powder it would take to blow him or her up recognizes the collaborative nature of book writing. In that context, this book reflects ideas and topics concerning which a set of colleagues and I have argued and published over the past two decades or so. In particular, Axel Meyer and I wrote a review of data indicating the widespread occurrence of reticulate evolution among primates, including *H. sapiens*. This, along with a 2004 review for the journal *Molecular Ecology*—where I was allowed to explore the effect of genetic exchange on the evolution of organisms with which humans interact—provided a solid conceptual basis for this book.

I also wish to thank Professors Peter Holland and Paul Harvey of the Department of Zoology and Professor Dame Jessica Rawson of Merton College for providing a Research Fellowship at Oxford University during which time I began the writing of this book. Similarly, I must thank Professor Wyatt Anderson of the University of Georgia who facilitated my work at Oxford by assuming some of my duties in the Department of Genetics. I have much gratitude also for Eleanor Kuntz, Jacob Moorad, Rebecca Okashah, Eileen Roy, and Natasha Sherman for reading and critiquing earlier drafts of the chapters. Theirs was definitely a labor of kindness toward the author. Ian Sherman, my editor and friend, has provided continual support during this project. I want to thank him for cheerfully answering the same questions more than once. Ian's assistant, Helen Eaton, also gave great guidance as this project developed. During the period of writing, I was supported financially by the National Science Foundation grant, DEB-0345123.

One of my favorite authors, John Piper, wrote in the preface to his book *Desiring God* (2003; Multnomah Books) that he relied on his wife like gravity and oxygen. I can think of no better expression of my undying gratitude for the person who knows me the best, and loves me anyway. Thank you Frances. Like the two previous books, I dedicate this to you and our children, Brian and Jenny.

Contents

CHAPTER 1

Reticulate evolution: an introduction

It is impossible to doubt that there are new species produced by hybrid generation.

(Linnaeus 1760)

...studies of some species complexes have indeed been couched in terms of a web-of-life, rather than a tree-of-life metaphor...while others were designed with an appreciation of both metaphors.

(Arnold 2006)

...there is growing evidence that lateral gene transfer has played an integral role in the evolution of bacterial genomes, and in the diversification and speciation of the enterics and other bacteria.

(Ochman *et al.* 2000)

It is suggested that the chief effect of hybridization in this genus in eastern North America...is to increase variability in the parental species.

(Anderson 1936)

The only data sets from which we might construct a universal hierarchy including prokaryotes, the sequences of genes, often disagree and can seldom be proven to agree. Hierarchical structure can always be imposed on or extracted from such data sets by algorithms designed to do so, but at its base the universal TOL rests on an unproven assumption about pattern that, given what we know about process, is unlikely to be broadly true.

(Doolittle and Bapteste 2007)

1.1 Reticulate evolution and the development of the web-of-life metaphor

The goal of this book is to provide a framework for understanding the evolutionary effects that are generated by the exchange of genes between organisms belonging to divergent lineages. In particular, I will examine the contribution of **reticulate evolution** to the origin and development of (1) our own species, (2) primates in general, (3) the organisms on which humans depend for food, clothing, and so on, and (4) the disease carrying and causing vectors and pathogens with which we battle for survival. Through the development of this framework, I hope to illustrate the conclusion that the tree-of-life metaphor is insufficient both in terms of predicting and explaining evolutionary patterns and process. As my colleagues and I have argued earlier (Arnold and Larson 2004; Arnold 2006), evolutionary diversification is best illustrated not as a bifurcating, ever-diverging, tree-like structure, but rather as a web made up of genetic interactions between different strands (i.e., lineages).

Studies in the field of evolutionary biology are often characterized as being either pattern or process focused. Those that examine the relationships among evolutionary lineages are considered pattern based while studies of the factors that lead to evolutionary change are said to be process oriented. Yet this dichotomy does not capture the underlying complexity. For example, analyses designed to provide phylogenetic resolution almost always can be used to at least generate, if not also test, hypotheses concerning the processes that contributed to the resolved pattern. Similarly, the resolution of processes—for instance, those factors that lead to some measure of reproductive isolation—are often used as data sets for determining evolutionary relatedness (i.e., pattern). In this chapter, I emphasize the process side of this dichotomy to provide the conceptual and terminological basis for the subsequent discussions. However, in the following chapters both process- and pattern-based studies and findings will be reviewed and discussed.

The restrictive nature of the pattern–process characterization of evolutionary studies is illustrative of the similarly limiting metaphor known as the "tree of life." In this case, the proposal that all life can be represented as a branching, evolutionary "tree" (Darwin 1859) led understandably to the construction of algorithms that allowed only the delimitation of dichotomously diverging representations (e.g., Swofford 1998). It is thus inevitable that with such a constraining assumption—that is, evolution proceeds primarily, or exclusively, by a process of divergence—phylogenetic representations would of course resolve into "trees." However, over the last several decades, some microbiologists interested in evolutionary pattern have argued that the accuracy of such representations, especially for prokaryotes, should be reevaluated (e.g., Doolittle *et al.* 1996, 2003; Zhaxybayeva *et al.* 2004). The reevaluation has resulted in the recognition of widespread reticulate evolution in the prokaryotic clade (Ochman *et al.* 2005). Furthermore, in terms of process-oriented effects, the widespread exchange of genes through **lateral or horizontal transfer** has resulted in the transfer and *de novo* origin of adaptations, allowing the recipient, prokaryotic species to exploit novel environmental settings (Lawrence and Ochman 1998). Indeed, **adaptive trait transfer** (Arnold 2006) forms part of the basis for the ecological breadth of numerous prokaryotes that cause devastating pathologies in human populations (e.g., Faruque *et al.* 2007).

Given the recognition of (1) the great diversity generated by the horizontal transfer of genetic elements in prokaryotes (and indeed in viral lineages as well; e.g., Heeney *et al.* 2006) and (2) the role of **natural hybridization** and **introgressive hybridization** (i.e., "**introgression**"; Anderson and Hubricht 1938) in the origin and evolution of plant lineages (Anderson 1949; Anderson and Stebbins 1954; Grant 1981; Arnold 1997, 2006), it is surprising that the tree-of-life metaphor has also been assumed the best descriptor by many evolutionary botanists (Grant 1981). In fact, entire clades are now known to rest on reticulate events. For example, the formation of **allopolyploid species** in flowering plants is one of the most important evolutionary processes and is estimated to underlie clades containing 50% or more of all angiosperms (Stebbins 1947, 1950; Grant 1981; Soltis and Soltis 1993; Masterson 1994). Similarly, introgressive hybridization is now recognized as one of the major outcomes from hybridization between related plant taxa (Arnold 1997, 2006). As mentioned above, genetic exchange events between bacterial and viral lineages may lead to the origin and/or transfer of adaptive traits. Such is also the case for introgression between different plant taxa (Anderson 1949; Arnold 1992; Heiser 1951; Kim and Rieseberg 1999, 2001; Martin *et al.* 2005, 2006; Whitney *et al.* 2006).

Introgression via sexual reproduction is not the only avenue by which genetic exchange events take place among plant lineages. Similar to prokaryotes, plant clades reflect the signatures of lateral transfer events as well. For example, the lateral transfer of plant mitochondrial DNA (i.e., mtDNA) appears frequent. One mechanism for mtDNA transfers (similar to what has been demonstrated for animal lineages: for example, see Houck *et al.* 1991; Kidwell 1993) has involved genetic exchange between host plants and their parasitic plant associates. Davis *et al.* (2005) detected an example of this form of horizontal transfer involving a parasitic flowering plant and a fern species. In particular, phylogenetic analyses using three mitochondrial gene regions positioned the rattlesnake fern (*Botrychium virginianum*) with other fern species. However, two other mtDNA gene elements indicated a closer evolutionary relationship with the largely parasitic angiosperm order that includes sandalwoods and mistletoes. Davis *et al.* (2005) concluded, "These discordant phylogenetic placements suggest that part of the genome in *B. virginianum* was acquired by horizontal gene transfer" (see Box 1.1).

Box 1.1 Genetic Exchange and Discordant Phylogenies

Phylogenetic discordance can be caused by the exchange of genes by the differential transfer of genetic elements between divergent evolutionary lineages. Discordance would thus arise if genetic loci that have been exchanged and those that have not been exchanged were utilized in constructing separate phylogenies. Panel 1 and 2 indicate the results of the phylogenies derived from exchanged and nonexchanged loci, respectively. Both are accurate reflections of evolutionary processes and the underlying evolutionary relatedness of the loci and thus the organisms under investigation. In one case (Panel 1), the close evolutionary relationship between different lineages has arisen more recently through reticulation, while the relationships defined in Panel 2 reflect more ancient associations. However, both reflect descent from a common ancestor for the specific loci used in the analysis. The detection of "phylogenetic discordance" (i.e., the disagreement between the resolved phylogenies) reflects the transfer of Locus 1, but not Locus 2 and is thus a signature of reticulate evolution.

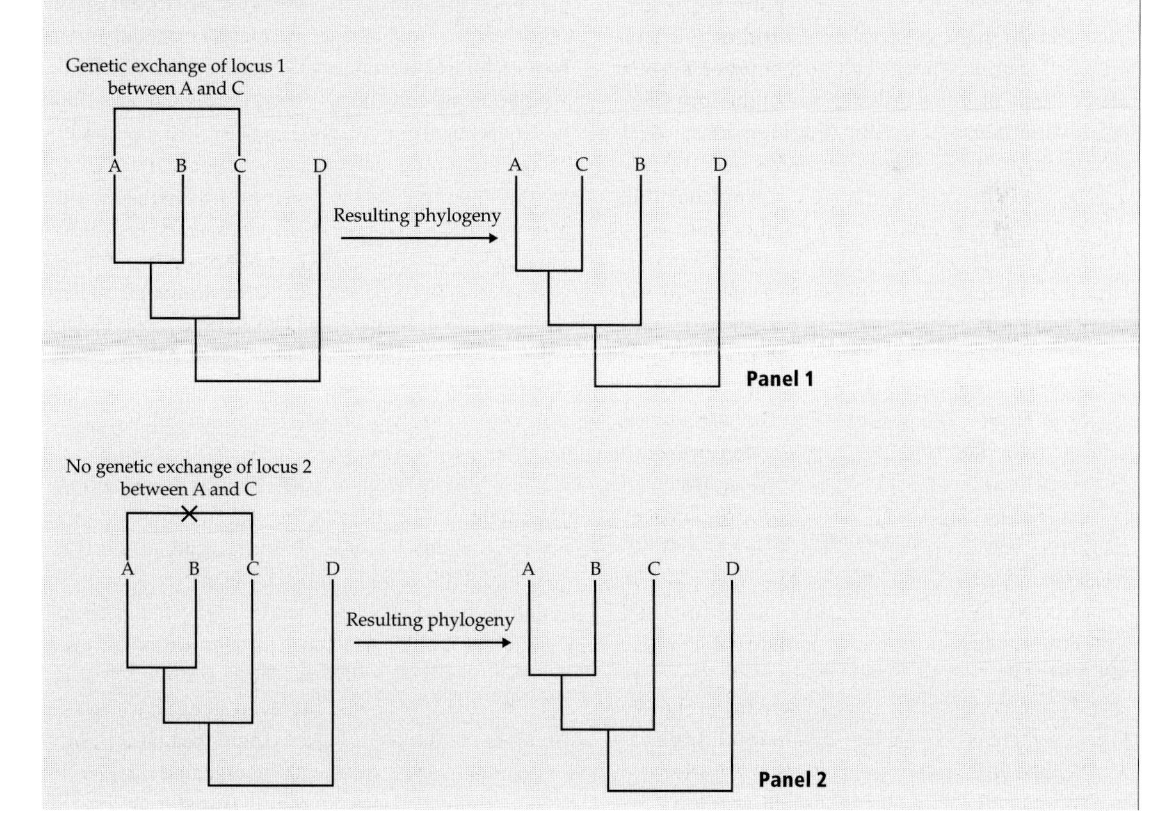

The above, brief, summary indicates that prokaryotic, viral, and plant evolution are often better represented by the metaphor of a web of life rather than a tree of life (Arnold 2006). Indeed, reticulate evolutionary patterns and processes are observed throughout these major clades. However, animal taxa also bear the genetic footprints of the processes of lateral transfer and introgression. For example, Tarlinton *et al.* (2006) detected the recent and ongoing invasion (i.e., lateral transfer) of the koala genome by an exogenous retrovirus that is transitioning into an endogenous viral element. In particular, they noted that

The finding that some isolated koala populations have not yet incorporated KoRV into their genomes, combined with its high level of activity and variability in individual koalas, suggests that KoRV is a virus in transition between an exogenous and endogenous element…and…provides an attractive model for studying the evolutionary event in which a retrovirus invades a mammalian genome. (Tarlinton *et al.* 2006)

In addition to genetic exchange via lateral transfer, animal clades are now understood to reflect frequent gene transfers through introgressive hybridization (Dowling and DeMarais 1993; Arnold 1997, 2006). For example, in a recent analysis of Z-chromosome loci from the hybridizing swallowtail butterfly species, *Papilio glaucus* and *Papilio canadensis*, Putnam *et al.* (2007) detected extremely discordant estimates of time since divergence. This led to the conclusion that the DNA sequence variation of the Z-chromosomes carried by these species had been structured by long-term introgressive hybridization, but with different regions of the chromosomes having been exchanged at different frequencies. In particular, Putnam *et al.* (2007) inferred that "the Z chromosome is a mosaic of regions that differ in the extent of historical gene flow, potentially due to isolating barriers that prevent the introgression of species-specific traits that result in hybrid incompatibilities."

In the following chapters, I will discuss numerous examples of the reticulate evolution of microorganism, plant, and animal species that have positive or deleterious effects on humans. Furthermore, I will describe in detail the available evidence indicating that *Homo sapiens* (and many other extant and extinct primates) evolved in the face of lateral transfers and introgressive hybridization. However, though the purpose of this book is to illustrate why the "web-of-life" metaphor

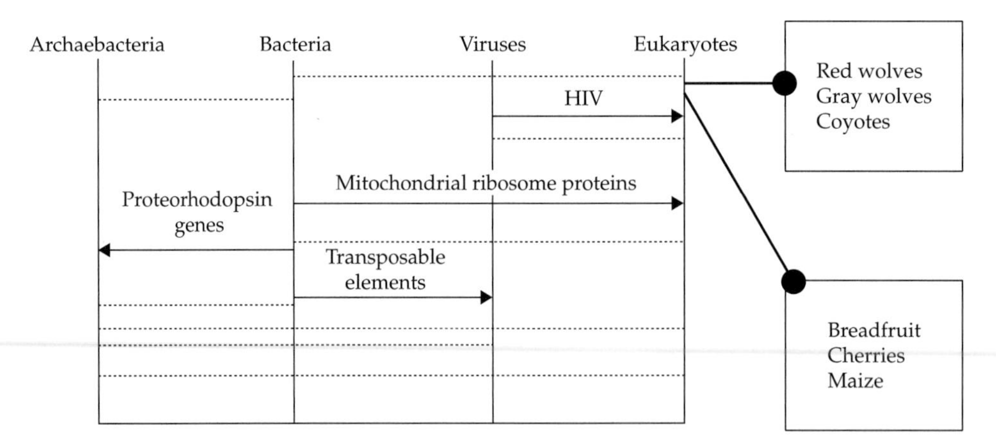

Figure 1.1 Schematic representation of the "web of life." The bold lines interconnecting the various lineages indicate representative, known lateral exchanges between different domains of life. The insets represent representative, known introgression events deriving from sexual reproduction between divergent lineages. The dashed lines reflect a small number of the additional exchange events known to have occurred. (References for the exchange events: "HIV," see discussion in Chapter 8; "mitochondrial ribosome proteins," see Bonen and Calixte 2006; "proteorhodopsin genes," see Frigaard *et al.* 2006; "transposable elements," see Filée *et al.* 2007; "red wolves, gray wolves, coyotes, breadfruit, cherries, maize," see discussion in Chapters 5–7.)

(Figure 1.1; Arnold 2006) best illustrates our own species' evolutionary trajectory, and that of taxa associated/related to *H. sapiens*, it is also useful to describe its efficacy in defining and predicting the patterns and processes in species not closely related or directly interacting with our own. In the remainder of this chapter, I will thus introduce the evolutionary effects possible from reticulate processes using several prokaryotic, plant, and animal clades.

1.2 Examples of evolutionary consequences from introgressive hybridization: animals

1.2.1 Evolution of adaptations and hybrid speciation: Darwin's finches

Darwin's finches are famous, not only because of their use as an evolutionary paradigm by numerous evolutionary biologists, but also because they represent one of the best examples of real-time measures of evolutionary change. As such, these species have been used to illustrate evolutionary and ecological processes including natural selection, character displacement, adaptation, competition, speciation, and adaptive radiations (e.g., Lack 1947; Schluter 1984; Petren *et al.* 2005; Abzhanov *et al.* 2006; Grant and Grant 2006; Huber *et al.* 2007). Most significant for the current discussion, however, has been the recognition of the central role of introgressive hybridization in the evolution of this species complex. In particular, Peter and Rosemary Grant and their colleagues have demonstrated the efficacy of introgression to feed genetic variation into animal populations resulting in the transfer of the material necessary for natural selection and adaptation.

I have argued previously (Arnold 2006) that, given what we know now of the evolution of this group, Darwin almost certainly was describing the effect on phenotype from repeated bouts of hybridization when he stated the following in *The Voyage of the Beagle*:

The most curious fact is the perfect gradation in the size of the beaks in the different species of Geospiza, from one as large as that of a hawfinch to that of a chaffinch, and... even to that of a warbler... Seeing this gradation and diversity of structure in one small, intimately related group of birds, one might really fancy that from an original paucity of birds in this archipelago, one species had been taken and modified for different ends. (Darwin 1845, pp. 401–402)

Darwin did not hold to a web-of-life paradigm, but rather emphasized (1) the role of natural selection alone in molding genetic and morphological variation and (2) the assumption that hybrid (i.e., in his terminology "mongrel") individuals would possess a lower fitness relative to their parents (Darwin 1859, pp. 276–277). Yet, his gift for noticing biological details provided evidence of crossbreeding as reflected in his statement that "instead of there being only one intermediate species... there are no less than six species with insensibly graduated beaks" (Darwin 1845, p. 402). This type of gradation of one species into another is expected, given the recombination between the genes underlying the morphological traits.

Darwin's completely understandable emphasis on a model of a bifurcating tree of life prevented his inferring a role for reticulation in the birds that were to become his namesake. No such limitation was reflected when Lowe (1936) reconsidered the pattern of morphological variability in this species complex. Instead, he concluded:

in the Finches of the Galápagos we are faced with a *swarm of hybridization segregates* which remind us... of the "plant" swarms described by Cockayne and Lotsy in New Zealand forests as the result of natural crossings... I think it was William Bateson who always maintained that the Finches... could only be explained on the assumption that they were segregates of a cross between ancestral forms. (Lowe 1936, pp. 320–321)

However, a decade later David Lack in his classic book, *Darwin's Finches* (Lack 1947), returned the emphasis on the evolution of this species complex to a strict Darwinian model. In particular, he explained the morphological variation and gradations from one form to another as being due to "the intermediate nature of their ecological requirements and not to a hybrid origin..." (Lack 1947, p. 100). This led him to the more general conclusion that "hybridization has not played an important part in the origin of new forms of Darwin's finches" (Lack 1947, p. 100).

Though others since Lack have considered the role of introgression as one of many possible evolutionary mechanisms affecting the diversification of the Darwin's finches, Grant (1993) accurately surmised, "Since 1947, and prior to the study reported here, hybridization in the Galápagos has been neither neglected nor satisfactorily demonstrated." It took the detailed, long-term analyses of these finch species by the Grants to confirm once and for all the combined effects of introgressive hybridization and natural selection in the evolution of *Geospiza* species. Their initial description of rare hybridization events leading to introgression between various Darwin's finch species (Grant and Grant 1992) laid the foundation for an understanding of the cause and significance of the diversity detected by Darwin and others. In the context of testing the applicability of the web-of-life metaphor for this group, two of the most important observations made by Peter and Rosemary Grant *et al.* were (1) the episodic nature of the impact of introgression events and (2) the fluctuating fitness estimates of hybrid and parental genotypes.

From 1976 to 1982 pairings of *Geospiza fortis* and *Geospiza scandens* and *Geospiza fortis* and *Geospiza fuliginosa* resulted in 1 and 32 fledgling(s), respectively (Grant and Grant 1993). Of these F_1 hybrids, two of the *G. fortis* × *G. fuliginosa* F_1s survived to breed (but not until after 1983). The sole *G. fortis* × *G. scandens* F_1 died without reproducing (Grant and Grant 1993). In contrast to their reduced fitness before 1983, between 1983 and 1991 the production of fledglings produced by hybrid genotypes exceeded the number necessary for them to replace themselves. During this same period, *G. fortis* and *G. fuliginosa* were not able to maintain their class sizes (Grant and Grant 1993). Grant and Grant (1992) concluded, "In the period 1983–1991 finches bred in 6 of 9 years. Those that hybridized were at no obvious disadvantage. They bred as many times as conspecific pairs and produced clutches of similar size...".

The observation of temporal variation in fitness estimates for hybrid (and nonhybrid) offspring was inferred to have been a consequence of an El Niño event during 1982–1983 that produced a record level (i.e., ~1400 mm) of rainfall (Grant and Grant 1993). This extreme climate fluctuation resulted in a radical ecological transition. However, the extraordinary environmental fluctuations were not restricted to the El Niño event of 1983. Negligible rainfall in the years 1985 and 1988 bracketed another very large rainfall total in 1987 (Grant and Grant 1993). The ecology of the Galápagos islands, and thus the environmental setting experienced by the hybrid/parental finches was perturbed repeatedly by this series of climatic disturbances. Grant and Grant (1993) described the perturbation in the following manner: "Changes in plant communities caused changes in the granivorous finch populations" and "is consistent with the survival advantage experienced by hybrids." The elevated fitness of hybrids was directly attributable to an increase in the abundance of small seeds (Figure 1.2; Grant and Grant 1993; Grant and Grant 1996). Thus, one effect of introgression between the finch species was the transfer/origin of adaptations leading to an elevated fitness of hybrid individuals due to their ability to utilize the abundant, small seeds.

Along with the apparent transfer of traits allowing hybrids to survive under the fluctuating conditions after 1982–1983, an overall pattern of morphological and genetic convergence took place between certain species (Grant *et al.* 2004; see below). Once again, this conclusion is reminiscent of the observations made by Darwin and others—that is, that there was noticeable gradation between forms. Attempts to derive phylogenetic trees from DNA data have also detected patterns consistent with past and ongoing introgression. For example, Freeland and Boag (1999) drew the following conclusion from an analysis of sequence variation in the mtDNA control region and the nuclear ribosomal internal transcribed spacer region of Darwin's finch species, "The differentiation of the ground finch species based on morphological data is not reflected in... DNA sequence phylogenies... We suggest that the absence of species-specific lineages can be attributed to ongoing hybridization involving all six species of *Geospiza*" (Freeland and Boag 1999). Sato *et al.* (1999) arrived

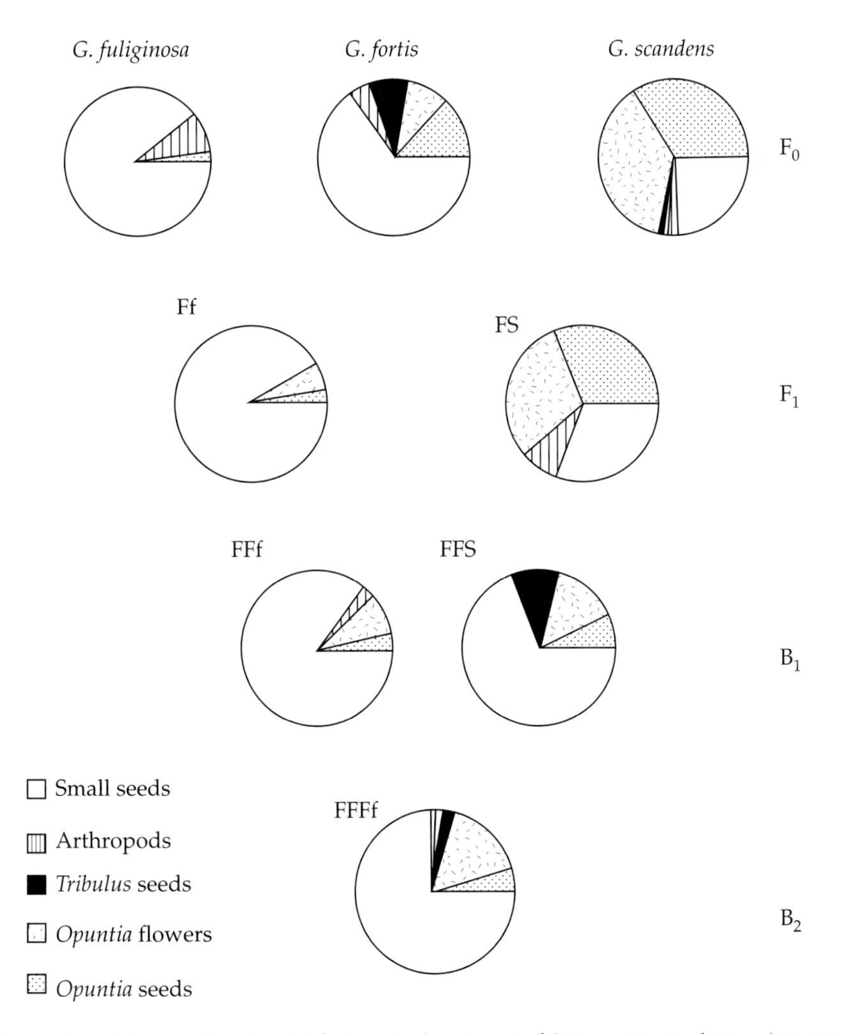

Figure 1.2 Differences in diets between three Darwin's finch species (i.e., *Geospiza fuliginosa*, *Geospiza fortis*, and *Geospiza scandens*) and three generations of hybrids. The three hybrid generations are: (1) *G. fortis* × *G. fuliginosa* (Ff) and *G. fortis* × *G. scandens* (FS) F_1 hybrids; (2) first generation backcrosses (B_1) formed from Ff × *G. fortis* (FFf) and FS × *G. fortis* (FFS); (3) second generation backcross (B_2) also involving *G. fortis* (FFFf).The change in relative abundance of different classes of seeds—caused by an extreme environmental fluctuation—is reflected in the distribution of the diets of the various parental and hybrid generations. Most importantly, this distribution reflected a major transition in the fitness of the parental and hybrid birds (Grant and Grant 1996).

at a similar conclusion from an analysis of additional mtDNA sequences. Specifically, their sequence information failed to place the ground and tree finch species into monophyletic clades. Instead, they found that "The inter- and intraspecies genetic distances overlap and on the phylogenetic trees, individuals representing different morphologically identified species are intermingled..." (Figure 1.3; Sato *et al.* 1999).

Recent estimates of the genetic similarities between sympatric and allopatric populations of pairs of Darwin's finch species have supported the inference of introgression postulated by Freeland and Boag (1999) and Sato *et al.* (1999). Specifically, Grant *et al.* (2005) found that species were more similar genetically to a sympatric relative than to allopatric populations of that same relative. Like the lack of a phylogenetic signal resulting in

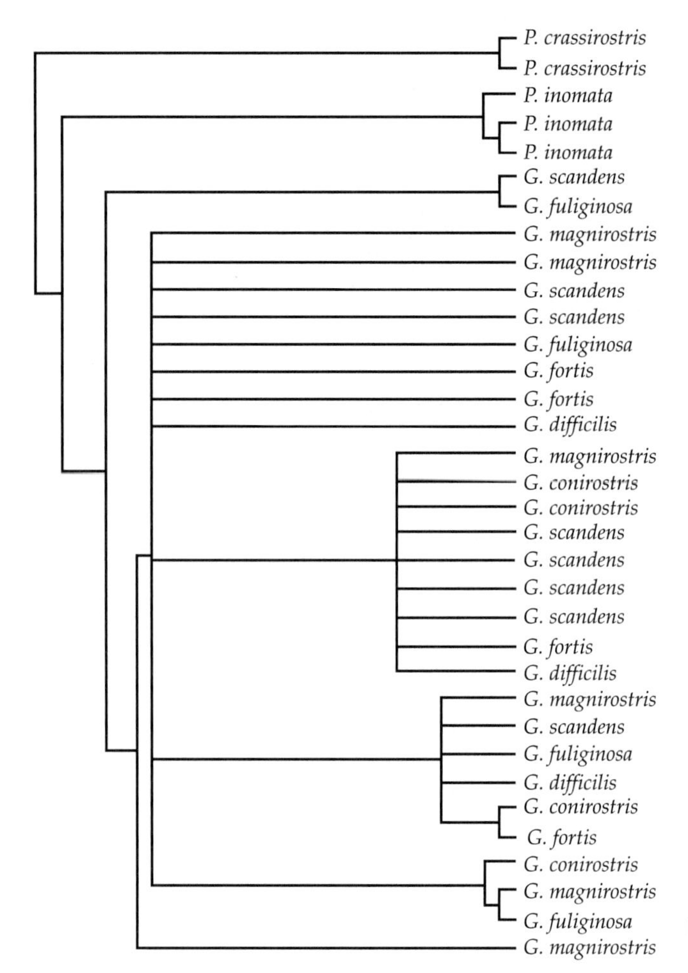

Figure 1.3 Maximum parsimony phylogeny for Darwin's finch species (Sato *et al.* 1999).

nonmonophyletic groupings of tree and ground finch species, the observation of greater similarity between sympatric, rather than allopatric, populations of different species is attributable to introgressive hybridization (Grant *et al.* 2005).

It is apparent from the above studies that introgressive hybridization has greatly affected the evolution of *Geospiza* species. Yet, past and ongoing hybridization are likely to have had varying impacts on the trajectories of different species. This hypothesis has indeed been supported by the pattern of genetic, phenotypic, and adaptive change in *G. fortis* and *G. scandens* on the Galápagos island of Daphne Major over a 30-year period. Specifically, Grant *et al.* (2004, 2005) observed (1) an increase in heterozygosity in *G. scandens*, but not in *G. fortis*, (2) high frequency introgression into *G. scandens* of alleles found in only *G. fortis* before the 1982 El Niño, (3) a significant increase in F_1 and backcross 1 hybrid individuals in "*G. scandens*-like," but not "*G. fortis*-like" samples between 1982 and 2002, and (4) a marked genetic convergence between the two species, but with the convergence explained by an asymmetric increase of similarity of the *G. scandens* samples to *G. fortis* (Grant *et al.* 2004).

The unidirectional pattern of change, resulting in *G. scandens* being drawn toward the genetic and phenotypic pattern of *G. fortis* can also be illustrated by comparisons of genetic variation and morphological character change during this same time period (Grant *et al.* 2005). For example,

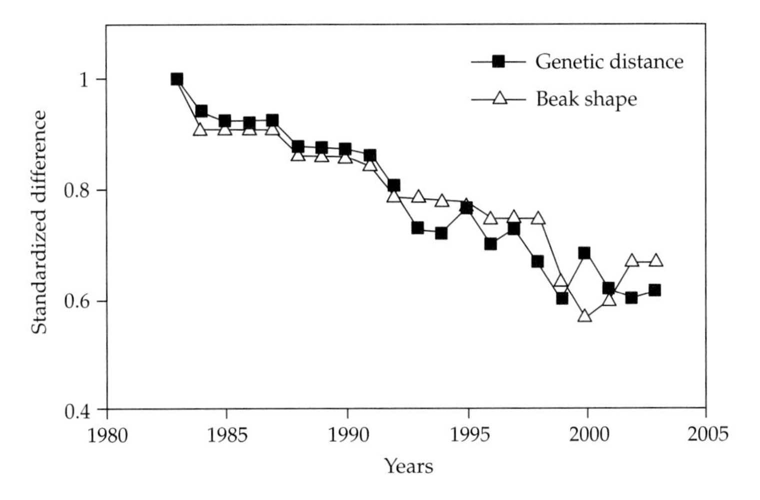

Figure 1.4 Convergence over ~20 years in overall genetic identity and beak shape of Darwin's finch species, *G. fortis* and *G. scandens*. The pattern of convergence was attributed to the dual actions of introgression and natural selection favoring a *G. fortis*-like morphotype (Grant *et al.* 2005).

both beak size and beak shape were affected by the introgression between these two species. Specifically, *G. scandens* became significantly more like *G. fortis* in terms of beak characteristics (Grant *et al.* 2004). Overall then, the transfer of genetic material between *G. scandens* and *G. fortis* affected the former species much more than the latter (Figure 1.4; Grant *et al.* 2005). It is once again important to reflect that the detection of clines in morphospace by Darwin, Lowe, and Lack, and more recently by the Grants and their colleagues, is easily explainable given past and ongoing introgressive hybridization. The importance of such genetic exchange is reflected by the fitness/adaptive consequences. As discussed above, the fitness differential between hybrid genotypes before and after the El Niño event of 1982 was due apparently to ecological selection mediated by a shift in habitat. In the same way, the introgression-mediated convergence of *G. scandens* and *G. fortis* (Figure 1.4) likely reflects the transfer of adaptations from the latter into the former species. This would result in selection favoring those hybrid/*G. scandens* individuals that approach a *G. fortis* type (Figure 1.4; Grant *et al.* 2004, 2005).

Selection leading to the convergence of *G. scandens* and *G. fortis* is not the only outcome of introgression posited for the Darwin's finch species complex. In fact, numerous papers have addressed the likelihood that introgressive hybridization has underlain the **adaptive radiation** of the entire clade. For example, Lowe (1936) and Freeland and Boag (1999) argued that the pattern of morphological variation among the finch species was the result of hybridization, with the latter authors concluding, "Hybridization has apparently played a role in the adaptive radiation of Darwin's finches." Most recently, Seehausen (2004) used the Darwin's finch clade as an example that supported his **hybrid swarm model of adaptive radiation**. Consistent with Seehausen's (2004) use of the Darwin's finches as a paradigm of introgression-mediated adaptive radiation, is the observation that "Hybridization may enhance fitness to different degrees by counteracting the effects of inbreeding depression, by other additive and non-additive genetic effects, and by producing phenotypes well suited to exploit particular ecological conditions" (Grant *et al.* 2003). It is important to note, however, that Grant *et al.* (2005) doubted that "hybridization was necessary for any part of the adaptive radiation of Darwin's finches."

Notwithstanding the role (or lack thereof) of introgression in the adaptive radiation of the entire clade, its effect on adaptive evolution within this clade is now well established. Introgressive hybridization has thus contributed to the ecological and evolutionary trajectories of certain species (e.g., *G. scandens* on Daphne Major). This outcome has occurred due to the transfer of genes for adaptations to environmental settings (Grant *et al.* 2004). As Petren *et al.* (2005) observed, far from constraining phenotypic divergence, introgression has actually enhanced genetic and phenotypic variation thereby facilitating evolution via natural selection. Finally, the episodic occurrence of introgression > ecological selection > genetic/phenotypic transformation reflects recurring **hybrid speciation**. In this regard then, though it is uncertain whether the adaptive radiation of *Geospiza* was underlain by introgressive hybridization, this process appears to have played a central role in the formation and evolution of individual Darwin's finch species.

1.2.2 Evolution of adaptations and hybrid speciation: African cichlids

As Clabaut *et al.* (2007) expressed so well, "The cichlids of East Africa are renowned as one of the most spectacular examples of adaptive radiation. They provide a unique opportunity to investigate the relationships between ecology, morphological diversity, and phylogeny in producing such remarkable diversity." The analyses by Clabaut *et al.* (2007) were designed to test the hypothesis that ecological selection had affected the adaptive radiation of the cichlid clade in Lake Tanganyika. They concluded that such selection had indeed played a significant role in this explosive diversification. Genner *et al.* (2007b) also tested for factors associated with the adaptive radiation of this extraordinarily diverse clade. However, in contrast to the most prevalently held view that the radiations occurred recently and within the current lake basins (e.g., within the Lake Tanganyika basin), these workers inferred much more ancient dates for the origin of some lineages. This inference led to the following conclusions:

> dates derived from Gondwanan fragmentation indicate that ancestors of every major tribe entered the lake independently and that molecular diversity within some tribes began to accumulate around the time of colonization, or indeed in 2 cases (Trematocarini and Ectodini) possibly before colonization . . . The Gondwanan estimates also suggest that the Haplochromini may have been split into several lineages already prior to the formation of deepwater conditions in Lake Tanganyika. (Genner *et al.* 2007b)

One implication of an earlier date of origin for the cichlid lineages followed by their invasion of the current rift lakes is that there was an increased opportunity for introgressive hybridization to contribute to levels of genetic and phenotypic variation. In support of such a significant evolutionary role for introgression is the common inference of genetic admixing between cichlid lineages (Figure 1.5; Rüber *et al.* 2001; Salzburger *et al.* 2002; Smith *et al.* 2003; Hey *et al.* 2004; Won *et al.* 2005), in spite of strong reproductive barriers (e.g., see Genner *et al.* 2007a). For example, Seehausen and his colleagues have produced data that are consistent with both a contemporary role for introgression and hybrid speciation in the cichlids and as a major catalyst for the adaptive radiation of the entire species complex (Seehausen *et al.* 2002; Seehausen 2004; Joyce *et al.* 2005). In particular, Seehausen (2004) applied his hybrid swarm model to explain the adaptive radiation of the rift lake cichlids. Specifically, he suggested that the cichlid adaptive radiation may have begun as a **syngameon.** Consistent with this conclusion is the recent findings by Samonte *et al.* (2007) that the species flock found in Lake Victoria is characterized by high levels of interspecific gene flow and low levels of genetic differentiation. Indeed, these authors suggested that their data reflected the presence of a single cichlid genus rather than the multitude of genera normally assigned to this flock (Samonte *et al.* 2007).

Numerous subsequent analyses have provided support for Seehausen's hypothesis that introgressive hybridization has played a significant role in cichlid evolution. For example, Koblmüller *et al.* (2007a) obtained phylogenetic and population genetic information for gastropod-shell-breeding species. Data for their inferences came from both mtDNA and nuclear sequences. These authors argued that the unique ecological setting—that is,

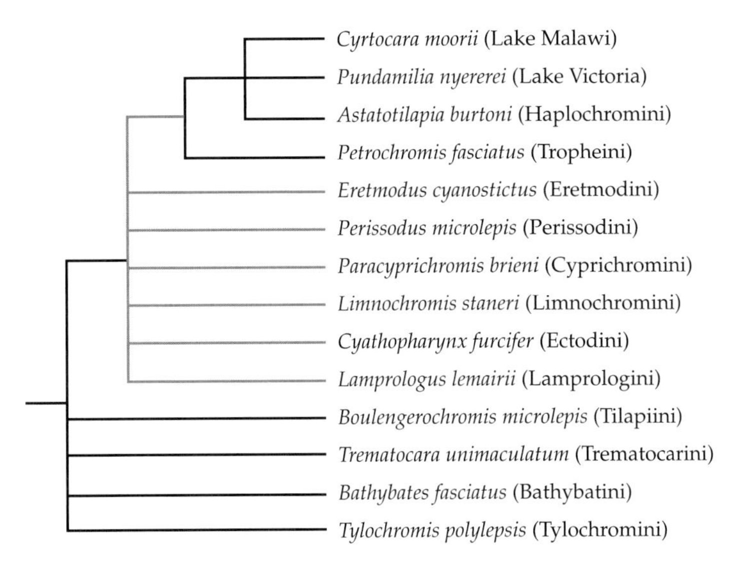

Figure 1.5 Phylogenetic tree for the 12 tribes of Lake Tanganyika cichlids. The tree was constructed using the patterns of insertion of transposable elements. The gray portions of the tree have been inferred to reflect the retention of ancestral polymorphisms (i.e., incomplete lineage sorting; Takahashi *et al.* 2001). However, these patterns have most often been interpreted as evidence for gene exchange through introgressive hybridization (Seehausen 2006).

living and breeding in empty gastropod shells—shared by these cichlid taxa would facilitate natural hybridization (Koblmüller *et al.* 2007a). Consistent with this hypothesis was the finding of incongruence between phylogenetic trees derived from the alternate genetic data sets. In fact, their phylogenetic results led these authors to infer that *Lamprologus meleagris, Lamprologus speciosus, Neolamprologus wauthioni* and *Neolamprologus multifasciatus* were hybrid species (Koblmüller *et al.* 2007a). Furthermore, different samples of two of the putative species in this group were placed into different clades (i.e., *Altolamprologus calvus* and *Lepidiolamprologus* sp. "meeli-boulengeri") suggestive of introgressive hybridization resulting in the sharing of mtDNA or nuclear loci between divergent lineages. In addition to the phylogenetic patterns indicative of past introgression and hybrid speciation, Koblmüller *et al.* (2007a) also collected population genetic data indicating contemporary genetic exchange. Specifically, mtDNA sequence information indicated that the maternal parents for a set of putative hybrid individuals belonged to the *Neolamprologus brevis/Neolamprologus calliuris* clade, while the nuclear loci in these animals suggested a paternal contribution from either *Lamprologus callipterus* or *Neolamprologus fasciatus* (Koblmüller *et al.* 2007a). In total, these findings suggest a significant role for reticulate evolution in the origin and diversification of this clade of cichlids (Figure 1.6; Koblmüller *et al.* 2007a).

Like Koblmüller *et al.* (2007a), Day *et al.* (2007) examined cichlids belonging to the tribe Lamprologini. Also, as in the results of the former study, those detected by the latter authors included the placement of closely related taxa into divergent portions of the phylogenies constructed. Furthermore, Day *et al.* (2007) also detected the nonmonophyly of members of a given taxon. In particular, species belonging to *Neolamprologus, Lamprologus, Julidiochromis,* and *Telmatochromis* were not uniformly resolved into their respective genera. These results, those discussed above, and those from earlier studies (Salzburger *et al.* 2002; Schelly *et al.* 2006), all support the hypothesis that this tribe of Lake Tanganyikan cichlids has been impacted greatly by introgressive hybridization leading to the diversification of hybrid lineages (i.e., species). Another example from a Lake Tanganyikan assemblage, in this case involving

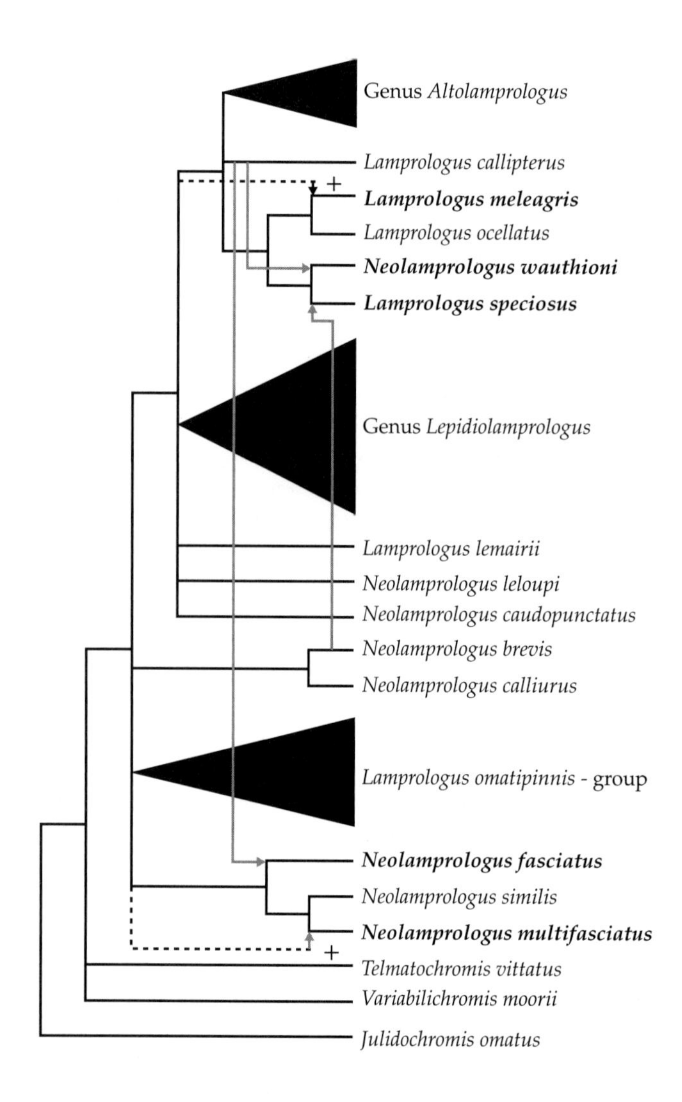

Figure 1.6 Reticulate relationships among species of gastropod-shell-breeding cichlids from Lake Tanganyika. Taxa names that are bolded reflect hypothesized hybrid species. Lines with arrows indicate the direction of mtDNA introgression into these species. Dashed lines indicate extinct lineages (Koblmüller *et al.* 2007a).

the tribe Perissodini, is also illustrative of the evolutionary effects possible from introgressive hybridization (Koblmüller *et al.* 2007b). These cichlid species utilize scales scraped from other fish as their food source. Unlike their analysis of the Lamprologini, Koblmüller *et al.* (2007b) concluded that **incomplete lineage sorting** rather than introgressive hybridization might explain the discordant patterns found in the Perissodini. Yet, given the widespread occurrence of past and contemporary introgression in cichlids in general, it would seem most likely that the numerous inconsistencies between the phylogenies constructed from the nuclear and mtDNA sequences (Koblmüller *et al.* 2007b) reflect at least some role for genetic exchange.

Recently, Seehausen (2006) reflected further on the potential for introgressive hybridization to act as a generator for adaptive radiations. The initial admixture of many divergent lineages was suggested to have the potential to produce founding populations and subsequent derivative species "enriched in adaptive variation at a large number of quantitative trait loci..." with "much adaptive genetic potential. Such enriched populations may possess an increased propensity to undergo rapid diversification if opportunity arises again" (Seehausen 2006). As indicated above, this hypothesis is consistent with numerous data sets. Reticulate evolution thus marks both the initial adaptive radiation of the African cichlid clade, as well as its ongoing diversification.

1.3 Examples of evolutionary consequences from introgressive hybridization: plants

1.3.1 Evolution of adaptations and hybrid speciation: Louisiana Irises

A major outcome of reticulate evolution, by definition, is the derivation of lineages with novel evolutionary and ecological trajectories (Anderson and Stebbins 1954; Grant 1981; Arnold 1997, 2006). For example, it has been estimated that a majority of flowering plant species derive from reticulate events (see Arnold 1997 and 2006 for discussions). In contrast, there are many fewer references to hybrid speciation events in the zoological literature, though this may be due more to definitional confusion than to a lack of the process (Arnold 2006).

Two categories have been defined to encompass many of the derivatives of hybrid speciation events: homoploid (i.e., diploid derivatives) and polyploid. Though the most common process in plants is thought to involve whole genome duplication events (i.e., polyploidy—Stebbins 1959; Soltis and Soltis 1993; Arnold 1997, 2006), numerous cases of homoploid hybrid diversification have been identified (Grant 1981; Abbott 1992; Rieseberg and Wendel 1993; Arnold 1997, 2006; Rieseberg 1997). There is a growing list of well-supported examples of homoploid animal taxa as well (Wayne and Jenks 1991; DeMarais *et al.* 1992; Salzburger *et al.* 2002; Tosi *et al.* 2003; Salazar *et al.* 2005; Schwarz *et al.* 2005; Meyer *et al.* 2006).

One example of homoploid hybrid speciation inferred for a plant taxon comes from the species complex commonly referred to as the Louisiana Irises. Randolph (1966) examined morphological and chromosomal characteristics (Randolph 1966) of natural Louisiana Iris populations and from these data concluded that *Iris nelsonii* was the product of reticulate evolution. A unique aspect of Randolph's hypothesis for the formation of *I. nelsonii* species was that it derived from hybridization among three species: *Iris fulva, Iris hexagona,* and *Iris brevicaulis* (Randolph 1966; Randolph *et al.* 1967). In the early 1990s, our group utilized a combination of isozyme, chloroplast DNA (i.e., cpDNA), and randomly amplified polymorphic DNA (i.e., RAPD) markers diagnostic for the three putative parents of *I. nelsonii* to test for any contribution to the origin of this species. The hypothesis that *I. nelsonii* was the product of a three-species interaction, was tested and supported by these molecular analyses (Arnold *et al.* 1990, 1991; Arnold 1993). *I. nelsonii* individuals were found to possess a combination of the nuclear and cpDNA markers that were diagnostic for *I. fulva, I. brevicaulis,* and *I. hexagona* (Arnold 1993). These studies led to the following question (Arnold 1993): "What then are the attributes that characterize *I. nelsonii* as a stabilized hybrid species?" The answer arrived at was that the attributes included "the population level pattern of genetic variation... distinctive ecological preference... marker chromosomes, and a characteristic morphology... The definition of *I. nelsonii* as a novel evolutionary lineage, as with any other species, depends upon a number of genetic and ecological components" (Arnold 1993).

Similar to the examples from the Darwin's finches and African cichlids, the Louisiana Iris species complex not only exemplifies hybrid speciation, but also the evolution of adaptations via introgressive hybridization. This conclusion reflects the outcome of tests of a longstanding hypothesis first proposed by Edgar Anderson in his book, *Introgressive Hybridization*. In particular, Anderson (1949; using data from Riley 1938) considered the morphological variation in natural populations of *I. fulva, I. hexagona,* and their hybrids, described the process of introgressive hybridization, and then highlighted some its potential evolutionary consequences in the Louisiana Irises. One hypothesis proposed by Anderson (1949, p. 62) was that a very small amount of introgression might be of enormous evolutionary potential.

Recent studies involving both natural and experimental hybrid populations of Louisiana Irises have supported the above conclusion. For example, Cornman *et al.* (2004) inferred both the spatial distribution of naturally occurring hybrid plants and the paternal contribution to their genotypes. The observations of (1) spatially structured genotypes and (2) the recruitment into the population of only a limited subset of possible genotypes were consistent with a higher fitness of the recruits resulting from adaptive introgression (Cornman *et al.* 2004). Furthermore, Cornman *et al.* (2004) argued for the outcome of evolutionary novelty arising from hybridization between the various Louisiana Iris species. Thus, differential selection that favored or disfavored different hybrids was inferred to have resulted in "the establishment of recombinant lineages that are more fit than the parental types in some habitats" (Cornman *et al.* 2004).

Recent quantitative trait locus (i.e., QTL) mapping experiments involving the Louisiana Iris species, *I. fulva* and *I. brevicaulis*, have provided further support for the hypothesis that adaptive evolution in this species complex has been affected by reticulation. Bouck *et al.* (2005) detected genomic regions (in reciprocal backcross (BC_1) individuals) that introgressed at significantly lower- or higher-than-expected levels. The detection of regions with significantly increased frequencies of introgression is consistent with the hypothesis of gene transfer that leads to the transfer of adaptations and thus the elevated fitness of some hybrid genotypes in certain habitats. This conclusion was supported when Martin *et al.* (2005) defined QTLs (Figure 1.7) associated with the phenotype of long-term survivorship in the same greenhouse environment utilized by Bouck *et al.* (2005). The greenhouse environment reflected a water-limited habitat for some hybrid genotypes. In particular, though *I. brevicaulis* is often found in dryer, greenhouse-like, natural environments, *I. fulva* plants most often occur in water-saturated soils (Viosca 1935, Cruzan and Arnold 1993; Johnston *et al.* 2001). The habitat associations for these two species lead to a prediction of higher mortality in the backcrosses toward "wet adapted" *I. fulva* relative to those toward "dry adapted" *I. brevicaulis*. Martin *et al.* (2005) did indeed find this pattern, with *I. fulva* backcrosses demonstrating twice the frequency of mortality as *I. brevicaulis* backcross plants. In addition, four QTLs in the *I. fulva* hybrids were significantly associated with survivorship (Martin *et al.* 2005). Three of the four QTLs, as expected, were associated with the introgression of alleles from *I. brevicaulis* (i.e., dry adapted). However, the fourth QTL reflected homozygosity of the recurrent (i.e., wet adapted *I. fulva*) parent's alleles (Martin *et al.* 2005). This latter result indicates the origin of a novel adaptive potential resulting from combining genes from divergent lineages (Arnold 1997, 2006).

Following from the above studies, Martin *et al.* (2006), transplanted the same genotypes into natural settings. This latter analysis was also designed to determine the genetic architecture (using QTL analyses) of survivorship. Unlike the relatively dry settings for the experiments of Bouck *et al.* (2005) and Martin *et al.* (2005), the third experimental analysis resulted in the exposure of the hybrid genotypes to a > 3 month flood. In contrast to selection favoring the dry alleles found in *I. brevicaulis*, the flooded environment was predicted to result largely in positive selection for alleles from wet-adapted *I. fulva*. Overall, the results from the flood event reflected this prediction. First, the rank survivorship of the various classes was *I. fulva* > backcrosses to *I. fulva* > backcrosses to *I. brevicaulis* > *I. brevicaulis* (Martin *et al.* 2006). Second, the frequency of survivorship of the *I. brevicaulis* backcross hybrids was increased by the presence of introgressed *I. fulva* alleles. Third, the fitness (as reflected by survivorship) of *I. fulva* backcross hybrids was affected by two epistatically interacting QTL (Figure 1.7; Martin *et al.* 2006). The two QTLs that affected survivorship in the *I. fulva* BC_1 hybrids were located on two of the same linkage groups that contained QTLs associated with survivorship in the dry environment (Martin *et al.* 2005). Intriguingly, the effects of the same genomic regions under the two different environments were found to be in opposite directions. Thus, introgression of alleles lowered survivorship in the dry habitat, but increased survivorship in the flooded environment (Martin *et al.* 2006).

The analyses by Bouck *et al.* (2005) and Martin *et al.* (2005, 2006) support Anderson's (1949) concept that "A trickle of genes so slight as to be without any practical taxonomic result might still

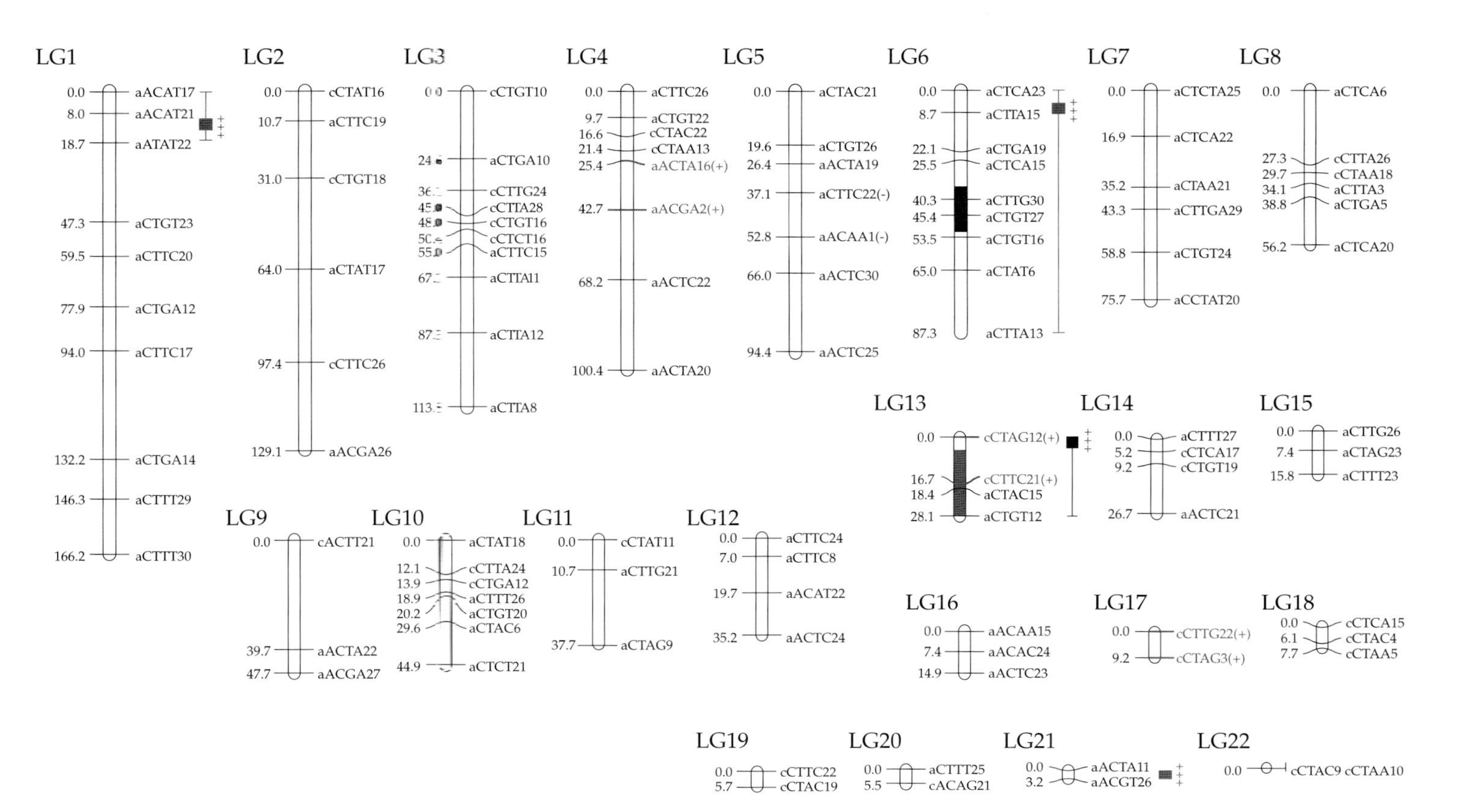

Figure 1.7 Linkage map of dominant *Iris brevicaulis* IRRE retrotransposon display markers (Kentner *et al.* 2003) in experimental backcross hybrids toward *Iris fulva*. Markers whose text is in italics reveal significant transmission ratio distortion (Bouck *et al.* 2005). Significant QTLs for survival in greenhouse conditions are denoted (with 2-lod confidence intervals) to the right of the marker names (Martin *et al.* 2005). QTLs for survival in an extended flood (Martin *et al.* 2006) are denoted by hatched and dotted bar segments (2-lod confidence intervals) on the linkage groups. The "hatched" segments represent regions where introgressed (hybrid/heterozygous) regions are favored, while the "dotted" segments represent regions where recurrent (parental/homozygous) regions are favored.

be many times more important than mutation…" Indeed, this trickle can apparently lead to not only an increase in genetic variation, but also to the origin and transfer of the underlying architecture supporting adaptations.

1.3.2 Introgression and hybrid speciation: *Arabidopsis*

Meinke *et al.* (1998) emphasized the fundamental importance of utilizing *Arabidopsis thaliana* as a model system for plant biology. Specifically, they observed that *A. thaliana* "is a small plant in the mustard family that has become the model system of choice for research in plant biology. Significant advances in understanding plant growth and development have been made by focusing on the molecular genetics of this simple angiosperm." Furthermore, these authors emphasized the general utility of knowledge concerning the genetic structure of this species and argued that "The current visibility of *Arabidopsis* research reflects the growing realization among biologists that this simple angiosperm can serve as a convenient model not only for plant biology but also for addressing fundamental questions of biological structure and function common to all eukaryotes." Though not emphasized by Meinke *et al.* (1998), *A. thaliana* and its relatives can also provide insights into the processes associated with reticulate evolution, including introgression and the origin of hybrid taxa. A number of studies have detected the signatures of past, and ongoing, introgressive hybridization resulting in significant genetic exchange and the evolution of hybrid species and subspecies.

In their paper, "Poorly known relatives of *Arabidopsis thaliana*," Clauss and Koch (2006) discussed the power of applying the genomic tools available for *A. thaliana* to various closely related species to test important evolutionary hypotheses. In particular, they suggested that the application of knowledge gleaned from studying this model organism to other species within this complex would help to elucidate "adaptive evolution of ecologically important traits and genomewide processes, such as polyploidy, speciation and reticulate evolution…". However, these authors emphasized that the extent to which reticulation had affected evolution within the genus had only recently begun to be appreciated (Figure 1.8; Clauss and Koch 2006). Yet, like other clades of organisms (Arnold 2006), the laundry list of *Arabidopsis* taxa affected by genetic exchange is known to be long. For example, allopolyploidy has resulted in the formation of several well-recognized lineages. Two of these are *Arabidopsis lyrata* ssp. *kamchatica* and *Arabidopsis suecica* (O'Kane *et al.* 1996; O'Kane and Al-Shehbaz 1997; Säll *et al.* 2003; Clauss and Koch 2006). The former taxon has been proposed to be a derivative from hybridization between *A. lyrata* and *Arabidopsis halleri* ssp. *gemmifera* (Clauss and Koch 2006), whereas *A. thaliana* and *Arabidopsis arenosa* have been identified as the parental lineages for *A. suecica* (e.g., see O'Kane *et al.* 1996 and Säll *et al.* 2003). Molecular dating of this latter allopolyploid has suggested a relatively recent origin (between 12,000 and 300,000 years ago), followed by population expansion as the ice shields retreated (Jakobsson *et al.* 2006).

Though allopolyploid lineage formation has occurred repeatedly in the genus *Arabidopsis* this is not the sole, or the most frequent, result of genetic exchange between divergent lineages. Past and contemporary introgressive hybridization, resulting in shared genetic variation, has thus been inferred for numerous lineages. For example, Koch and Matschinger (2007) carried out an analysis of both cpDNA and ribosomal DNA (rDNA) internal transcribed sequence (i.e., ITS) variation to determine relationships among the following *Arabidopsis* species: *halleri, arenosa, lyrata, croatica, cebennensis, pedemontana,* and *thaliana*. Though Koch and Matschinger (2007) argued for the effect of retained ancestral polymorphisms to explain some of the discordances found between the cpDNA and ITS phylogenies, they noted that reticulate evolution was also a likely contributor to patterns of overlapping genetic variation. Indeed, Clauss and Koch (2006), apparently reflecting on this same data set, came to a similar conclusion in that they argued for an effect from both past and contemporary reticulations and retention of ancestral polymorphism in structuring the genetic variation in present-day populations of *A. thaliana* and its relatives (Figure 1.8). In regard to the effects from natural hybridization, they concluded

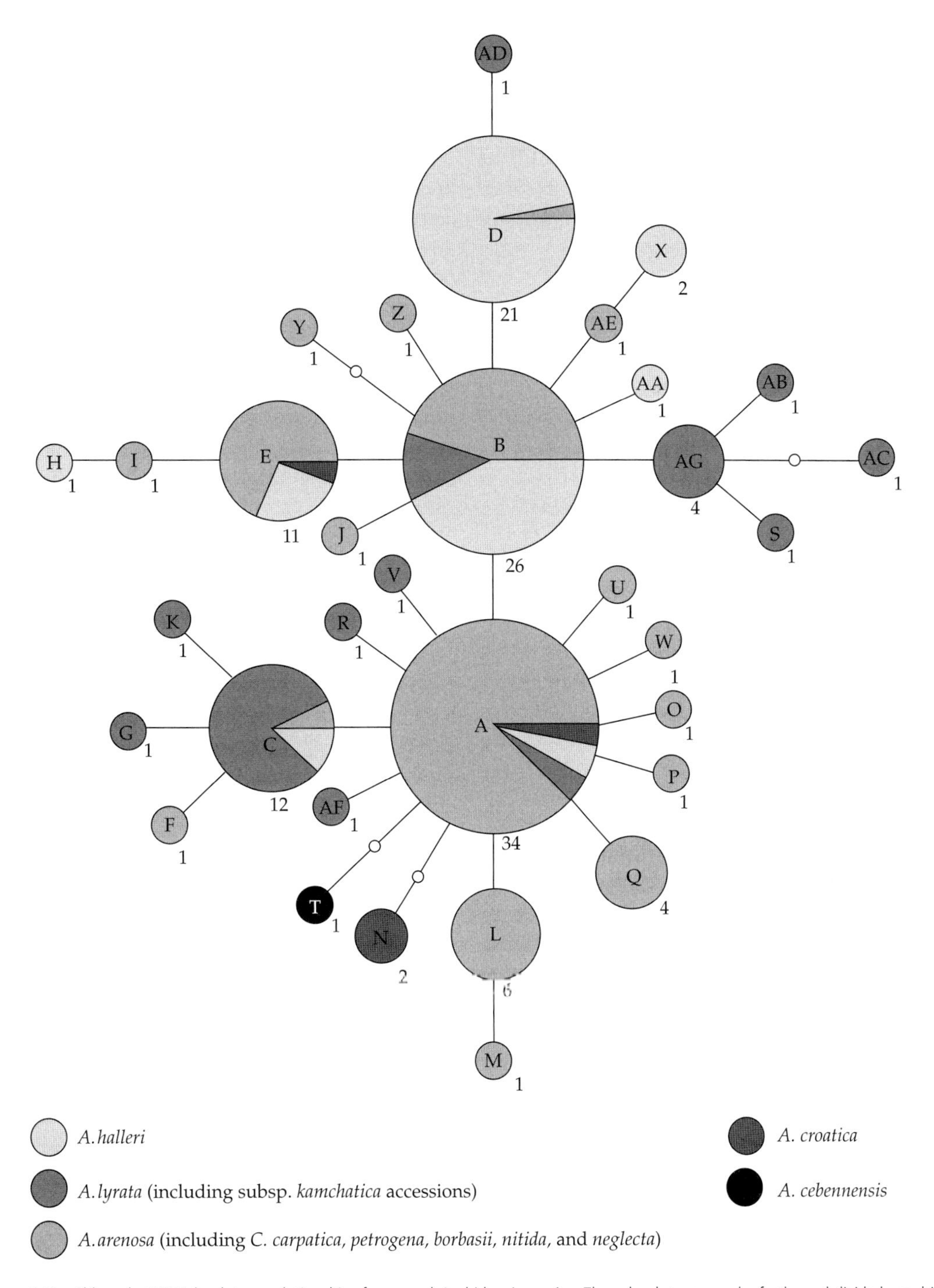

Figure 1.8 Chloroplast DNA haplotype relationships for several *Arabidopsis* species. These haplotypes can be further subdivided, resulting in 145 unique haplotypes in total (indicated by numbers below each circle). The network of relationships demonstrates that haplotypes from the inner part of the network, and thus inferred to be older (A, B, C, and E) are shared between all three-species groups. Several evolutionary explanations can account for this pattern of haplotype sharing, however, the observation of widespread reticulate evolution among *Arabidopsis* species argues for a contribution from introgressive hybridization (Clauss and Koch 2006).

that "hybridization between *Arabidopsis* species in some geographical regions is, or has been, common, and we expect this process of reticulate evolution to affect the patterns of molecular, karyological and morphological diversity..." (Clauss and Koch 2006). Similar to finches, cichlids, and irises, the evolutionary history of the model plant genus, *Arabidopsis*, has included reticulation.

1.4 Examples of the evolutionary consequences from horizontal gene transfer: prokaryotes

1.4.1 Transfer of adaptive machinery: bacteria and archaebacteria

The horizontal transfer of genomic material among prokaryotic and even eukaryotic organisms is now well documented (e.g., Bergthorsson *et al.* 2003; Doolittle and Bapteste 2007; Sorek *et al.* 2007). Furthermore, many exchange events reflect the origin or transfer of adaptations to novel (at least for the recipient organisms) environments (see Arnold 2006 for a review). Numerous instances of adaptive trait origin and/or transfer—for example, involving human pathogens—will be discussed in the following chapters. A recent report by Frigaard *et al.* (2006) gives a clear illustration of the adaptive potential of lateral gene transfer. In this study, the phylogenetic and ecological distribution of genes associated with light-mediated metabolic functions (Figure 1.9) led to an inference of an adaptive, horizontal transfer/acquisition in a nonpathogenic organism.

In a 1992 paper, DeLong described the detection of rDNA sequences characteristic of archaebacteria (i.e., Archaea) in a previously unknown environment for these organisms, that is, oxygenated coastal surface waters. The usual settings from which Archaea had been isolated included such extreme environments as anaerobic, hydrothermal, or those high in saline (DeLong 1992). The detection of lineages of Archaea in environments containing eubacteria suggested that these two types of prokaryotes would be in competition for common resources (DeLong 1992). Frigaard *et al.* (2006) reflected this hypothesis when they stated, "Planktonic bacteria, Archaea, and Eukarya reside and compete in the ocean's photic zone under the pervasive influence of light." This hypothesis leads to a series of expectations, one of which is that in order for Archaea and Eubacteria to compete for light-mediated resources, they must both possess genes controlling the utilization of photic energy. Significantly, genes encoding proteins that allow the harvesting of light in a variety of environments, including those of the oxygenated, near surface zones, have been isolated from Archaebacteria and Eubacteria (e.g., Béjà *et al.* 2000, 2001; Balashov *et al.* 2005; Frigaard *et al.* 2006). In particular, the class of photoproteins known as proteorhodopsins has been detected (Béjà *et al.* 2000). Proteorhodopsins exhibit various functions including that of light collection (Balashov *et al.* 2005).

In the context of the known functional characteristics of proteorhodopsins, it is also significant that Frigaard *et al.* (2006) detected signatures of lateral transfer between "planktonic Bacteria and Archaea." Specifically, these authors found that (1) there were unique associations between the proteorhodopsin genes and rRNA genes consistent with recent acquisition and (2) the proteorhodopsin genes isolated from euryarchaeotes were present in those isolates taken from photic, but not subphotic,

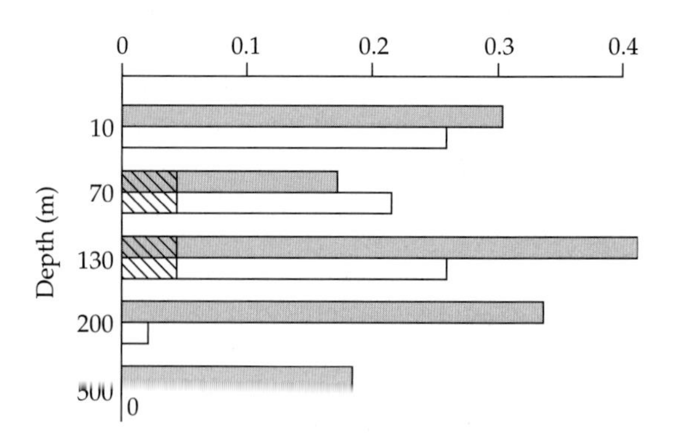

Figure 1.9 Bars indicate the fraction of clones at each depth interval that contain (1) the euryarchaeal SSU rRNA gene (gray), (2) the archaeal-like proteorhodopsin gene (white), or (3) both of these genes (hatched). The high frequency of both genes in the photic portion of the water column (i.e., 200 m and above), but the absence of the light-utilizing archaeal-like proteorhodopsin gene in the nonphotic zone is consistent with the adaptive acquisition of the latter gene through horizontal exchange (Frigaard *et al.* 2006).

regions of the water column (Figure 1.9). Frigaard *et al.* (2006) surmised from this latter observation that the organisms in the light-limited zones would likely gain no benefit from such genetic architecture and the transfer of the proteorhodopsin gene, if it occurred in these regions, would be unlikely to be retained. Frigaard *et al.* (2006) also hypothesized that the limited number of additional genes (as few as two) needed to have a functional light-utilizing system would facilitate the acquisition of this adaptation via lateral transfer by organisms in the photic zone. Finally, Frigaard *et al.* (2006), by considering the biology of the organisms involved and the phylogenetic and spatial/environmental distribution of the proteorhodopsin genes, concluded that "lateral gene dispersal mechanisms, coupled with strong selection for proteorhodopsin in the light, have contributed to the distribution of these photoproteins among various members of all three of life's domains."

1.4.2 Horizontal transfer and species distributions: thermophylic bacteria

Nesbø *et al.* (2006) have argued that some of the most hotly debated concepts in the field of prokaryotic biology relate to the definition of the prokaryotic species. In particular, citing Fenchel (2003) and Finlay and Fenchel (2004), they provided two questions that summarize two of these contentious issues: "What are prokaryotic species?" and "Are such species cosmopolitan in their distribution?" Encapsulated within these two questions are a series of fundamentally important basic and applied scientific issues. For example, being able to accurately define bacterial species is necessary for the application of control measures. Also, as with any organismic group, defining species facilitates an understanding of evolutionary diversification in the face of gene flow—that is, within the conceptual framework of the web of life (Arnold 2006). Indeed, the data provided by Nesbø *et al.* (2006) in their discussion of the prokaryotic species concept and the biogeography of this class of organism also highlight the process of reticulate evolution of prokaryotic lineages.

The prokaryotes analyzed by Nesbø *et al.* (2006) were members of the hyperthermophilic genus *Thermotoga*, which these authors defined as "obligately anaerobic heterotrophs, with optimal growth between 66° and 80°...". The various strains and species examined in this study were definable on the basis of physiology, DNA sequence variation, the ecological settings in which they occurred, and their geographical distributions (Nesbø *et al.* 2006). Genetically, the four lineages included in this analysis demonstrated less than 96% similarity for their average gene sequences. Given that bacterial species have been routinely defined as containing members with >97% sequence identity (Rosselló-Mora and Amann 2001), the organisms included in this study represented well-differentiated clades. Notwithstanding the distinctiveness of the *Thermotoga* isolates, sequence analysis of these lineages revealed widespread genetic exchange. A consequence of this exchange was reflected in discordance between phylogenies constructed from either rDNA or other genomic sequences (Figure 1.10; Nesbø *et al.* 2006). The alternate topologies of the trees constructed from the different sequences of the *Thermotoga* isolates was directly attributable to lateral exchange events resulting in high levels of similarity between distantly related strains for the non-rDNA regions (Figure 1.10). Likewise, phylogenetic discordance involving *Thermotoga maritima* and *Thermotoga neapolitana* was attributable to an exchange of an 88-kb fragment from the former to the latter species (Nesbø *et al.* 2006).

The findings of Nesbø *et al.* (2006), like those for all of the other examples discussed above, indicate a pervasive evolutionary effect from reticulate events. In regard to the questions posed, the authors concluded that there was no single species concept that could reflect adequately the widespread recombination among ecologically, physiologically, and geographically distinct lineages. This lack of a descriptive and predictive species concept for prokaryotes—due to lateral exchange—also indicated the untenable nature of using a species nomenclature to test for the distribution of certain taxa. Nesbø *et al.* (2006) proposed that biogeographical analyses, instead, should be designed to test for "the global distribution of genes and their alleles and their patterns of divergence and dispersal."

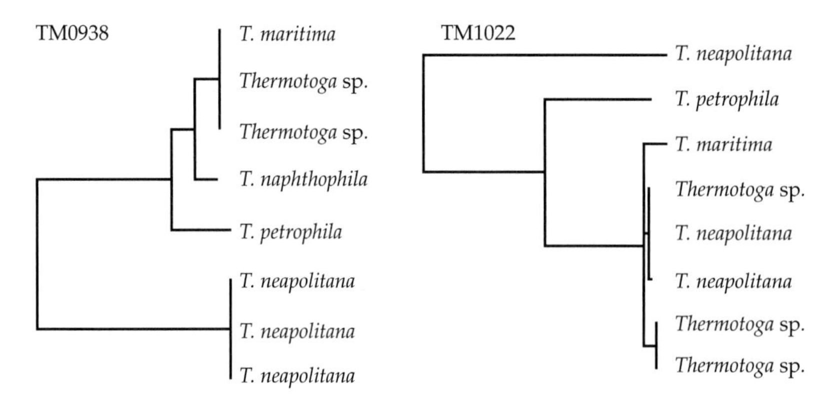

Figure 1.10 Two phylogenetic trees derived from two different genes (left panel, TM0938; right panel, TM1022) isolated from the same lineages of the hyperthermophilic bacterial genus, *Thermotoga*. Note the different placements of the various bacterial lineages, indicating the effects of separate transfer events of the two gene sequences between different members of this genus (Nesbø *et al.* 2006).

1.5 Summary and conclusions

This tree-of-life notion of evolution attained near-iconic status in the mid-20th century with the modern neo-Darwinian synthesis in biology. But over the past 15 years, new discoveries have led many evolutionary biologists to conclude that the concept is seriously misleading and, in the case of some evolutionary developments, just plain wrong. Evolution, they say, is better seen as a tangled web. (Arnold and Larson 2004)

In this way, my colleague Ed Larson and I tried to capture the ingrained, and in some ways inadequate, nature of the tree-of-life metaphor while at the same time reflecting the explanatory and predictive power of the metaphor known as the web of life. The examples discussed above are illustrative of both of these conclusions. In the following chapters I will turn my attention to highlighting cases from primates (including our own species), and those organisms that humans eat, play with, wear, or from which we contract diseases. I will also discuss many lineages that we have "created" intentionally or accidentally for our own purposes through the various avenues of genetic exchange. In this way, I hope to illustrate how our own and related species have been impacted by web-of-life processes. Just as irises, prokaryotes, fish, and bird complexes reflect hybrid speciation and adaptive evolution, so does our own lineage (and related lineages) appear to bear the imprint of reticulate evolutionary change.

CHAPTER 2

Reticulate evolution: nonhominine primates

...study of the presence of hybrids in fragmented and intact forest tracts will reveal whether human-induced forest fragmentation has instigated hybridization by confining members of both species to small areas and limiting access to conspecific mates.

(Cortés-Ortiz *et al.* 2007)

...certain hybrid crosses [i.e., those involving a black lemur (*Eulemur macaco macaco*) parent] usually yield sterile offspring, while others can yield fertile offspring between different parental species or subspecies ....

(Horvath and Willard 2007)

The mitochondrial paraphyly of Ethiopian hamadryas and anubis (*P. anubis*) baboons suggests an extensive and complex history of sex-specific introgression.

(Wildman *et al.* 2004)

Although sympatric hybridization occurs in the absence of human disturbance, and may even have been a creative force in cercopithecine evolution, anthropogenic habitat fragmentation may increase its incidence.

(Detwiler *et al.* 2005)

...the evidence that orangutans have been impacted by reticulate evolution comes from discordant results from different molecular studies.

(Arnold and Meyer 2006)

2.1 Reticulate evolution in New World nonhominines

The use of analogy is prevalent in studies of evolutionary patterns and processes. Darwin used this approach repeatedly in *The Origin* to illustrate evolutionary mechanisms. For example, he used numerous cases from the animal-breeding literature of his day to indicate the strength of human-mediated selection. He then used this analogy to argue for the role of *natural* selection as a major causal factor in evolutionary change. In the same way, instances of hybridization, introgression, and hybrid speciation in primate groups that are more or less distantly related to our own species (Figure 2.1) are appropriate analogies for understanding the likelihood of reticulate evolution in the clade that includes *Homo sapiens*. In this chapter I will begin constructing the analogy using primate groups from Central and South America. I will then continue by considering Old World primate clades, including lemurs, lorises, langurs, baboons, guenons, mangabeys, macaques, gibbons, and orangutans.

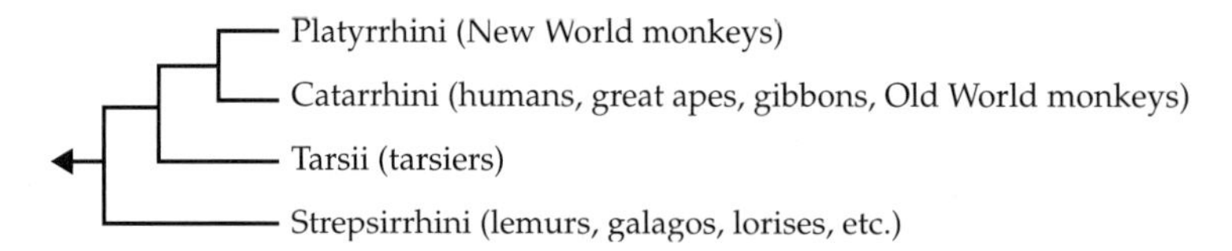

Figure 2.1 Phylogenetic relationships between New World and Old World primates (from Tree of Life Web Project, University of Arizona, http://tolweb.org/tree/phylogeny.html).

2.1.1 Introgressive hybridization: howler monkeys

The Neotropical, or Platyrrhine, monkeys belong to an assemblage marked by morphologically, behaviorally, and genetically diverse taxa (Figure 2.2). It may also be that the Platyrrhine complex reflects a lower frequency of natural hybridization than do their Old World counterparts (Cortés-Ortiz *et al.* 2007). Yet, numerous instances consistent with hybridization and introgression have been identified for Neotropical species (e.g., see Arnold and Meyer 2006 for a review). In the present section (and the following two sections), the role played by introgressive hybridization in the evolutionary history of Platyrrhines will be illustrated from studies of genetic and morphological diversity among howler monkeys, spider monkeys, marmosets, and tamarins. These four species complexes demonstrate the characteristic mosaicism (both genetic and morphological) common to all instances of genetic exchange through introgression. These primate taxonomic groups reflect the transfer of genetic elements and thereby the origin of novel, hybrid, evolutionary units (Arnold 2006).

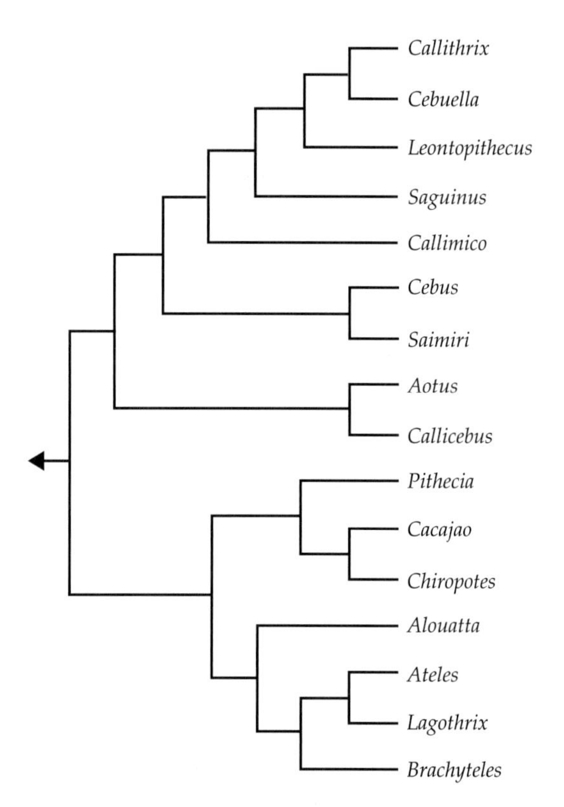

Figure 2.2 Phylogenetic relationships among the New World primate genera (from Tree of Life Web Project, University of Arizona, http://tolweb.org/tree/phylogeny.html).

The genus *Alouatta* (i.e., howler monkeys) has a geographic range throughout both Meso- and South America, with 10 and 19 recognized species and subspecies, respectively. An initial survey of mtDNA sequence variation by Cortés-Ortiz *et al.* (2003) resolved phylogenetic trees containing reciprocal monophyly for the species from each of the two major geographic subdivisions. The divergence time between these two clades, estimated from the molecular data, ranged from 6.6 to 6.8 million years ago (mya). However, as in other groups of Platyrrhine species—and primates in general—ancient and present-day areas of overlap, leading to introgression, have evidently occurred in the howler monkey clade.

The above conclusion is supported by phylogenies typified by individuals grouping not with members of their own species, but instead with members of other species (Figure 2.3; Cortés-Ortiz *et al.* 2003). Gene trees based on mitochondrial and nuclear sequences demonstrate the para- and

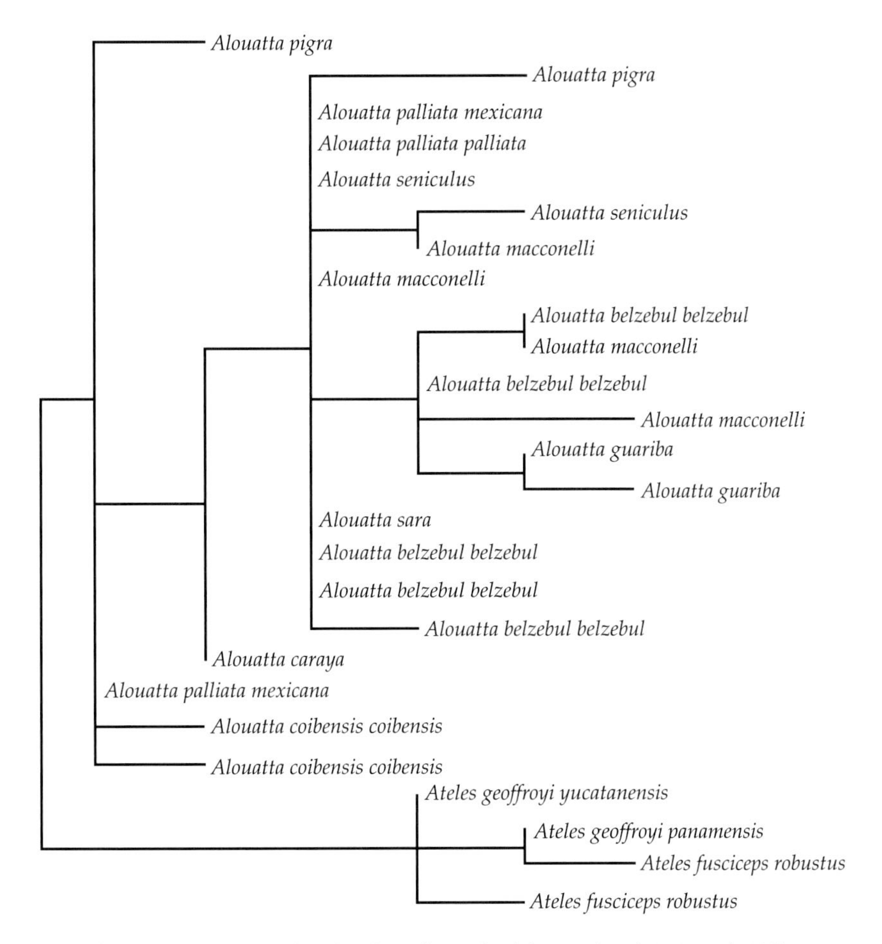

Figure 2.3 Phylogeny of Howler monkey species based on the nuclear *Calmodulin* gene (Cortés-Ortiz *et al.* 2003).

polyphyly expected if past exchange led to the introgression of portions of the nuclear genome, or the entire mitochondrial genome, between various subspecies and species (Cortes-Ortiz *et al.* 2003). For example, Cortés-Ortiz *et al.* (2003), in a phylogeny constructed from mtDNA sequences, detected a paraphyletic association involving the geographically widespread *Alouatta palliata* and the much more restricted *Alouatta coibensis.* In the case of the results shown in Figure 2.3 (Cortés-Ortiz *et al.* 2003), a lack of sufficient sequence variation to resolve species placement might also explain the nonmonophyletic relationships. However, the inference of a role for introgression in the admixing of the *Alouatta* lineages is supported by the detection of present-day howler monkey hybrid zones between *A. palliata* and *A. pigra* (Cortés-Ortiz *et al.* 2003, 2007). Though an alternate explanation for the variation present in these regions is parapatric divergence into two new forms (i.e., *A. palliata* and *A. pigra*), Cortés-Ortiz *et al.* (2003) argued that the formation of mixed troops containing putative hybrid individuals was consistent with introgression. They followed up on this observation by examining genetic variation at mtDNA, microsatellite, and Y-chromosome loci (Cortés-Ortiz *et al.* 2007). This latter analysis inferred a hybrid status for 13 of 36 individuals. Cortés-Ortiz *et al.* (2007) detected a pattern of variation suggesting that: (1) there is a lack of hybrid production between female *palliata* and male *pigra*; (2) there are infertile or inviable male hybrids produced from the reciprocal cross; and (3) that hybrid females carrying the mtDNA haplotype of *A. pigra* and hybrid males carrying the Y-chromosome marker from this species predominate in producing further hybrid generations.

Aguiar *et al.* (2008) have reported morphological variation in Brazilian troops of *Alouatta* that suggest the occurrence of introgression between South American howler monkey species as well. First, Aguiar *et al.* (2007) examined morphological traits of individuals from eight separate troops. Five and two of the troops were characterized by morphological traits of *Alouatta caraya* and *Alouatta clamitans*, respectively. The eighth group of howler monkeys contained two adult males and two adult females possessing *A. caraya* morphological traits, but also two adult females and a subadult male that were typified by a combination of the pelage patterns of the two species (Aguiar *et al.* 2007). These findings were consistent with previously reported morphological variation in specimens collected in the 1940s from the same region of Brazil (Aguiar *et al.* 2007). In a second analysis, Aguiar *et al.* (2008) recorded even more extensive evidence for introgressive hybridization between these two species. Specifically, of 11 groups examined, only 4 did not contain morphological hybrids (Aguiar *et al.* 2008). The "rediscovery" (Aguiar *et al.* 2007) of hybrids between *A. caraya* and *A. clamitans* thus reflects a significant impact from ongoing introgression between these South American howler monkey species.

2.1.2 Introgressive hybridization: spider monkeys

Spider monkeys, genus *Ateles*, belong to a clade that also includes the howler (*Alouatta*), woolly (*Lagothrix*), and muriqui (*Brachyteles*) lineages (Meireles *et al.* 1999; Collins and Dubach 2000; Celeira de Lima *et al.* 2007). Furthermore, both phylogenetic and population-level analyses have repeatedly detected patterns of variation indicative of past and ongoing introgressive hybridization between spider monkey lineages. For example, Rossan and Baerg (1977) reported the production of an F_1 hybrid between a captive pair of *Ateles geoffroyi panamensis* and *Ateles fusciceps robustus*. The significance of this finding was that the morphological admixture reflected in an artificial hybrid offspring was also detected in individuals collected from an area of sympathy between these taxa in Panama (Rosin and Berg 1977).

The detection of contemporary hybrid zones between spider monkey taxa also helps to account for the frequent phylogenetic and population genetic discordances defined by numerous workers. This result was indicated by Collins and Dubach (2000) when they stated "The phylogenetic relationships of similar haplotypes do not match their geographic distribution…significant gene flow must have occurred at some time in the past to link such geographically separate haplotypes." In a subsequent study that combined both mtDNA and nuclear DNA (i.e., from the gene aldolase) data, Collins and Dubach (2001) found general agreement between phylogenies derived from the alternate DNA data sets. However, they also detected phylogenetic discordances consistent with past and ongoing introgression. These discordances were reflected by (1) lineages in the mtDNA and nuclear phylogenies showing divergent sister group relationships and (2) unresolved relationships for some taxa that were resolved by sequence variation in the alternate data set. Similarly, Nieves *et al.* (2005), using both chromosome structure and mtDNA COII sequence variation, concluded that introgressive hybridization had contributed substantially to the evolutionary trajectory of spider monkey taxa. In particular, they argued for the effect of introgression leading to the fixation of chromosomal rearrangements that act as **postzygotic reproductive barriers** (Nieves *et al.* 2005). As with all examples of introgressive hybridization, these events would have the potential to produce adaptive changes as well (Arnold 1997, 2006).

In phylogenetic analysis of the woolly spider and muriqui lineages, Collins (2004) concluded that evolutionary relationships among these taxa were best represented as an unresolved trichotomy. This conclusion came from Collins' (2004) observation that three loci used to define phylogenetic relationships among these taxa resulted in conflicting trees (Figure 2.4; Collins 2004). Specifically, members of the three genera were alternately inferred to be sister taxa depending on whether the trees were based on mtDNA COII, mtDNA D-loop, or nuclear aldolase sequences (Figure 2.4). Celeira de Lima *et al.* (2007) disagreed with Collins' conclusion that a trichotomy best reflected the evolutionary history of taxa that included spider monkeys. Yet, their data also reflected discordance among the phylogenies based on different genomic regions; only four of the eight data sets supported the same phylogenetic arrangement (Celeira de Lima *et al.* 2007).

Axel Meyer and I (Arnold and Meyer 2006) argued that the above patterns (i.e., discordance in population genetic and phylogenetic analyses) are stereotypical for instances of genetic exchange. In particular, we related the discordance found in phylogenetic reconstructions using different portions of the genome to the concept of a "semi-permeable boundary" between the hybridizing forms (Key, 1968; Harrison, 1986). We concluded that "As Key (1968) argued when applying the term 'semi-permeable' to hybridizing taxa, different portions of the genome are expected to introgress at varying rates due to the action of selection, drift, etc., resulting in mosaic hybrid genomes constituted with markers from both taxa" (Arnold and Meyer 2006). If phylogenetic analyses are based on different regions of the

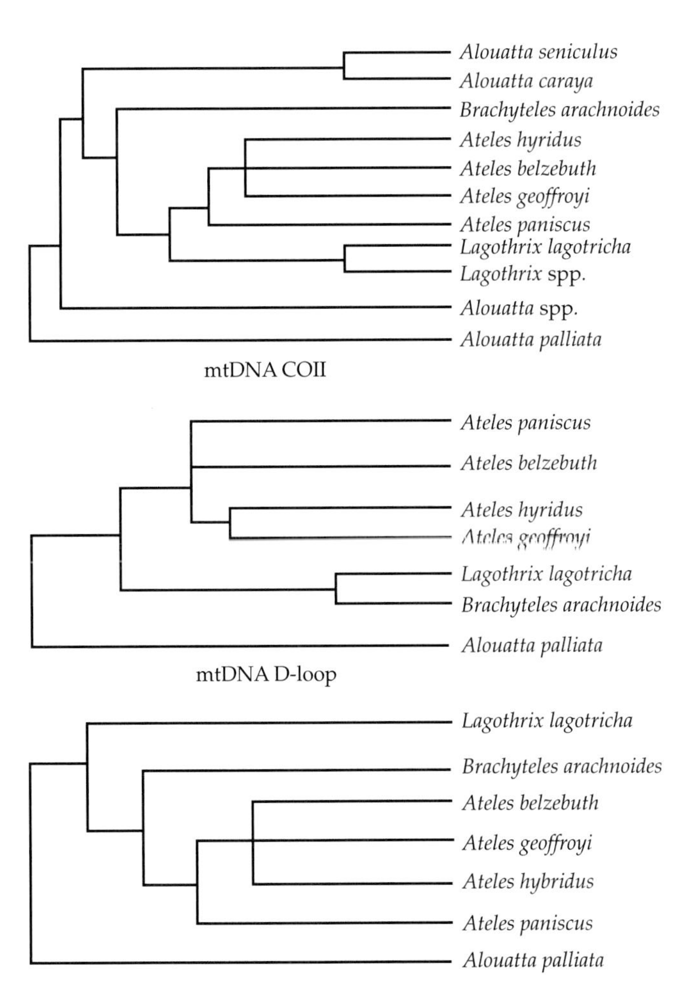

Figure 2.4 Phylogenetic trees derived from three different genomic sequences for various taxa belonging to the subfamily Atelinae. Included are species of howler (*Alouatta*), woolly (*Lagothrix*), muriqui (*Brachyteles*), and spider (*Ateles*) monkeys. Note the alternate placement of the same taxa depending on the sequences (i.e., the two mtDNA and one nuclear regions) used for phylogenetic reconstruction (Collins 2004).

genomes of related taxa, it is likely that loci would be chosen that had been differentially impacted by introgression and thus they would resolve discordant phylogenetic and population genetic patterns. This would seem to be the case for the results reported for the spider monkeys and their sister taxa.

2.1.3 Introgressive hybridization: marmosets and tamarins

The final two examples illustrating reticulate evolution in New World primates involve the related networks of marmosets and tamarins. Data for both of these assemblages reveal signatures of variation leading to an inference of past and present introgressive hybridization. Many of the species examined thus demonstrate a mosaic of characters that apparently derived from multiple lineages. For example, Tagliaro *et al.* (1997) stated the following concerning a series of marmoset taxa: "Morphological species that did not form monophyletic groups included *Callithrix mauesi*, *C. penicillata*, and *C. kuhli*" These same authors also spoke of the "confused picture regarding the *C. penicillata*, *C. kuhli*, *C. jacchus* group" (Tagliaro *et al.* 1997). These observations reflect the high degree of uncertainty relating to the phylogenetic placement of many of the marmoset species belonging to the South American genus *Callithrix*. Phylogenetic admixture is indicated by the placement of different individuals belonging to the same morphological species into more than one clade (Tagliaro *et al.* 1997).

As with the examples given in Chapter 1, and the other New World primate taxa discussed above, a likely explanation for the phylogenetic placement of members of a single species into different, well-supported clades is introgression leading to a reticulate, phylogenetic signal. For the marmosets, this inference is supported by the observation of parapatric distributions and hybrid zones between various combinations of species (Tagliaro *et al.* 1997; Marroig *et al.* 2004). In addition to the hypothesis of introgressive hybridization among various marmoset species, it has also been hypothesized that one of the taxa, *C. kuhli*, is a hybrid derivative from crosses between *C. geoffroyi* and *C. penicillata*. Consistent with this hypothesis is the finding that mtDNA variation in *C. kuhli* placed individuals into two clades separated by haplotypes found in *C. penicillata* and *C. jacchus* (Tagliaro *et al.* 1997). This paraphyletic pattern supports a hypothesis of hybrid origin for these *C. kuhli* individuals. However, it also suggests that the more likely progenitors of *C. kuhli* were *C. penicillata* and *C. jacchus*, rather than *C. penicillata* and *C. geoffroyi*.

Tamarins (genus *Saguinus*) are phylogenetically closely allied to the clade containing the marmosets. Indeed, various taxonomic treatments (see Cropp *et al.* 1999 for a discussion) have placed members of this genus and those of marmosets within a common subfamily or family. As with marmosets, data for the tamarin clade provide additional support for the effect of reticulation on the evolution of New World primates. Specifically, mtDNA-based phylogenetic reconstructions found discordant patterns for several species. Cropp *et al.* (1999) resolved a phylogenetic arrangement that separated *Saguinus fuscicollis fuscus* away from other *fuscicollis* subspecies and placed it in a clade with *Saguinus nigricollis*. Similarly, *Saguinus fuscicollis lagonotus* was separated from other members of its own taxonomic group and placed into a clade with *Saguinus tripartitus*. It is important to note that Cropp *et al.* (1999) did not infer natural hybridization as the cause of discordance between the morphological and mtDNA data. However, if reticulate evolution within tamarins has occurred—as has been demonstrated for the closely related marmoset clade—the discordance between the morphological and mtDNA data would be expected.

2.2 Reticulate evolution in Old World nonhominines

Old World primate groups include the Prosimians, Tarsiers, and Catarrhines. In the following sections, I will consider examples from the Prosimians (in particular, the lemurs and lorises) and the nonhominine Catarrhines. Evidence of reticulate evolution in the hominine, Catarrhine taxa—including our own species—will be discussed in Chapters 3 and 4).

2.2.1 Introgressive hybridization in Lemuriformes

The most basal primate group, the Prosimians (i.e., Strepsirrhini; Figure 2.1), like the phylogenetically more derived assemblages, also possesses genetic and morphological signatures of reticulate relationships. The diverse array of lemurs ("all Prosimian primates endemic to the island of Madagascar and surrounding Comoro islands," Horvath and Willard 2007), appear no different in this respect. A role for reticulation in the evolution of this complex is consistent with the observation by Horvath and Willard (2007) that phylogenetic topologies, though generally well defined at the genus level, were disputed for specific and subspecific relationships (e.g., Pastorini *et al.* 2003; Roos *et al.* 2004; Yoder and Yang 2004). Phylogenetic analyses for lemurs detect footprints of weblike rather than treelike processes in primates (Arnold 2006).

"The discovery of hybrid zones between lemur populations should not be surprising given the high species richness and close proximity of many related taxa" (Wyner *et al.* 2002). This conclusion reflected not only the extensive taxonomic diversity in the Prosimian clade, but also the portion of the genetic diversity detected in this clade resulting from the process of introgressive hybridization. Wyner *et al.* (2002) were specifically referring to the introgression-affected genetic structure of lemurs, and in particular findings from their analysis of a hybrid zone between the three species, *Eulemur fulvus rufus, Eulemur albocollaris,* and *Eulemur collaris*. Members of *Eulemur* are recognized as having the capacity to form intersubspecific, as well as interspecific, hybrids even when their chromosome numbers are widely divergent (Horvath and Willard 2007). Wyner *et al.*'s (2002) survey of mtDNA and nuclear intron loci detected genotypes indicating interspecific admixtures; of the 21 individuals examined from the area of sympatry, 18 were designated as genotypic hybrids, and the remaining 3 possessed hybrid-like morphologies (Wyner *et al.* 2002). The nonconcordance detected between the phenotypic (i.e., hybrid-like) and genotypic (i.e., *E. albocollaris*) character sets present in the latter three individuals is consistent with the transfer of genes underlying the morphological characteristics of *E. fulvus* into *E. albocollaris*, in the absence of mtDNA or nuclear intron introgression. A study of intersex body mass and canine size dimorphism, in concert with estimates of testes volume, carried out for *E. f. rufus*, *E. albocollaris*, and presumptive hybrid individuals also detected variation reflective of mosaics of morphological characteristics (Johnson *et al.* 2005). Support for the conclusion that the individuals of *Eulemur* examined by Wyner *et al.* (2002) were the products of reticulation comes also from an earlier analysis by Rabarivola *et al.* (1991). In this earlier study, *Eulemur macaco macaco* and *Eulemur macaco flavifrons* were the taxa examined. Rabarivola *et al.* (1991) collected only morphological data, yet they were able to use diagnostic morphological characters to identify an extensive hybrid zone between these two lineages (Rabarivola *et al.* 1991).

Disagreements between patterns of morphological and molecular variation, or between molecular data sets for different loci, are expected for advanced generation hybrid genotypes in which recombination and selection have changed associations between loci (Arnold 1997; Burke and Arnold 2001). Such genomic discontinuities, resulting in discordance between phylogenetic placement and morphological/taxonomic assignments for members of *Eulemur*, were also detected in a survey of mtDNA variability for most members of the lemur clade. In particular, Pastorini *et al.* (2003) found nonmonophyletic distributions involving numerous members of the *E. fulvus* complex (Figure 2.5). Similarly, these authors reported paraphyletic associations among members of the genus *Hapalemur*, specifically involving *Hapalemur griseus* subspecies.

Interestingly, the last example of probable reticulate evolution among lemur lineages comes from a study in which the authors concluded that incomplete lineage sorting had been the major source for the detected phylogenetic discordances. Heckman *et al.* (2007) stated:

> Due to the effects of incomplete lineage sorting, systematic methods can become inappropriate and uninformative at the boundaries of intra- and inter-specific divergence...At this level, phylogenies are inadequate in discerning relationships

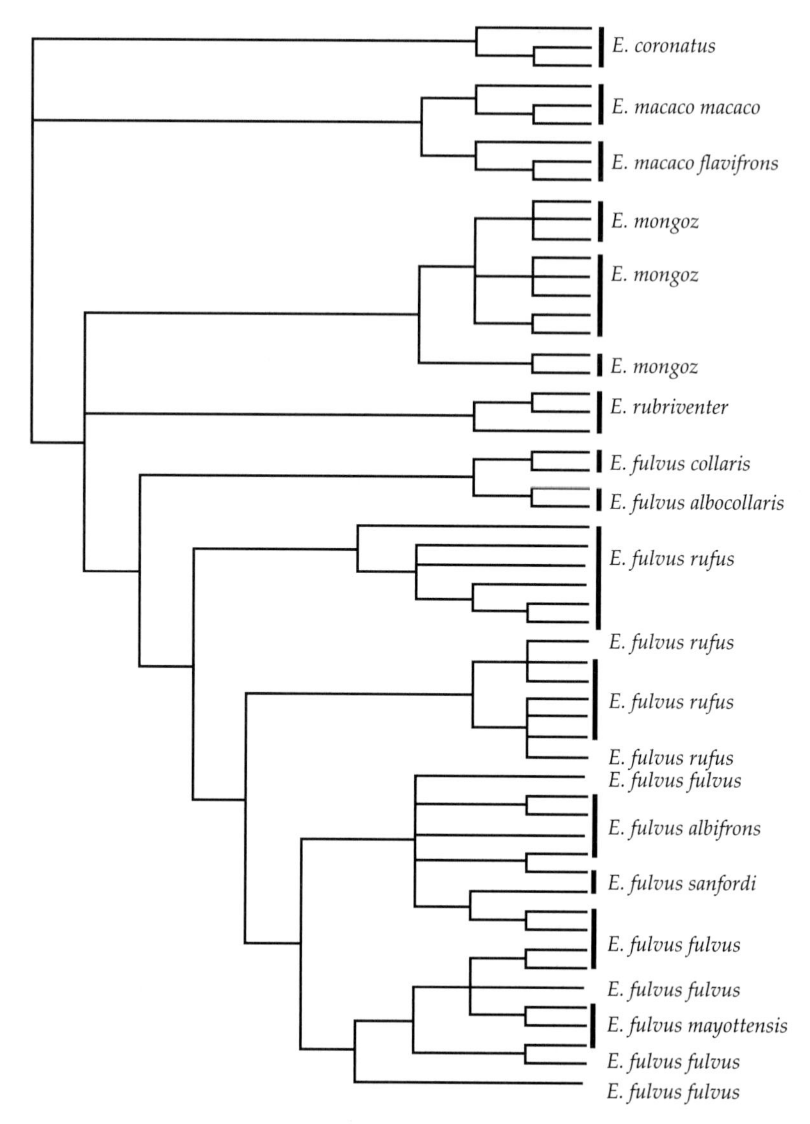

Figure 2.5 Phylogenetic inferences for relationships within the lemur genus, *Eulemur.* Note the paraphyletic placement of taxa belonging to the *Eulemur fulvus* complex (Pastorini *et al.* 2003).

between taxonomic groups. When lineage sorting is incomplete, a dichotomous branching pattern is no longer applicable... as insufficient numbers of lineages have sorted and ancestral polymorphisms are retained. Nuclear DNA has a relatively larger effective population size than mtDNA; therefore, the point in the phylogeny when a single lineage has reached fixation at a nuclear locus will theoretically be at a higher taxonomic level (e.g., at or above the species level) than for mitochondrial loci. In this circumstance, depicting the pattern of evolution as reticulated, or net-like, may better reflect the relationships among individual alleles, as alternative interconnections between point mutations may be necessary.

The significance of this quote, and why I have chosen to include it nearly in its entirety, is that it reflects precisely the predictions of reticulations due to genetic exchange. Also, the observation of frequent introgressive hybridization in the lemur clade makes this inference even less likely. Specifically, the patterns detected by Heckman *et al.* (2007)—that is, discordance between the evolutionary relationships defined by phylogenies derived from DNA sequence information versus the taxonomy of the lineages analyzed—reflect those seen in studies of lemurs undergoing contemporaneous (and likely ancient) introgressive hybridization. Though these authors are correct in suggesting a possible role for incomplete lineage sorting in affecting the phylogenetic patterns, the evolutionary biology of lemurs, *sensu lato*, argues against this being the sole, or the most significant, factor producing the weblike relationships.

2.2.2 Introgressive hybridization in slow lorises

"Slow lorises (*Nycticebus* spp.) are small and sluggish primates designated as endangered according to IUCN." This quote by Chen *et al.* (2006) reflects one of the endearing qualities of this Strepsirrhine assemblage, and its threatened status. Unfortunately for conservation applications, information concerning the systematics and evolution of this genus is extremely limited. Owing to a lack of strong morphological differentiation, the number of species and subspecies recognized for this group has varied. Most recently, it has been suggested that there are four species, *Nycticebus coucang*, *Nycticebus pygmaeus*, *Nycticebus bengalensis*, and *Nycticebus javanicus* (sometimes considered a subspecies of *N. coucang*). Furthermore, *N. coucang* has been divided into two subspecies—*N. coucang coucang* and *N. coucang menagensis* (see Chen *et al.* 2006 for a discussion and references concerning systematic treatments).

The systematic relationships of slow lorises, based on the limited morphological variation, have recently been tested through the use of mtDNA sequence information. Chen *et al.* (2006) collected both D-loop and cytochrome b sequences for 22 individuals that represented each of the *Nycticebus* species and subspecies. This analysis resolved most of the recognized lineages. However, significant phylogenetic discordance was detected, with some individuals of *N. coucang coucang* and *N. bengalensis* sharing a closer evolutionary relationship with each other than to members of their own species. Introgressive hybridization between these two taxa in a region of sympatry in southern Thailand was suggested as the potential cause of this admixture (Chen *et al.* 2006). This hypothesis of reticulate evolution was supported by a recent comparison of mtDNA sequence variation found within *N. bengalensis* and the pygmy slow loris, *N. pygmaeus*. In this analysis, Pan *et al.* (2007) genotyped 119 *N. pygmaeus* and 21 *N. bengalensis* individuals. In spite of the much smaller sample size for the latter species, its nucleotide diversity was 1.38%, compared to only 0.19% for *N. pygmaeus*. Several factors were suggested to explain this paradoxical observation, one being introgression between *N. bengalensis* and *N. coucang coucang*. Specifically, Pan *et al.* (2007) concluded that (1) the observation of phylogenetic admixture by Chen *et al.* (2006) and (2) the detection of significantly higher nucleotide diversities in the smaller sample of Bengal slow lorises could be explained by introgression from *N. coucang coucang* into *N. bengalensis*.

2.2.3 Introgressive hybridization in langurs and leaf monkeys

Asian leaf monkeys, or langurs, along with their sister clade of African colobus monkeys are well defined as having a unique fermentation system. These species possess both a complex foregut, in which leaf material undergoes bacterial fermentation, along with a true stomach where high levels of bacteriolytic lysozyme are expressed (Messier and Stewart 1997). Notwithstanding their unique digestive system, the langurs, like other primates, appear to have been impacted by reticulate processes. In this regard, Rosenblum *et al.* (1997) made the following observations: (1) "Within the Javan *Trachypithecus auratus*, our analysis does not support the distinction of two subspecies"; and (2) "*T. auratus* and *T. cristatus* are not internally monophyletic with respect to each other." These conclusions came from a phylogenetic analysis of restriction fragment length polymorphisms in the mitochondrial *NADH* gene. Rosenblum *et al.* (1997) discovered phylogenetic admixture of the two *Trachypithecus* species.

In particular, the mtDNA haplotypes detected in this study clustered into groups not reflective of morphological similarity. Rosenblum *et al.* (1997) used the paraphyly observed for the mitochondrial haplotypes from *T. auratus* and *T. cristatus* to suggest that, for conservation efforts, this species pair should be considered only as "one large polymorphic, conservation unit."

Conservation issues notwithstanding, the nonconcordant results from the phenotypic and molecular data for the langur species can be interpreted as resulting from introgression of only a portion of the genetic markers. Specifically, introgression between these species has likely resulted in the transfer of the mitochondrial genomes, but not the genes underlying morphological traits. In fact, Rosenblum *et al.* (1997), though not arguing in favor of reticulation, suggested a behavioral pattern that would explain a high level of [maternally inherited] mtDNA introgression. They observed that, unlike many mammals and particularly primates, leaf monkey females sometimes transfer between groups. If this involved interspecific movement of females, introgressive hybridization of the mtDNA would result.

Consistent with the occurrence of introgressive hybridization among langurs, Ting *et al.* (2008) detected incongruent phylogenetic relationships among the various genera of leaf monkeys/langurs based on morphology and X-chromosomal sequences on the one hand and mtDNA sequences on the other. Specifically, these authors reported the alternative sister-clade relationships of *Semnopithecus/Trachypithecus* (morphological and X-chromosomal data) and *Presbytis/Trachypithecus* (mtDNA data). Though both incomplete lineage sorting and introgressive hybridization were discussed as explanations for the incongruence, Ting *et al.* (2008) reflected that genetic exchange "may have allowed for the exchange of mitochondrial or X-chromosomal alleles between langur and leaf monkey taxa, thus resulting in gene tree incongruence."

2.2.4 Introgressive hybridization and hybrid speciation in cercopithecines

The subfamily Cercopithecinae (within the family Cercopithecidae; Figure 2.6)—including macaques, mangabeys, mandrills, baboons, vervets, and guenons—provides a diverse array of examples of the evolutionary outcomes possible from reticulate processes. I will first consider this diversity from comparisons between the genera belonging to this assemblage. I will then examine data for taxa belonging to the genera *Papio* and *Macaca*.

Reflecting on the variety of evolutionary effects from reticulate processes within the Cercopithecinae, Detwiler *et al.* (2005) identified two categories of natural hybridization. The first was termed "parapatric hybridization" and defined as "natural hybridization at interfaces between differentiated, parapatric populations assigned to the same, or very closely related, species" (Detwiler *et al.* 2005). One name applied to such closely related, primate lineages that are morphologically distinct, but not completely reproductively isolated is "allotaxa" (Jolly 2001). In the case of parapatric hybridization between allotaxa, the hybrid zones formed were discovered in ecotones or river headwaters. Figure 2.7 illustrates the parapatric distribution of one group of cercopithecines, the baboon allotaxa. As discussed below, this pattern of geographical association between the lineages belonging to the genus *Papio* (i.e., baboons) has resulted in frequent introgressive hybridization (Jolly 2001).

The second class of hybridization identified between members of the Cercopithecinae has

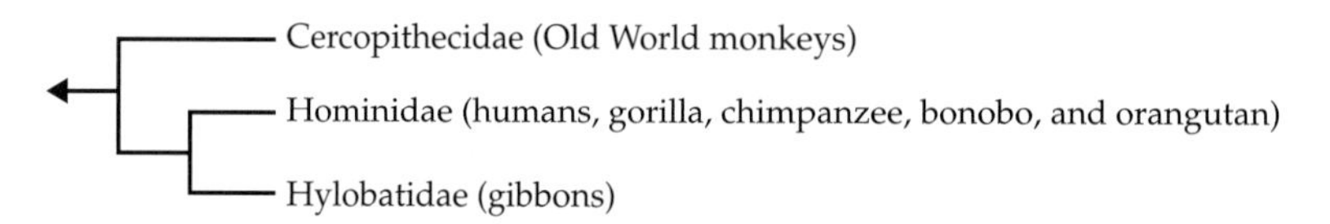

Figure 2.6 Phylogenetic relationships of Old World monkeys, hominids, and hylobatids (from Tree of Life Web Project, University of Arizona, http://tolweb.org/tree/phylogeny.html).

been assigned the term "sympatric hybridization" (Detwiler *et al.* 2005). This type of hybridization, as the name implies, involves species with broadly overlapping geographic ranges. Furthermore, this class of genetic exchange typically occurs between ecologically distinct taxa. Its frequency has been reported as lower than the instances of parapatric hybridization, and "In monkeys it is usually sporadic and ephemeral, and often occurs under unusual environmental circumstances" (Detwiler *et al.* 2005). These authors suggested that one category of unique environmental condition that might facilitate sympatric hybridization occurred when natural or human-mediated barriers restricted dispersal of monkeys and thus access to conspecific mates. Indeed, this latter situation was viewed as a potential catalyst for the loss of a rare form through the genetic assimilation by a more abundant or prolific taxon. However, Detwiler *et al.* (2005) also concluded that rare, sympatric introgression might have resulted in the production of novel lineages and/or adaptations. Support for the origin of evolutionary novelty through hybridization between cercopithecine lineages was garnered from an analysis of experimental hybrids between the olive (*Papio anubis*) and yellow (*Papio cynocephalus*) baboons. Specifically, Ackermann *et al.* (2006) detected heterosis and novel morphologies in F_1 (i.e., olive × yellow) and B_1 (i.e., olive × F_1) progeny. When compared to museum specimens from known, natural hybrid zones, similar levels and types of variation were observed (Ackermann *et al.* 2006). These results not only indicated a method for detecting introgression in fossil specimens (Ackermann *et al.* 2006), but also supported the hypothesis that reticulate processes can lead

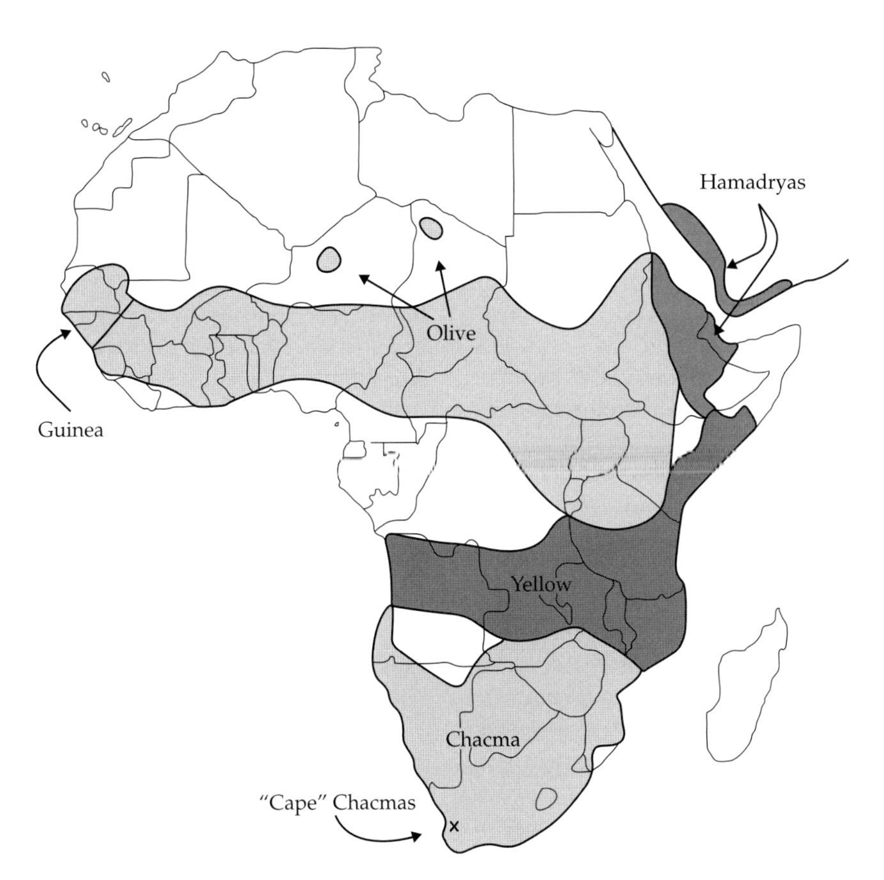

Figure 2.7 Distribution of five *Papio* (i.e., baboon) allotaxa. The parapatric distribution of these forms has yielded numerous instances of hybrid zones and introgressive hybridization (Newman *et al.* 2004).

to the derivation of new evolutionary trajectories (Arnold 1997, 2006).

Whether the outcomes of introgression between divergent forms fall more within the category of evolutionary novelty or conservation concern (Detwiler *et al.* 2005), it is evident that reticulation is a common process in the subfamily Cercopithecinae. For example, hybrid zone formation, introgression, and the founding of hybrid taxa have been demonstrated for the baboon and macaque clades. Furthermore, the tribe Papionini, which also includes the geladas (*Theropithecus*), drills and mandrills (*Mandrillus*), and mangabeys (*Cercocebus, Lophocebus*), reflects population genetic and phylogenetic signatures consistent with reticulate evolution (Shotake *et al.* 1977; Shotake 1981; Jolly *et al.* 1997; Harris and Disotell 1998; Szmulewicz *et al.* 1999; Alberts and Altmann 2001; Dirks *et al.* 2002; Newman *et al.* 2004). One piece of evidence that supports this conclusion comes from tests for recombination in mtDNA. Some of the strongest signals of recombination in animal mtDNA have come from baboon, macaque, and mandrill taxa (Piganeau *et al.* 2004). The recombinant molecules involved sequences so divergent that Piganeau *et al.* (2004) suggested their origin was from hybridization between different subspecies or species.

Phylogenetic inferences for the tribe Papionini, and the identification and analyses of contemporary hybrid zones, are consistent with the widespread effect from reticulate evolution. In order to provide a phylogenetic structure for the papionin genera, Harris and Disotell (1998) defined evolutionary relationships using five, unlinked, nuclear loci. Consistent with past reticulations, these authors detected a paraphyletic placement of species from the two mangabey genera, *Cercocebus* and *Lophocebus*, into clades containing other members of the papionin tribe. Similarly, a phylogeny based on mtDNA sequences had previously revealed widespread paraphyly for the two mangabey genera (Disotell, 1994). Also, the phylogenies based on (1) the mtDNA sequences and (2) some of the nuclear gene loci revealed discordances (Harris and Disotell, 1998) in the placement of *Papio* (baboon) and *Theropithecus* (gelada); these two genera resolved as sister lineages in some trees, but not in others.

To explain all of the above phylogenetic discordances among the papionin genera, Harris and Disotell (1998) invoked either incomplete lineage sorting or reticulation. They also observed that (1) "It is significant that hybridization has been reported in wild and captive populations of *Papio* and *Theropithecus*" and (2) "Hybridization between papionin species with subsequent introgression... may lead genera to falsely appear as sister taxa." These data all point to the conclusion that the evolutionary trajectory of papionin taxa involved repeated bouts of introgressive hybridization. This introgression apparently affected even the progenitors of what are now recognized as different genera.

Papio

As described above, phylogenetic and phylogeographic analyses have demonstrated the effect of introgression on the various papionin genera. Well-defined cases of reticulation have also come from studies within the genera *Papio* and *Macaca*. For example, Wildman *et al.* (2004) and Winney *et al.* (2004) carried out phylogenetic and phylogeographic analyses, respectively, for species belonging to *Papio*. The phylogenetic assessment by Wildman *et al.* (2004) detected paraphyly involving samples of *P. anubis* and *P. hamadryas*. Specifically, some of the *P. anubis* and *P. hamadryas* populations from Ethiopia formed a clade distinct from samples of their own species from Ethiopia, Arabia, or Kenya/Tanzania. This finding led to an inference of "an extensive and complex history of sex-specific introgression" (i.e., with the bias involving introgression of the maternally inherited mtDNA) between *P. anubis* and *P. hamadryas* (Wildman *et al.* 2004). Such an inference is consistent with the biological attributes of *P. hamadryas*; this species demonstrates the rare, life history trait in which females (rather than males) disperse (Hapke *et al.* 2001).

A similar conclusion to that drawn by Wildman *et al.* (2004) also resulted from the phylogeographic analyses of Winney *et al.* (2004). In their analysis, mtDNA sequence information was collected from Saudi Arabian and Eritrean baboons. Not only did these data elucidate the probable dispersal history of *P. hamadryas* between Arabia and Africa, but also, like an earlier study of only Eritrean baboon populations (Hapke *et al.* 2001), detected a paraphyletic arrangement of *P. hamadryas* and *P. anubis* (referred to as *Papio hamadryas anubis* by Winney *et al.* 2004). Once again, reticulate processes were inferred to have affected the evolutionary history of these two baboon taxa. Indeed, even the origin of *P. anubis* has been suggested to have involved reticulation, in this case hybrid speciation. Specifically, Jolly (2001) reviewed genetic and morphological evidence that supported numerous rounds of introgressive hybridization among the various *Papio* species. He further asserted that the phenotype of *P. anubis* reflected an admixture of traits from "northern" and "southern" lineages of papionins and thus indicated a reticulate origin for this taxon (Jolly 2001).

In addition to past genetic exchange events, reticulate evolution within the papionin tribe is also affecting the genetic structure of contemporary populations. This is seen clearly in two well-defined instances of ongoing introgressive hybridization involving three species of baboons. The best-documented case involves a zone of overlap and introgression in the Awash National Park, Ethiopia. This hybrid zone involves the olive (*P. anubis*) and hamadryas (*P. hamadryas*) baboons. The hybrid zone occurs where gallery forest (i.e., *P. anubis* habitat) along the Awash River extends into a region of semidesert (i.e., *P. hamadryas* habitat; Detwiler *et al.* 2005). Along the forest-to-semidesert gradient, there is clinal changeover from *P. anubis* individuals to recombinant phenotypes that contain an increasingly greater proportion of *P. hamadryas*-like characteristics (Detwiler *et al.* 2005). Species-specific morphology, blood proteins, Alu repeat elements and mitochondrial DNA markers indicate that hybridization in the Awash National Park between *P. hamadryas* and *P. anubis* has resulted in bidirectional, but asymmetric, introgression with gene flow proceeding largely from the former into the latter species (Shotake *et al.* 1977; Szmulewicz *et al.* 1999; Jolly, 2001).

The introgression detected between *P. anubis* and *P. hamadryas* has resulted in not only genetic but likewise behavioral novelty (Bergman *et al.* 2008). Bergman and Beehner (2004) detected unique behavioral characteristics within this hybrid zone, involving social organization, social structure, and mating system. The behavior of the two parental species were described in the following manner:

> anubis consists of large, cohesive multimale, multifemale groups with no permanent substructuring...while hamadryas live in a multilevel society characterized by tightly-bonded unimale, multifemale groups (one-male units or OMUs) that aggregate at multiple levels into clans, bands, and troops...Mating in anubis societies is promiscuous, with male competition for access to fertile females mediated by a strict dominance hierarchy...while hamadryas have a harem polygynous mating system...In anubis groups, the strongest bonds among group members are within matrilines of natal females...while the strongest bonds in hamadryas society are between a male and the females within the OMU.

In contrast, to the parental species, the hybrid individuals possessed unique combinations and expressions of the *P. anubis* and *P. hamadryas* behavioral/social characteristics (Bergman and Beehner 2004). Specifically, they demonstrated an overall pattern of multimale/female group organization that lacked the cohesiveness seen in the parental taxa. Furthermore, though OMUs were present and characterized by members being more *P. hamadryas*-like, the multilevel society structure characteristic of *P. hamadryas* was not detected. The observed mating system was biased toward the *P. hamadryas* mode, but with some small frequency of promiscuity as well (Bergman and Beehner 2004). Finally, estimates of intersexual bonding detected an intermediate expression between that seen in the two species (Bergman and Beehner 2004). These patterns support the hypothesis that recombination between the divergent genomes has shuffled the underlying genetic architecture of the two species thus leading to a hybrid population exhibiting novel behavioral attributes. Notwithstanding their unique behavioral attributes, hybrid individuals were found to have some of the highest estimated fitness values. Specifically, Bergman *et al.* (2008) documented—over a nine-year period—the high reproductive success of hybrid males in the Awash hybrid zone.

The second hybrid zone has been identified from the environs of Amboseli, Kenya. Similar to the Awash zone of hybridization, the Amboseli area of overlap includes *P. anubis*, but with a third species, *Papio cynocephalus* (i.e., yellow baboon). Though both the Ethiopian and Kenyan hybrid zones are characterized by numerous recombinant genotypes, the olive/*hamadryas* (i.e., Awash zone) phenotypic structure demonstrates a bimodality (i.e., hybrids are morphologically olive-like or *hamadryas*-like) whereas the olive/yellow hybrid zone (i.e., Kenyan) reflects a continuous morphological distribution from the yellow to olive phenotypes (Alberts and Altmann, 2001). Consistent with the continuous phenotypic distribution was the finding that mitochondrial haplotype variation did not separate phylogenetically into the expected *P. anubis* and *P. cynocephalus* clades. This once again reflected phylogenetic discordance caused by extensive introgression (Newman *et al.* 2004). The frequency of hybridization and introgression appears to be increasing in the Amboseli zone of overlap (Figure 2.8), most likely due to increased immigration of *P. anubis* and hybrid individuals fleeing from human-mediated habitat destruction (Alberts and Altmann, 2001).

Macaca

Similar to other cercopithecines, the genus *Macaca* (i.e., macaques) provides examples of numerous signatures of reticulate events, including discordant phylogenetic inferences and contemporary areas of geographic overlap containing individuals defined as morphological and genetic recombinants. As important as the estimate of the role of hybridization is for an understanding of the evolutionary history of *Macaca*, it is also important from the perspective of this groups use as a biomedical, model system. For example, the Rhesus macaque genome has now been sequenced and is being used in comparisons to other primate genomes to test biomedical hypotheses (Rhesus Macaque Genome Sequencing and Analysis Consortium 2007). Furthermore, macaques continue to be used in clinical trials with the hope of developing control and treatment regimes to ameliorate pandemics (e.g., influenza; Kobasa *et al.* 2007).

Smith *et al.* (2006) have identified potential problems with applying this model system to medical applications. In particular, they noted a high frequency of misidentification of the country of origin leading to an uncertainty concerning which evolutionary lineages of Rhesus macaques were

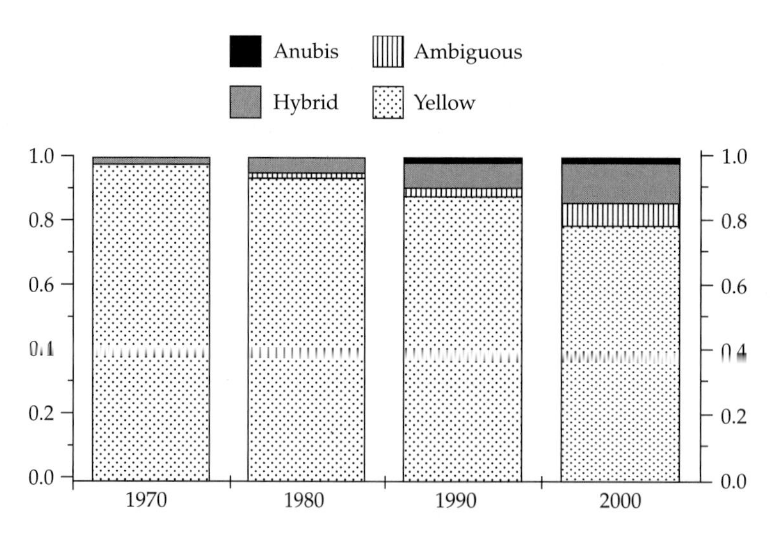

Figure 2.8 Graph illustrating the increase over time in the proportion of *P. anubis* ("Anubis") and hybrids in the Kenyan hybrid zone between *P. anubis* and *P. cynocephalus* ("Yellow"). Each bar reflects census information gathered on the year indicated on the X-axis (Alberts and Altmann 2001).

present in studies of human diseases. Because of this uncertainty, hybridization (in captivity) between the divergent lineages was likely (Smith *et al.* 2006). Given this, conclusions drawn from the effects on macaques by exposure to human diseases—as a predictor of human responses—might be compromised due to the use of admixed genomes. Smith *et al.* (2006) concluded that it would become of increasing importance "to confirm the country of origin and suspected mixed ancestry in domestically bred rhesus macaques to maintain the integrity of animal models for the study of human disease." Indeed, Tosi and Coke (2007) have demonstrated that *Macaca fascicularis* from the island of Mauritius, a common source of macaque biomedical research colonies, consisted of individuals which possessed the Y-chromosome lineage and mtDNA genetic markers from different macaque lineages. Tosi and Coke (2007) concluded that the Mauritian samples were products of introgressive hybridization. They also suggested that, given the hybrid derivation of these individuals, the genes controlling biomedically important traits might have originated from either, or both, of the divergent genomes present. In addition, the genes from the two parental lineages could easily have different effects on the disease pathology. This latter conclusion is supported by the observation that rhesus macaques from the Chinese lineage, after being exposed to the simian immunodeficiency virus, develop AIDS-like symptoms more slowly than the divergent rhesus macaque individuals from India (Ling 2002). Because hybrids often combine parental traits and/or demonstrate novelty in their phenotypic, physiological, and so forth characteristics (Arnold 1997, 2006), using hybrid macaques to understand primate responses to diseases such as HIV could be problematic.

In addition to their use as a primate model for biomedical research, macaques have also provided exemplars for understanding processes affecting evolution within their own genus as well as within other primate genera. For example, limits to hybridization within *Macaca*, either from **prezygotic** or postzygotic reproductive barriers have been estimated. Fujita *et al.* (1997) documented a partial, prezygotic reproductive barrier between Sulawesi macaques based on mate choice. These authors found that male individuals responded much more strongly to photos of animals from their own species than they did to individuals from other species. Females did not respond as strongly as did the males (Fujita *et al.* 1997). Though this assortative behavior would limit gene flow, it was not absolute and does not reflect a complete barrier to hybridization and introgression. Similarly, Gotoh *et al.* (2001) examined a potential contributor to hybrid fitness and thus postzygotic reproductive isolation. In this case, hematological (e.g., red blood cell counts) and parasite load estimates were collected from naturally occurring *Macaca hecki* and hybrids between *M. hecki* and *Macaca tonkeana*, from Sulawesi (Gotoh *et al.* 2001). If the hybrid individuals consistently possessed either lower hematological counts or higher parasite loads, this would suggest a lower fitness relative to the parental species and therefore reflect a component contributing to postzygotic reproductive isolation. However, no consistent pattern was found (Gotoh *et al.* 2001). This suggests an inference commonly arrived at for hybrids—they are not necessarily less fit than the parental lineages from which they are derived (Arnold and Hodges 1995; Arnold 1997).

The genus *Macaca* has also been utilized as a model system to infer processes affecting the evolutionary trajectory of other members of the primate assemblage. For example, a hybrid derivative between a Rhesus macaque (*Macaca mulatta*) and a baboon (*P. hamadryas*), known as a Rheboon, was used to decipher the cytogenetic differences between *Macaca* and *Papio* (Moore *et al.* 1999). Though this F_1 hybrid animal lacked sperm, cytological analyses detected no differences between the macaque and baboon chromosomes. On the basis of these results, Moore *et al.* (1999) proposed a single nomenclature for the chromosomes belonging to the two parents. These results further suggest that karyotypic evolution is not the basis of the postzygotic reproductive isolation between the baboon and macaque lineages. Instead, genetic divergence likely underlies the observed hybrid sterility.

A final example of how macaques have been used to inform other scientific investigations, again with regard to primate evolution, involves predictions concerning the degree of morphological divergence expected between reproductively isolated taxa. Specifically, Schillaci *et al.* (2005) used morphological data from *Macaca maurus, M. tonkeana,* and their natural hybrids to determine the amount of morphological divergence between these two species and their hybrids. They then compared their estimates of (1) growth allometry and (2) the shape and form of the craniofacial characteristics, during macaque development, to data for *H. sapiens* and *Homo neanderthalensis.* Specifically, Schillaci *et al.* (2005) tested the hypothesis that the degree of morphological distinctiveness of *H. sapiens* and *H. neanderthalensis* was indicative of reproductive isolation. Schillaci *et al.* (2005) concluded that the morphological divergence between *H. sapiens* and *H. neanderthalensis,* compared to that found for the introgressing macaque species, did not indicate that the *Homo* taxa would have been completely reproductively isolated.

As stated above, the evidence for a significant role of reticulate processes in the evolution of the genus *Macaca* reflects both ancient (mainly detected by phylogenetic analyses) and recent (mainly detected by analyses of contemporary hybrid zones) genetic exchange. In terms of phylogenetic analyses, studies examining the entire genus as well as those focusing on specific assemblages within *Macaca,* have been reported. Morales and Melnick (1998) utilized mtDNA sequence information to infer phylogenetic relationships among the 19 recognized *Macaca* species. In addition to conclusions concerning the direction of evolution within the genus (e.g., that taxa possessing more specialized genitalia evolved from less derived taxa), Morales and Melnick (1998) defined patterns consistent with genetic exchange between lineages. In particular, they noted the paraphyletic arrangement of members of the *M. fascicularis* species group (i.e., *M. fascicularis, M. mulatta, Macaca cyclopis,* and *Macaca fuscata*). Their phylogenetic assignments detected the paraphyletic separation of samples of the following *Macaca* species as well: *assamensis, nemestrina, maura, hecki, nigrescens, nigra, tonkeana,* and *maura* (Morales and Melnick 1998).

The hypothesis that many of the paraphyletic distributions detected by Morales and Melnick (1998) resulted from introgressive hybridization has been supported by additional phylogenetic and population genetic analyses. With regard to phylogenetic surveys, Evans *et al.* (1999, 2003), Tosi *et al.* (2000, 2002, 2003), Smith and McDonough (2005), Chu *et al.* (2007), Tosi and Coke (2007), and Ziegler *et al.* (2007) examined relationships within and among various macaque species. By using male-specific (Y-chromosome), female-specific (mtDNA), and biparental (autosomal) genetic markers these authors were able to define cases of introgression—both ancient and contemporary. Specifically, the studies revealed the following indicators of reticulate evolution:

1. mtDNA sequences of *M. nemestrina* demonstrated paraphyletic associations with other species of *Macaca* from Sulawesi (Evans *et al.* 1999).
2. Monophyletic or paraphyletic associations of macaque species were resolved when either Y-chromosome or mtDNA sequences, respectively, were utilized (Tosi *et al.* 2000)—indicative of Y-chromosome introgression between *M. fascicularis* and *M. mulatta* (Tosi *et al.* 2002).
3. mtDNA paraphyly for *M. nemestrina, M. tonkeana,* and *M. hecki* (Evans *et al.* 2003; Ziegler *et al.* 2007).
4. Paraphyly in phylogenies based on Y-chromosome, mtDNA, and biparental markers (Tosi *et al.* 2003).
5. Admixtures of divergent mtDNA lineages in southern Chinese and Vietnamese populations of *M. mulatta* (Chu *et al.* 2007).
6. Phylogenetic clustering of some samples of *M. mulatta* from India in a clade with Burmese samples of this species, separate from other Indian individuals (Smith and McDonough 2005).
7. Y-chromosome introgression of "continental" lineages into Sumatran *M. fascicularis* (Tosi and Coke 2007).

One of the conclusions arrived at in several of these studies was that the higher frequency of either male-specific or biparental marker introgression, relative to the female-specific mtDNA markers, was likely caused by the female philopatry (i.e., lack of female dispersal from their natal sites), coupled with obligate male dispersal (Tosi et al. 2000). Dispersing males of one lineage would thus carry both male-specific (i.e., Y-chromosome) and biparental (i.e., located on autosomal linkage groups) loci into new regions leading to the genetic transfer of these alleles, while the maternal-specific (i.e., mtDNA) markers would remain behind with the nondispersing female macaques.

In addition to detecting past reticulations, several phylogenetic surveys can be used to infer cases of hybrid speciation. The allocation of some reticulate events into the category of "hybrid speciation" while assigning others (like those discussed above) to the category of introgressive hybridization can be somewhat arbitrary and misleading. I have argued previously that the most significant question is not whether a newly formed lineage is deemed to be deserving of a taxonomic label, but instead whether it reflects evolutionary diversification (Arnold 2006, p. 124). However, for the present discussion I will highlight two examples that have been placed into the category of ancient reticulations resulting in new taxa—that is, hybrid speciation or hybrid lineage formation. These two examples come on the one hand from studies by Tosi *et al.* (2000, 2003) and on the other by Chakraborty *et al.* (2007). From their analyses of paternal, maternal, and biparental molecular markers, Tosi *et al.* (2000, 2003) concluded that *Macaca arctoides* had derived from ancient hybridization between individuals belonging to what they referred to as a proto-*M. assamensis/thibetana* lineage and a proto-*M. fascicularis* lineage. Tosi *et al.* (2003) also concluded that instances of reticulate evolution, such as that resulting in the origin of *M. arctoides*, would likely remain undetected if only a single molecular marker system were utilized.

Similar to the analyses of Tosi *et al.*, that of Chakraborty *et al.* (2007) also reflected the analysis of multiple loci. Chakraborty *et al.* (2007) were interested in defining the evolutionary relationships and history of the newly described species, *Macaca munzala*. This species, described in 2005, was discovered in northeast India in the state of Arunachal Pradesh, and was assigned to the *M. sinica* species complex (Chakraborty *et al.* 2007). The analyses of *M. munzala*, in combination with other macaque species, utilized both mtDNA and Y-chromosome sequences. This approach facilitated the diagnosis of the relative contributions of male- and female-mediated genetic exchange. Figure 2.9 illustrates the phylogenetic trees derived from the two marker systems. The inference of a reticulate origin for this species comes from a comparison of the arrangements of the Y-chromosome and mtDNA phylogenies. Specifically, the mtDNA sequence variation reflects a sister relationship between *M. munzala* and *Macaca radiata*. In contrast, *M. munzala* is placed in a clade containing *Macaca thibetana* and *M. assamensis*, but not *M. radiata* based on Y-chromosome sequences (Figure 2.9). Morphological analyses, however, only indicate a close similarity between the new species and *M. thibetana/M. assamensis* (Chakraborty *et al.* 2007). Taken together, these findings suggested the occurrence of ancient, male-mediated introgression from an *M. assamensis/M. thibetana*-like lineage into the ancestor of *M. munzala* (Chakraborty *et al.* 2007).

In addition to the extensive literature on phylogenetic signatures of reticulate evolution in *Macaca*, population-level surveys also reveal patterns of variation indicating genetic exchange. There are two (related) classes of investigation that have detected these patterns of variation. A study reported by Hernandez *et al.* (2007) illustrates the first of these approaches. The purpose of this study was to determine the demographic history and the strength of association between unlinked loci (i.e., linkage disequilibrium; Hernandez *et al.* 2007) in macaque individuals. These researchers examined autosomal DNA loci from 9 and 38 captive animals whose parents were wild-collected Chinese and Indian rhesus macaques, respectively. In the present context—that is, testing for patterns consistent with introgressive hybridization—it is significant that 1 out of 9 (11%) "Chinese" macaques and 8 of the 38 (21%) "Indian" macaques reflected admixtures of genes from both lineages (Hernandez *et al.* 2007). Furthermore, the estimates of linkage disequilibrium for the samples indicated a greater degree of association between loci in the "Indian" samples. Hernandez *et al.* (2007) suggested recent introgression between individuals from the Indian lineage with Burmese animals (see above discussion and Smith and McDonough 2005) to explain this finding. Hernandez *et al.*'s (2007) results reflect

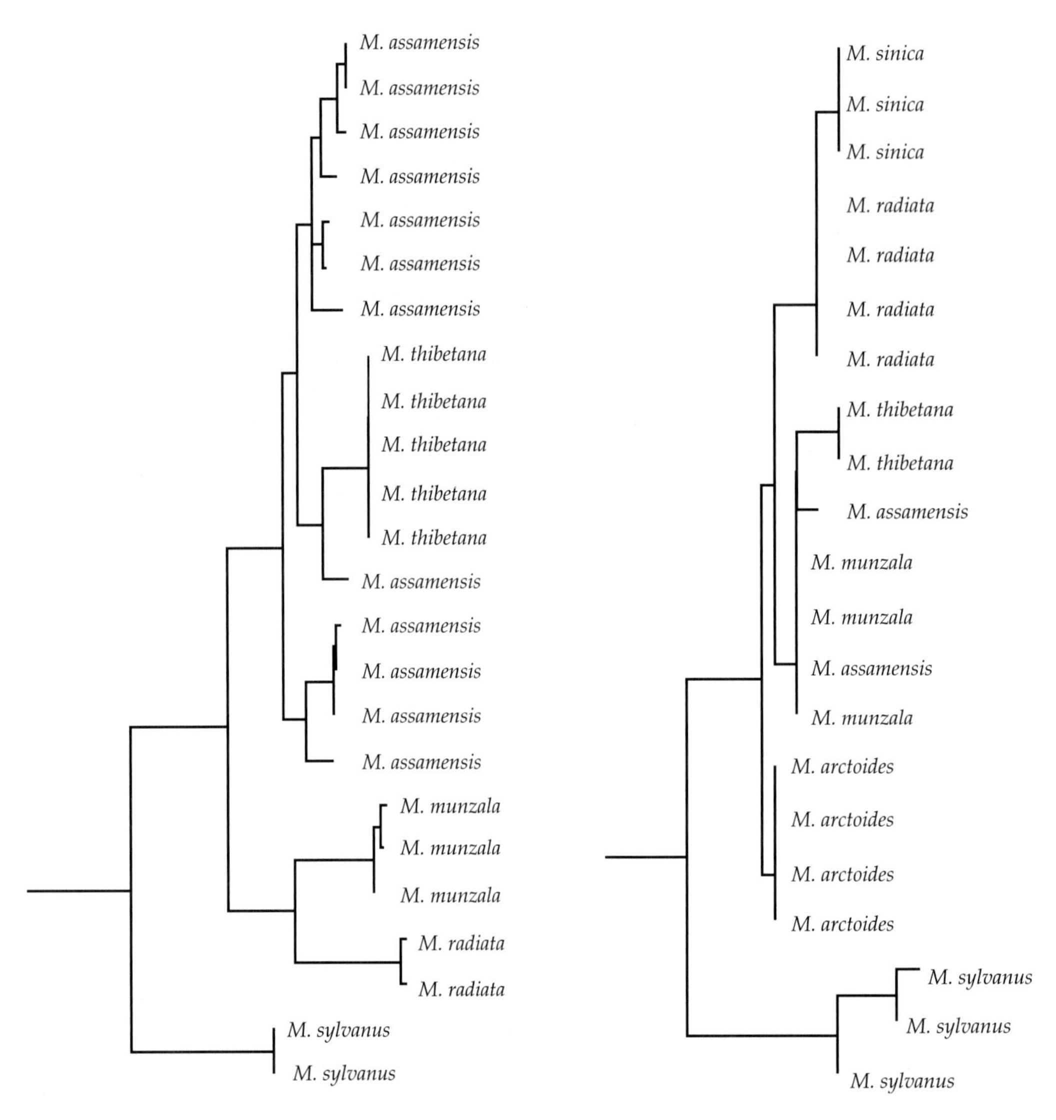

Figure 2.9 Phylogenetic relationships among species from the *M. sinica* complex, including the newly described taxon, *M. munzala*. The left-hand phylogeny derives from mtDNA sequence variation while the right-hand phylogeny was determined using Y-chromosome-specific sequences (Chakraborty *et al.* 2007).

a significant role for reticulate processes in the contemporary, genetic structure of Indian and Chinese rhesus macaque populations.

A second population-level approach for discerning the affect of reticulate events in macaque evolution has involved the analysis of contemporary areas of geographic overlap. In these regions of overlap, characteristics (morphological, molecular, etc.) that are taxon specific have been examined to test for the presence of ongoing introgressive hybridization. The first three studies that I will discuss involved the use of morphology to infer introgression between different macaque species. In the first two of these studies, Bynum and her colleagues (Bynum *et al.* 1997; Bynum 2002) defined a set of morphological traits (e.g., crown hair orientation, cheek whisker contrast, throat contrast) that were diagnostic for *M. tonkeana* and *M. hecki*, two species that had been reported to hybridize in Central Sulawesi, Indonesia. Analysis of the morphological traits in individuals from this region supported the previous report of hybridization. In particular, Bynum *et al.* (1997) and Bynum (2002) observed the following concerning this hybrid zone. First, there was significant clinal variation (ranging from *M. tonkeana*-like to *M. hecki*-like) for the diagnostic morphological characteristics, and the clines for the individual traits were geographically coincident. Second, the clinal shift occurred over a distance of 1,500–2,000 m for both males and females. However, adult males possessed a greater degree of *M. tonkeana*-like characteristics across this zone indicating asymmetric introgression of this species' traits. Third, the center of the morphological clines was located on a road that crossed the isthmus of Central Sulawesi. It was suggested that continued habitat disturbance associated with this road would likely affect the evolution of the hybrid zone. Fourth, museum specimens from individuals collected from this region in 1916 reflected the presence of this hybrid zone by that time.

Hamada *et al.* (2006) described a third example of morphological variation in a region of sympatry between macaque taxa. In this case, the zone of overlap was in northeastern Thailand and involved *M. mulatta* (i.e., Rhesus macaques) and *M. fascicularis* (long-tailed macaques). By comparing body size, body proportion, and pelage color for individuals from a Rhesus-like troop within the zone of overlap, to wild and captive reared individuals of the two species, Hamada *et al.* (2006) detected admixture of morphological traits. Specifically, the female Rhesus-like individuals were 10–20% smaller and had a greater relative tail length than the Rhesus control animals (indicative of similarity to long-tailed macaques). However, these Rhesus-like individuals did possess the Rhesus bipartite pelage color pattern. Hamada *et al.* (2006) concluded that introgression of *M. fascicularis* traits, into a largely *M. mulatta* genetic background, accounted for the morphological variation in this macaque troop.

My final two examples of reticulate evolution in *Macaca* involved the examination of DNA sequence variation rather than morphology. In particular, Evans *et al.* (2001) and Kawamoto (2005) defined genetic interactions between *M. maura*/*M. tonkeana* in Sulawesi and *M. fuscata*/*M. cyclopis* in Japan, respectively. By examining mtDNA and nuclear (i.e., autosomal and Y-chromosome loci), Evans *et al.* (2001) detected high levels of recombination within the hybrid zone between *M. maura* and *M. tonkeana*. However, no significant introgression was detected outside the present-day region of overlap. The high degree of admixture within the zone, but lack of introgression outside this zone, suggested the following model. Continued immigration by parental individuals into the area of overlap provided a source for new hybrid formation, but selection against hybrid genotypes restricted introgression to the hybrid zone (Evans *et al.* 2001; see Barton and Hewitt 1985 for a discussion of this class of hybrid zone model).

The final example of macaque introgressive hybridization comes from a study reported by Kawamoto (2005). Their description of the genetic variation present in populations of macaques in Japan suggested that introgression followed by assortative mating favoring individuals with the introduced species' genotype had resulted in a hybrid zone between the native Japanese macaque species, *M. fuscata,* and the introduced Taiwanese taxon, *M. cyclopis*. Unlike the *M. maura*/*M. tonkeana* example, the likelihood of extensive introgression into the native species was considered so great that attempts to eliminate the hybrid population were begun in 2001 (Kawamoto 2005). Selection against hybrids, if present, did not appear to be restricting the introgression to the contact zone.

Cercopithecus

The final example of a cercopithecine assemblage reflecting reticulate evolution involves the guenon clade. Detwiler *et al.* (2005) list 24 and 10 instances of parapatric and sympatric hybridization among guenon species, respectively (see beginning of Section 2.2.3 for a description of these two classes of hybridization). Of these, they described in greater detail data from four instances of sympatric introgression. These included hybridization between *C. cephus* × *C. nictitans* (Lopé Reserve, Gabon), *C. mitis* × *C. ascanius*

(Kibale and Budongo Forests, Uganda), *C. mitis* × *C. wolfi* (Nyungwe Forest National Park, Rwanda), and *C. mitis* × *C. ascanius* (Gombe National Park, Tanzania). Similar patterns of morphological variation were described for the first three of these hybrid zones, leading to the inference of a common set of causal factors. In particular, these three examples were ascribed to instances in which conspecific mates were limited or absent and thus interspecific sexual reproduction was the most frequent, or only, option (Detwiler *et al.* 2005). Furthermore, the effects from the hybridization episodes were considered ephemeral as the following statement reflects: "Nevertheless, as at Budongo, hybridization apparently died out at Ngogo" (Detwiler *et al.* 2005). However, as pointed out by Anderson and Hubricht (1938)—the authors of the term "Introgressive Hybridization"—morphological data is not adequate to test for advanced generation introgression due to the diluting effects of recombination on polygenic, phenotypic traits. Indeed, Tosi *et al.* (2005) reflect this same conclusion for guenons when they stated "Although the frequency of hybridization is probably very low among sympatric guenon species…the amount of resultant genetic exchange between them remains unclear…because the progeny of repeated backcrosses can quickly take on the appearance of the parental types with which they mate." These authors argued for the utility of discrete molecular markers for testing hypotheses of introgression among guenon species.

Unlike the first three cases discussed by Detwiler *et al.* (2005), the fourth example (involving hybridization between *C. mitis* × *C. ascanius* in Gombe National Park) reflected extensive and persistent introgressive hybridization. Though the guenons in this park are now restricted to an island of habitat bordered on all sides by cultivated land, the origin of the hybrid zone occurred at a time when there was still an intact and widespread ecosystem favorable to guenons (Detwiler *et al.* 2005). Anthropogenic effects were therefore not hypothesized as causal in the initiation of hybridization. Furthermore, recombinant individuals are a high frequency and dynamic contributor to the guenon breeding population of Gombe. For example, in 1996, hybrid individuals were detected in every valley located in the northern two-thirds of the park (Detwiler *et al.* 2005).

Hybridization in the guenon assemblage—both parapatric and sympatric—is common and evolutionarily significant. Detwiler *et al.* (2005) expressed it well by stating, "Hybridization…is intrinsically neither good or bad…It is best treated as one of many natural evolutionary processes that have played an important role in shaping the biodiversity that conservation aims to protect."

2.2.5 Introgressive hybridization in the Hominidae and Hylobatidae

The hylobatids and hominids (Figure 2.6) include the final two examples of reticulate evolution to be discussed in this chapter, the gibbons, and orangutans, respectively. Both gibbon and orangutan taxa reflect the transfer of genetic material via introgression, once again as detected by phylogenetic and population genetic analyses. These taxa are the closest extant organisms to the assemblage that includes *Homo*. As such, the inferences drawn concerning reticulate evolution in gibbons and orangutans represent one of the strongest analogies for the potential significance of this process in our own species.

Gibbons

Phylogenetic discordances among species of gibbons, genus *Hylobates*, have been detected by the use of various molecular markers (e.g., compare results of Noda *et al.* 2001 with those of Hayashi *et al.* 1995 and Whittaker *et al.* 2007). That these discordances may be due to introgression is supported by the observation that hybridization between siamangs (genus *Symphalangus*) and *Hylobates* has been accomplished with captive animals (Myers and Shafer 1979). Noda *et al.*'s (2001) conclusion that "There are several competing hypotheses on the phylogenetic relationships of species within subgenus *Hylobates*…these gibbon species are closely related, hybridization is possible, and this may be the source of incongruent phylogenetic relationships" reflects both the closer evolutionary relationships *within* the gibbon clade and the likelihood of natural hybridization between sympatric forms.

Given the above, it comes as no surprise that introgressive hybridization has apparently led to the differential introgression of nuclear and/or cytoplasmic markers between species of *Hylobates*. As discussed in Section 2.1.2, this differential transfer is expected due to the semipermeable nature of genomes (Key 1968). As in all the other cases described in this and the previous chapter, the exchange of some, but not all, markers can result in different phylogenies reflecting varying arrangements of lineages. This is the case for phylogenetic studies of gibbons. In fact, even individuals belonging to different subgenera have been found to group together on the basis of some DNA sequences. For example, sequence variation in the mitochondrial *16S* rRNA gene resulted in the paraphyletic placement of one individual of *Hylobates pileatus*, subgenus *Hylobates*, with *Hylobates concolor*, subgenus *Nomascus* (Noda *et al.* 2001). Furthermore, Noda *et al.* (2001) found that within the subgenus *Hylobates*, individuals of *Hylobates klossii* were more closely related to individuals of *Hylobates agilis* or *Hylobates lar* than they were to gibbons with which they shared a much closer morphological resemblance. In contrast to the latter findings, Hayashi *et al.* (1995) reported a monophyletic relationship of individuals of *H. lar* and *H. klossii*. Because such discordance between phylogenies is the expected outcome from differential genetic exchange, Noda *et al.* (2001) concluded that introgressive hybridization "may be the source of incongruent phylogenetic relationships, depending on the individuals used from the same species."

Nonconcordant phylogenies for gibbons have also resulted from higher taxonomic comparisons as well. Specifically, the identity of the basal *Hylobates* taxa has varied between different analyses (Noda *et al.* 2001; Roos and Geissmann 2001). Given the apparent impact of reticulate processes on the evolutionary trajectory of more closely related species, this result seems likely to reflect the effect of introgression within the entire gibbon assemblage.

Monda *et al.* (2007) posed a similar hypothesis based on their analysis of mtDNA sequence variation within the IndoChinese concolor gibbons. In this regard, they detected paraphyly among four individuals of *Nomascus siki*, with two individuals each clustering with *Nomascus gabriellae* and *Nomascus leucogenys*, respectively (Monda *et al.* 2007). The *N. siki* specimens from which the DNA samples were obtained were not available for examination by Monda *et al.* (2007). Thus, the cause of the paraphyly could not be conclusively linked to introgression due to the possibility of specimen misidentification. However, given that hybridization within this group is known to occur (Monda *et al.* 2007), it is likely that reticulation has occurred.

Orangutans

Like gibbons, orangutans belong to the super family Hominidae. Orangutans are then further allied to the human/chimp/gorilla clade within the family Hominoidea (Figure 2.6). Orangutan lineages are often described as Bornean or Sumatran, reflecting their island of origin. The orangutans inhabiting these islands differ by two chromosomal rearrangements (involving autosome 2 and the Y-chromosome) and by the presence of a nucleolar organizing region on the distal portion of the long arm of the Y-chromosome (i.e., found in Sumatran but not Bornean individuals; Xu and Arnason 1996). In addition, numerous comparisons of DNA sequence variation between Bornean and Sumatran individuals detected equivalent, or higher, levels of divergence as found for other pairs of species (e.g., pygmy and common chimpanzees) or genera (e.g., *Homo* and *Pan*; Xu and Arnason 1996; Zhi *et al.* 1996). The genetic distances calculated from mitochondrial (RFLP and *16S* sequences) and nuclear (minisatellite) loci resulted in an estimated divergence time from a common ancestor of 1.5–1.7 mya. It was concluded that this divergence occurred "well before the two islands separated and long enough [ago] for species-level differentiation" (Zhi *et al.* 1996). Although many workers refer to the orangutan taxa as subspecies (i.e., Bornean, *Pongo pygmaeus pygmaeus*; Sumatran, *Pongo pygmaeus abelii*) the level of sequence divergence has led some investigators to designate them as separate species (e.g., *P. pygmaeus and P. abelii*; Xu and Arnason 1996).

In spite of their distinctiveness, *Pongo* (i.e., orangutan) lineages reflect the effect of past, and likely ongoing, genetic exchange. On the one hand, the analyses of Xu and Arnason (1996), Zhi *et al.* (1996), Noda *et al.* (2001), and Warren *et al.* (2001) resolved reciprocal monophyly for orangutans from the islands of Sumatra and Borneo. In contrast, Muir *et al.* (2000), Kanthaswamy and Smith (2002), and Kanthaswamy *et al.* (2006) detected introgression between the two islands (Figure 2.10). Muir *et al.* (2000) reported that "The Borneo populations, sampled widely from the island, form a compact collection of mtDNA haplotypes. Sumatra, in contrast, appears to include at least three very distinct lineages, including one of the Borneo lineages." They proposed a model of allopatric divergence and recurring introgression to explain these admixtures (Muir *et al.* 2000). Such a model is not only supported by geological data, but also by the observation that the Bornean and Sumatran taxa are capable of forming fertile offspring (Xu and Arnason, 1996). In regard to the cause of the observed introgression (Figure 2.10), Kanthaswamy and Smith (2002) and Kanthaswamy *et al.* (2006) suggested an alternative model; they argued that human-mediated transfer may explain some of the genetic exchange. They proposed that translocation of orangutans had led to some, but likely not all of the observed admixtures. As concluded for gibbons, and other primates, hybridization between divergent lineages of orangutans seems to have affected phylogenetic and population genetic variation.

2.3 Summary and conclusions

I began this chapter by arguing, as did Darwin and many others since, that the use of analogy

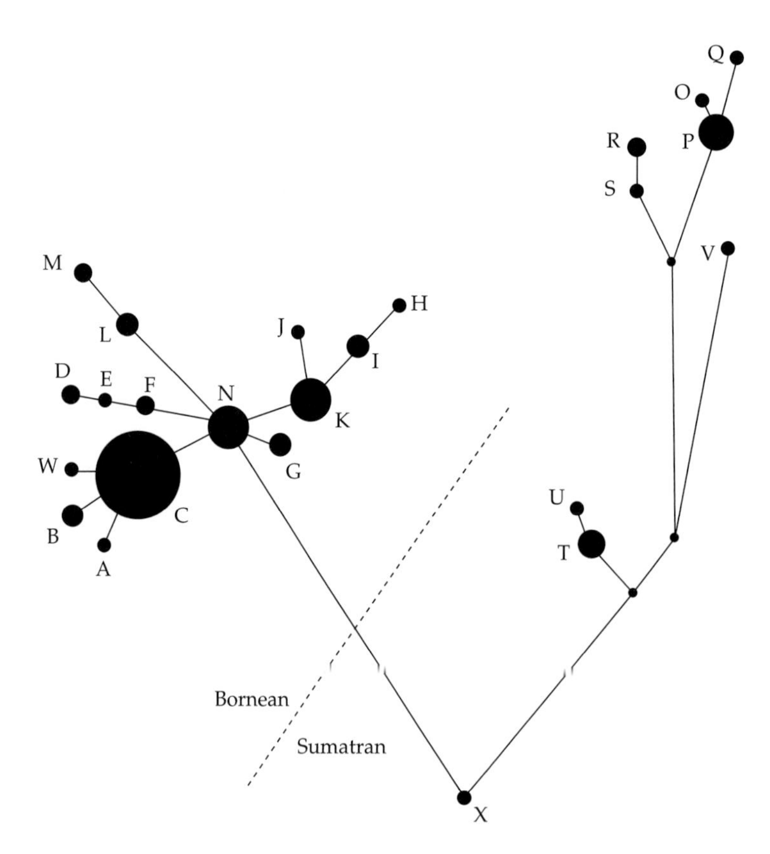

Figure 2.10 Network of relationships based on mtDNA sequences from Bornean and Sumatran orangutans. Note that the sequences "C", "N," and "W" within the Bornean cluster were found for animals collected from Sumatra, reflecting introgression between the lineages (Kanthaswamy *et al.* 2006).

can facilitate the construction and testing of evolutionary hypotheses. In the current context that involves testing for reticulate events in widely divergent organisms—as I did in Chapter 1—as well as in organisms belonging to the assemblage of interest, in this case primates. In the present chapter, I have highlighted the evidence for the important role of genetic exchange in the evolution of numerous species complexes of New and Old World primates. A common observation is that not only are contemporary zones of hybridization detectable, but ancient reticulations are as well. These exchange events have resulted in both mosaic genomes and also the formation of new evolutionary lineages. In some cases, these new lineages have been recognized taxonomically at the level of species, genus, and so on. The primate groups discussed in this chapter thus reflect what is common for all other major categories of life; reticulation is widespread and genetically, ecologically, and evolutionarily creative (Arnold 1997, 2006). As an analogy for (1) our own species, (2) our closest primate relatives, and (3) species with which we have to cope such as diseases, food sources, and so on, the previous examples suggest an important role for reticulate evolution. In the following chapters, I will explore this hypothesis.

CHAPTER 3

Reticulate evolution in the hominines

...not only chimpanzee subspecies, but also bonobos and chimpanzees, may have an intermixed genetic relationship.

(Kaessmann *et al.* 1999)

These new dates would make *A. africanus* and *A. boisei* synchronic taxa; hybridization between these two species in zones of ecological overlap could explain the persistence of *A. africanus* apomorphic features...in *A. robustus*.

(Holliday 2003)

Although the majority of regions in our genome are most closely related to chimpanzees and bonobos, a non-trivial fraction is more closely related to gorillas.

(Pääbo 2003)

...a prolonged period of gene flow between ancestral chimpanzee and gorilla seems plausible.

(Osada and Wu 2005)

These unexpected features would be explained if the human and chimpanzee lineages initially diverged, then later exchanged genes before separating permanently.

(Patterson *et al.* 2006)

Among different population-split scenarios, our data suggest a complex history for western and eastern gorillas including an initial population split at around 0.9–1.6 MYA and subsequent, primarily male-mediated gene flow until approximately 80,000–200,000 years ago.

(Thalmann *et al.* 2007)

3.1 Reticulate evolution within and among *Pan*, *Gorilla*, and *Homo* and their ancestors

In this and the following chapter I will discuss the evidence for (1) reticulations involving the ancestors of the network that includes gorillas, chimpanzees, and humans and (2) the assemblage that includes species of *Homo*, respectively. Drawing this distinction is somewhat arbitrary due to the continuum that is evolutionary change, especially when we are discussing intermittent genetic exchange during organismal diversification. This continuum will necessarily lead to some overlap in the studies discussed in both chapters, and thus some redundancy. Furthermore, unlike examples from the genus *Homo*, I will discuss examples of hybridization *within* the genera, *Pan* and *Gorilla* in this chapter rather than in Chapter 4.

There are three reasons for the topical structures of Chapters 3 and 4. First, this book is designed to use the genus *Homo* as the central exemplar for discussing reticulate evolution and the web-of-life metaphor. Because of this, I want to highlight repeatedly how the lineages that interacted to give rise to our species, along with our own lineage, have been impacted by reticulate events. Second, the evolution of the genus *Homo* has been marked by significant and unusual anatomical and behavioral shifts (e.g., incorporation of stone technology possibly leading to increased carnivory—de Heinzelin 1999; increase in cranial capacity—Conroy *et al.* 1998, Falk 1998) that may have contributed to its unique position in the ecology of the entire planet. In this light, and given the context of this book, it is important to consider how genetic exchange might have contributed to the origin and evolution of these attributes in, for example, *Australopithecus* as well as our own species. Third, some of the greatest interest, and most heated debates, surrounding the role of reticulate evolution in the development of the chimp/gorilla/human web has centered around the question of whether or not different lineages of *Homo* (1) coexisted in space and time and (2) hybridized (e.g., compare Zilhão *et al.* 2006 with Mellars *et al.* 2007 and Currat and Excoffier 2004 with Templeton 2007). Given these three goals/issues, the division of the set of closely related topics surrounding the genus *Homo* into Chapters 3 and 4 seems warranted.

3.2 Reticulate evolution in the hominine fossil record: analogies and *Australopithecus*

In his paper "A proper study for mankind: Analogies from the papionin monkeys and their implications for human evolution," Jolly (2001) argued that evolutionary pattern and process observed in the papionins was likely to have occurred in the hominins as well. For example, Jolly (2001) tested a diet-based hypothesis of hominin origins by examining various morphological characters in papionins—in particular the genus *Theropithecus*—associated with their processing of food items. From this use of analogy, he demonstrated that various morphological trends thought to be associated with adaptations for feeding in *Australopithecus africanus* were also found in *Theropithecus* (Jolly 2001). The parallel trends in these unrelated primates suggested to Jolly (2001) common causes, specifically the action of natural selection, due to shifts in diet and ecological setting.

Along with analogies like the above, Jolly (2001) also utilized conclusions drawn from papionin data to illustrate the potential for reticulation to have played a significant role in the evolution of fossil (and extant) hominins. In particular, he hypothesized that the numerous examples of hybridization between individuals belonging to different species and even genera of papionins suggested the plausibility of crosses between now extinct hominin taxa (e.g., *Paranthropus boisei* × *Kenyanthropus rudolfensis*, *Australopithecus afarensis* × *Kenyanthropus platyops*, *Ardipithecus ramidus kadabba* × proto-chimpanzee; Jolly 2001). In summarizing his hypothesis concerning hominin reticulate evolution, Jolly (2001) posited that, like extant papionins (and for that matter, numerous mammalian lineages; Arnold 1997, 2006), (1) spatial and temporal overlap in now extinct hominins was likely to have resulted in introgressive hybridization and (2) the resulting genetic exchange could have resulted in recognizable changes in fossil forms and their extant derivatives (e.g., increase in brain size in the hominin assemblage).

Similar to the conclusions drawn from analogy with the papionins, direct assessment of fossil forms has also suggested the occurrence of reticulation between now extinct hominins. In particular, Holliday (2003) reviewed evidence of introgressive hybridization and hybrid speciation involving australopithecines. Holliday's (2003) conclusions included inferences concerning the role of reticulation in the evolutionary trajectory of *Australopithecus boisei, Australopithecus robustus,* and *Australopithecus africanus*. For example, he argued that the close evolutionary relationship and the overlap in space and time of *A. boisei, A. africanus,* and *A. robustus* would have provided ample opportunity for hybridization. Indeed, Holliday (2003) concluded that introgressive hybridization between *A. boisei* and *A. robustus* was extremely likely due to their status as sister taxa and their broadly overlapping spatial and temporal distributions. A second hypothesis derived from the fossil data was that introgressive hybridization between *A. boisei* and *A. africanus* had resulted in the establishment of some of the characteristic morphological features of *A. robustus* (Holliday 2003). It was thus hypothesized that derived morphological features present in *A. africanus* were also present in *A. robustus* due to reticulation. This hypothesized, reticulate event had thus given rise to a stable hybrid derivative (i.e., *A. robustus*) of the type found in extant animal and plant species complexes (e.g., Rieseberg *et al.* 1990; DeMarais *et al.* 1992; Arnold 1993).

3.3 Reticulate evolution in the hominines: molecular evidence of introgressive hybridization and hybrid lineage formation

All subdisciplines of evolutionary biology have been challenged, and sometimes refreshed, by the collection and analysis of molecular data. Early studies utilizing protein electrophoresis, reassociation kinetics, and amino acid sequencing have given way to the analysis of the base pairs making up specific RNA and DNA molecules. At each stage, the development of molecular methodologies has provided tests (not always successfully; e.g., see Lewontin 1974, p. 189) of some of evolutionary biology's most revered hypotheses. Regarding the evolution of the chimpanzee/gorilla/human web, a comparison of proteins from humans and chimpanzees resulted in an estimate of 99% similarity of amino acid sequence from the two species (King and Wilson 1975). The high similarity in protein sequences in the light of the significant morphological, behavioral, and so forth differences between chimpanzees and humans resulted in King and Wilson's (1975) conclusion that "evolutionary changes in anatomy and way of life are more often based on changes in the mechanisms controlling the expression of genes than on sequence changes in proteins." Given this conclusion, they argued for the importance of mutations in regulatory loci as the basis of the evolutionary divergence between *Pan* and *Homo* (King and Wilson 1975).

Recent molecular analyses have confirmed that the high degree of protein sequence similarity is also reflected in DNA sequences. Various studies have provided estimates of DNA base pair similarities ranging from 95% to 99.4% (e.g., Britten 2002; Wildman *et al.* 2003). It is thus apparent that our genome is extremely similar to that of the chimpanzee (and gorilla; e.g., Caccone and Powell 1989). Such a high degree of sequence identity is normally associated with taxa belonging to the same genus, rather than different families. Findings from an analysis of nonsynonymous (i.e., those base pair positions at which mutations cause changes in the amino acid sequence of proteins) and synonymous DNA sequences supported such a systematic rearrangement of these taxa. Wildman *et al.* (2003) detected only 0.58% and 0.74% nonsynonymous base pair changes in comparisons of humans and chimpanzees and humans and gorillas, respectively. Synonymous substitution percentages for the same species pairs were also very low (chimpanzee/human = 1.63%; gorilla/human = 1.76%). Wildman *et al.*'s (2003) findings suggested the following systematic scheme: family Hominidae would include all apes (gibbons, orangutans humans, etc.) and the genus *Homo* would include gorillas, chimpanzees, and humans.

Additional molecular analyses of loci found on human, chimp, and gorilla Y-chromosomes have also yielded tests for the affects from such widely divergent processes as natural selection and cultural/religious practices on the levels and patterns of sequence variation (e.g., Burrows and Ryder 1997; Skorecki *et al.* 1997; Hammer and Zegura 2002). Furthermore, coordinated evolution of the X- and Y-chromosomes has been detected resulting in a detailed understanding of exchange events between these chromosomes during the evolution of the chimpanzee/human/gorilla assemblage (e.g., Rozen *et al.* 2003; Skaletsky *et al.* 2003). Finally, comparisons of whole genomes within and between chimpanzees, humans, and gorillas have identified (1) recombination in supposedly nonrecombining genomes (i.e., mtDNA; Awadalla *et al.* 1999) and (2) regions that may be associated with many of the traits that make our species unique from all other primates (see Enard and Pääbo 2004 and Gagneux 2004 for discussions).

As most of the examples discussed in this chapter and in Chapters 1 and 2, have already illustrated, the relationship between molecular data sets and hypothesis testing is nowhere more obvious than in studies that have intentionally, or accidentally, resulted in descriptions of reticulate evolution. I will continue to highlight other types of data that have provided rigorous tests for signatures of reticulate evolution. However, it is clear that the analyses of molecules—particularly DNA sequences—most often provide the crucial information for determining whether or not a particular biological example falls into the category of reticulate evolution. In the remainder of this chapter, I will explore examples of the use of molecular data sets to test for genetic exchange involving *Pan*, *Gorilla*, and *Homo*.

3.3.1 Evidence of *Homo* × *Gorilla* × *Pan* introgression and the origin of hybrid taxa

The landmark paper by Sarich and Wilson (1967), in which they examined similarities between the albumin blood serum proteins from a number of primates, resulted in a revolutionary (for that time at least) conclusion that we and our two closest living relatives evolved from a common ancestor *c*.5 million years ago (mya; Figure 3.1). Sarich and Wilson (1967) also concluded that the phylogenetic relationships within the chimp/human/gorilla assemblage were too close to determine which, if any, of the species pairs shared a more recent common ancestor. They emphasized this latter inference by illustrating the relationships among the three taxa as an unresolved trichotomy (Figure 3.1; Sarich and Wilson 1967). Though a debate ensued concerning both of these conclusions, after 40 years the recent sharing of a common ancestor has been repeatedly demonstrated and is now accepted as fact. However, the consensus from numerous data sets is that humans and chimpanzees shared a more recent common ancestor relative to gorillas (see Avise 1994, pp. 329–331).

Nearly every data set employed to test for the relationships among *Pan*, *Homo*, and *Gorilla*

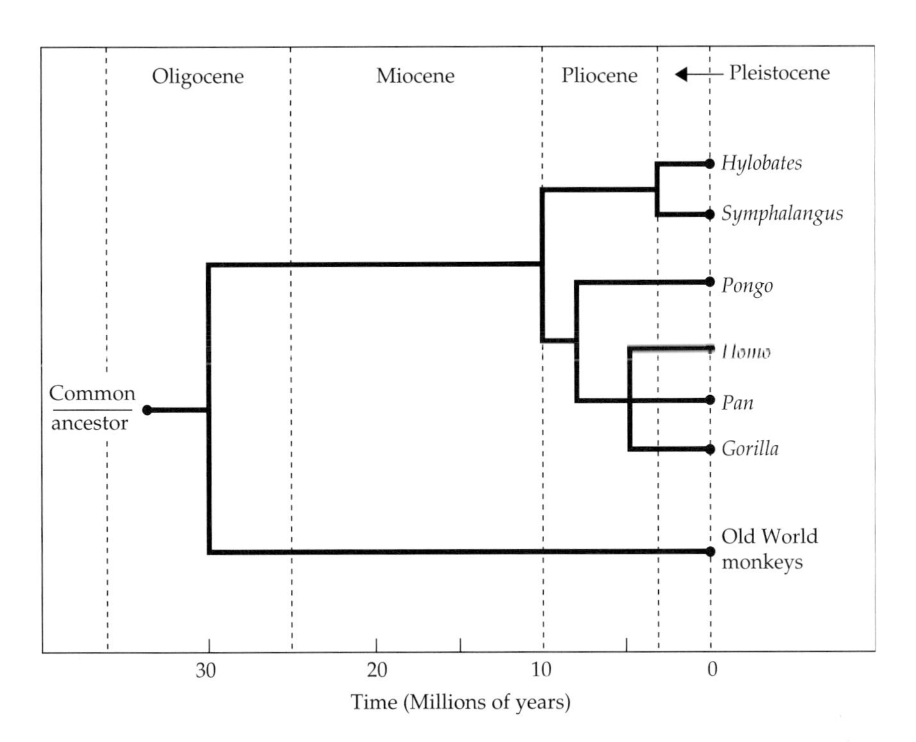

Figure 3.1 Phylogenetic tree reflecting relationships and times of divergence among various Old World primates. The depiction of the *Homo/Pan/Gorilla* clade as a trichotomy reflects the conclusion that their divergence from a common ancestor was essentially simultaneous (Sarich and Wilson 1967).

has resulted in some frequency of phylogenetic discordance. As I have indicated in the previous discussions, one widely supported explanation for this type of discordance is reticulation resulting in the differential transfer of loci (see Arnold 1997, 2006 for reviews and Box 1.1 for an illustration). In regard to introgression among the lineages that gave rise to *Pan, Gorilla,* and *Homo* it is significant that only *c.*5 million years has passed since divergence from a common ancestor. Prager and Wilson (1975) and Fitzpatrick (2004) have provided an estimate of 2–4 million years as the average duration necessary for the development of reproductive isolation between mammalian lineages. If this estimate applies to chimpanzees, gorillas, and humans, introgression was possible for 40–80% of the time period since their divergence from a common ancestor.

In fairness to some of the authors of the references cited below, I should state that not all workers in primate evolutionary biology infer reticulation as the cause of detected discordances. My goal will be to present a description of the findings including the inferences drawn. I will, however, argue that the data from individual studies, and the overall patterns from all of the relevant studies, point to the importance of introgression and hybrid lineage formation in *Homo, Pan* and *Gorilla*.

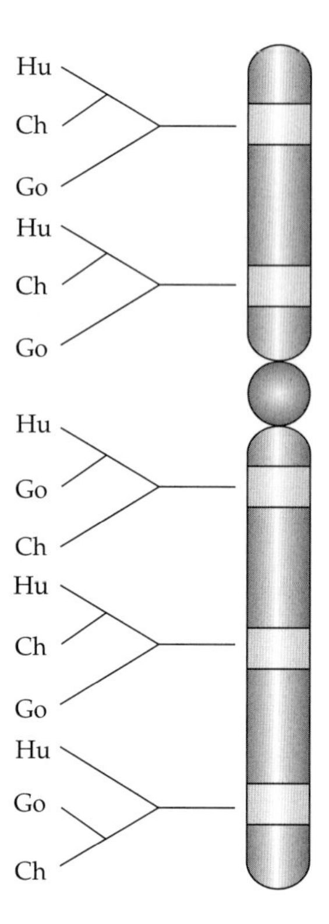

Figure 3.2 Illustration of the consequence of phylogenetic discordance due to alternate allelic sharing between chimpanzees ("Ch"), gorillas ("Go"), and humans ("Hu") at different sites along a single chromosome (Pääbo 2003). This pattern is predicted given differential introgression of loci.

Homo × *Gorilla*

Though the sequence divergence between *Gorilla* and *Homo* is slightly higher than for *Pan* and *Homo*, the frequency of base pair substitutions is small (see above). Furthermore, phylogenetic analyses of this assemblage have repeatedly detected sequence sharing at some loci indicating a closer relationship between humans and gorillas than between humans and chimpanzees (Figure 3.2). Pääbo (2003) reflected this observation in the quote at the start of this chapter. He concluded that variation carried over during the intervening and brief time period between the divergence of the gorilla lineage and the lineage that subsequently gave rise to chimpanzees and humans until the divergence of the human and chimpanzee lineages likely caused the sharing of similarities between humans and gorillas. In a similar vein, Deinard and Kidd (1999) concluded that the combination of the close temporal association of the origin of the three lineages coupled with high levels of ancestral polymorphisms would make it most accurate to represent the associations among *Pan, Homo,* and *Gorilla* as an unresolved trichotomy. Notwithstanding the conclusions of Pääbo (2003) and Deinard and Kidd (1999), the identification of a "non trivial" fraction of shared sequence identity between gorillas and humans (Pääbo 2003), relative to chimpanzees, along with the numerous examples of present-day hybridization between primates suggests an important evolutionary role for introgressive hybridization as these three lineages diverged.

One data set that illustrates the sharing of allelic variation between gorillas and humans comes

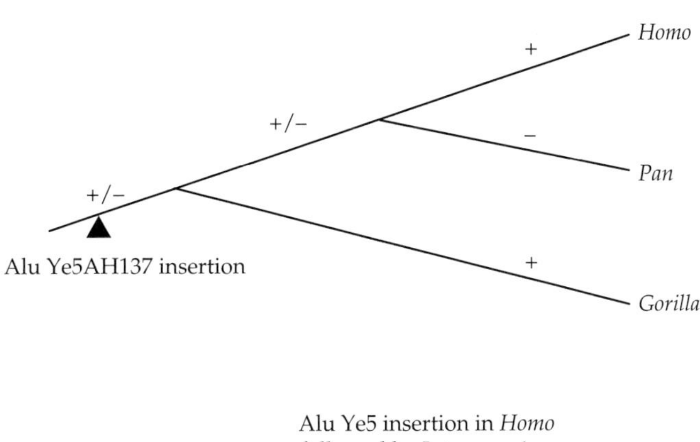

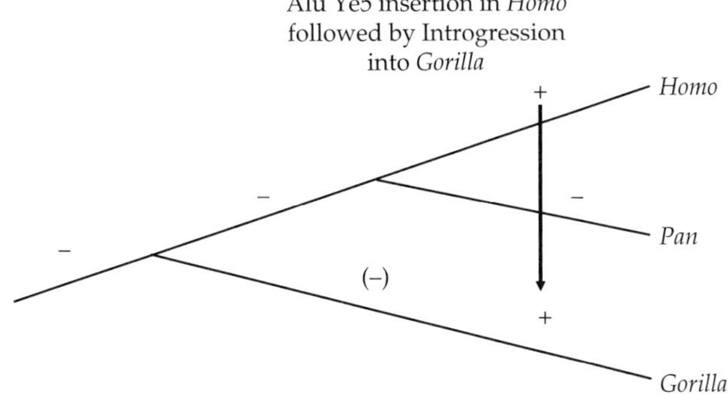

Figure 3.3 Two phylogenies that illustrate alternative explanations for the sharing of an Alu insertion between gorillas and humans. The upper phylogeny comes from Salem *et al.* (2003) and illustrates their conclusion that ancestral polymorphism for the presence (+) and absence (−) alleles explains the observed patterns. The bottom phylogeny indicates that the pattern can also be explained by reticulate evolution. The ancestral polymorphism hypothesis requires three separate events while the reticulate evolution hypothesis requires only two (see text for a discussion). Introgression of the insertion allele is illustrated as proceeding from humans into gorillas. The alternative scenario—the origin of the insertion in gorilla and its introgression into humans—is equally parsimonious.

from a study of the Alu family of transposable elements. Salem *et al.* (2003) analyzed Alu elements (i.e., the Alu Ye5 subfamily) that amplified during hominid evolution. They identified and compared the genomic insertion points of Ye5 in *Homo, Pan* (both common chimpanzees and bonobos), *Gorilla, Pongo* (orangutan), *Chlorocebus* (green monkey), and *Aotus* (owl monkey). The insertion of elements was not detected in either the green or owl monkey samples, thus identifying the expansion of Ye5 subsequent to the divergence of the hominids from other primates (Salem *et al.* 2003).

Overall, insertions of the Ye5 elements indicated a closer phylogenetic relationship of the two species of chimpanzee with humans, relative to gorillas. However, Salem *et al.* (2003) detected an Alu element insertion that was present in gorilla and human, but absent in both common chimpanzees and bonobos. These authors concluded that a polymorphism for insertion/lack of insertion in the ancestor of chimpanzee, gorilla, and humans best explained this finding (Figure 3.3). Yet, these authors cited a similar pattern found in other analyses in which transposable elements had been used to reconstruct phylogenies. In these other studies it was concluded that "interspecies hybridization" (quote from Salem *et al.* 2003) was the likely cause of shared insertion sites. Figure 3.3 includes the phylogeny of *Pan, Gorilla,* and *Homo* from Salem *et al.* (2003) indicating their inference of how shared ancestral polymorphism could have resulted in the present-day distribution of the Alu insertion shared between gorillas and humans. I have also included an alternative phylogeny indicating how the same distribution could have been obtained instead by a gain of the insertion in either the gorilla or human lineage followed by introgressive hybridization (Figure 3.3). In the scenario of Salem *et al.* (2003), the "lack of insertion" (i.e., "–"), allele would have been lost in both *Gorilla* and *Homo*, and the "presence of insertion" (i.e., "+"), allele would have been lost in *Pan*—three events are necessary to explain this hypothesis (Figure 3.3). In the alternative, reticulate, hypothesis only two events are necessary, the gain of the insertion allele and its replacement of the lack of insertion allele through introgression (Figure 3.3). Parsimony and the observation of numerous examples of introgression in this and other primate assemblages support the latter hypothesis.

Like Salem *et al.* (2003), O'hUigin *et al.* (2002) examined DNA sequence variation in primates including the human/chimpanzee/gorilla clade. Similarly, these latter workers detected conflicting patterns of sequence variation. Findings from this analysis were in accord with their introductory statement concerning previous attempts to resolve the phylogenetic relationships of *Pan, Gorilla,* and *Homo*:

Perhaps the best-known example of prolonged controversy... [regarding phylogenetic placement]... is the one involving the human species... The consensus approach identifies the chimpanzee as the nearest living relative of humans, but the evidence supporting this conclusion is not overwhelming... Inconsistency in the inferred patterns of shared-derived substitutions... is apparent both between and within loci of the three species comprising the trichotomy.

O'hUigin *et al.*'s (2002) study involved the analysis of segments of 51 loci (both autosomal and X-chromosome) from human, chimpanzee, gorilla, orangutan, macaque, and tamarin. As with other studies, O'hUigin *et al.* (2002) detected the greatest support (i.e., 52% of the informative base pair substitutions) for a phylogenetic arrangement in which humans and chimpanzees shared the most recent common ancestor. Yet, alternative phylogenetic arrangements were indicated by variation at other loci (as illustrated by Figure 3.2). In terms of the evidence for introgression between the lineages leading to gorilla and humans, it is significant that 16% of the informative sites supported a closer relationship between these two taxa, relative to the chimpanzee lineage (O'hUigin *et al.* 2002). Though some of these sites were inferred to be due to the sorting of ancestral polymorphism, many were not (O'hUigin *et al.* 2002) and were thus consistent with a model of evolution among humans, chimps, and gorillas that included reticulation.

Gorilla × *Pan*

Similar to the evidence for *Homo* × *Gorilla* genetic exchange, Pääbo's (2003) review of various data sets reflects the finding of shared characteristics between *Gorilla* and *Pan* as well. Sequence variation at certain loci indicated that chimpanzees and gorillas were more closely related to one another than either was to humans. Figure 3.2 illustrates the phylogenetic signature that would lend support to this conclusion. Similarly, O'hUigin *et al.*'s (2002) analysis discussed above also allowed a test for introgressive exchange of alleles between the lineages leading to gorillas and chimpanzees subsequent to their divergence from a common ancestor. In fact, nearly twice as many (i.e., 30%) of the informative base pair substitutions detected by O'hUigin *et al.* (2002) reflected shared variation between *Pan* and *Gorilla*, relative to *Homo* and *Gorilla*. Once again, these authors suggested that the sorting of ancestral polymorphisms could account for some, but not all, of the shared variation between chimpanzees and gorillas. Thus, a significant proportion of the allelic variation is likely due to introgression between the diverging lineages.

Osada and Wu (2005) tested the allopatric divergence model for the human/gorilla/chimpanzee assemblage. They designed an analysis of genic and intergenic sequences from chimpanzee, gorilla, and humans to assess whether or not the three taxa had diverged with or without interlineage gene flow. With regard to *Pan* × *Gorilla*, they argued that though the patterns of variation did not reject the null hypothesis of divergence without gene flow (i.e., the allopatric model), there was a signature that suggested past introgressive hybridization. Osada and Wu (2005) found that there was a high frequency of allelic sharing (as found in the statistically significant comparison between chimpanzees and humans). They concluded that the lack of a rejection of the allopatric model most likely reflected the limited number of available sequences from gorilla. They further argued that the shared biogeographic history of *Pan* and *Gorilla* would have facilitated the introgression between proto-gorilla and proto-chimpanzee individuals.

Homo × *Pan*

Of the three species combinations discussed in this section, to date, the one that has been explored in most detail is the chimpanzee–human pair. This is likely due to (1) the closer assumed phylogenetic relationship between *Pan* and *Homo*, but most importantly, (2) the availability of sequence information for large portions of the genomes of both species. In the not-to-distant future, with the availability of the entire genome sequence for gorilla, this picture should change.

Comparisons of the human and chimpanzee genomes have resulted in numerous opportunities to test the hypothesis that divergence of these lineages occurred in concert with ongoing introgressive hybridization. One study that generated an enormous amount of interest was that of Navarro and Barton (2003b). This study was not specifically designed to test for genetic exchange, but rather for the molecular evolutionary footprints expected if speciation occurred between lineages that (1) possessed different chromosomal rearrangements and (2) introgressed with one another as they diverged. These authors had previously derived a model that predicted greater sequence divergence in regions differing by chromosomal rearrangements than in collinear (i.e., nonrearranged chromosomal segments) segments (Navarro and Barton 2003a). This pattern was expected due to the possibility of gene exchange through recombination in hybrids in the nonrearranged, but not the rearranged portions of the chromosomes. Navarro and Barton (2003b) did indeed report elevated divergence between chimpanzee and human loci found in rearranged portions of the genome, relative to loci located in collinear regions. This resulted in the conclusion that "These patterns of divergence and polymorphism may be, at least in part, the molecular footprint of speciation events in the human and chimpanzee lineages" (Navarro and Barton 2003b). Furthermore, their data supported an exchange of genetic material between the proto-human and proto-chimpanzee lineages, especially involving the collinear genomic segments (Navarro and Barton 2003b; Rieseberg and Livingstone 2003).

Navarro and Barton's (2003b) use of the *Pan*/*Homo* genomic sequences to test the divergence-with-gene-flow model resulted in numerous response papers. Though some responses took issue with the conclusion that chimpanzees and humans evolved with some level of ongoing introgression (e.g., Zhang *et al.* 2004), most addressed the association between chromosome rearrangements and genic divergence (e.g., Hey 2003; Lu *et al.* 2003; Marquès-Bonet *et al.* 2004). Indeed, many of the conclusions contained in the responses can be summarized by Lu *et al.*'s (2003) statement, "We wish to emphasize that this comment does not contradict the elegant model of parapatric speciation driven by differential adaptation."

Additional studies that have found support for the model of human–chimpanzee divergence with gene flow (but see Barton 2006; Innan and Watanabe 2006) include the analysis of sequence variation by Osada and Wu (2005) discussed above in the context of gorilla × chimpanzee hybridization. For *Pan* and *Gorilla*, the relatively small amount of sequence data for the latter taxon resulted in the availability of only 76 coding sequences and 53 intergenic sequences for comparison (Osada and Wu 2005). This led to limited statistical power for testing inferences concerning past reticulation between proto-chimpanzees and proto-gorillas. In contrast, 345 coding and 143 intergenic sequences were available for comparisons between humans and chimpanzees. This allowed rigorous statistical tests for genetic exchange between the lineages leading to *Pan* and *Homo*.

Osada and Wu (2005) used several approaches to test the null hypothesis that chimpanzees and humans (and gorillas) diverged while separated from one another geographically (i.e., the allopatric model of divergence). These analyses included estimations of the time since divergence of alleles at the loci examined as well as congruence between allele associations and the phylogenetic relationships among *Pan, Gorilla,* and *Homo.* Using the divergence times (along with other estimated parameters), Osada and Wu (2005) found that the allopatric model was rejected, while the parapatric model (i.e., divergence with some level of gene flow between lineages) was not rejected. In addition, the phylogenetic inferences drawn from the sequence variation at individual coding and intergenic loci reflected support for different combinations of the three species (Table 3.1). Data from most of the loci supported a chimpanzee/human association relative to gorilla. However, the other two possible phylogenetic arrangements were also supported, albeit by a smaller percentage of the coding and intergenic sequence loci (Table 3.1; Osada and Wu 2005). This reflects not only support for a relatively larger amount of gene flow during divergence between proto-*Homo* and proto-*Pan,* but also for some level of gene flow between these two lineages and proto-*Gorilla.*

As I have already illustrated in organisms as diverse as archaebacteria and chimpanzees, different genomic segments can have vastly different evolutionary histories due to lateral transfers and introgressive hybridization. Patterson *et al.* (2006), in an analysis of sequence variation between humans, chimpanzees, gorillas, orangutans, and macaques, introduced their paper with a statement reflecting this conclusion: "The genetic divergence time between two species varies substantially across the genome, conveying important information about the timing and process of speciation." Patterson *et al.*'s (2006) analysis of 9.3 million base pairs between all five species, and 18.3 million base pairs between *Homo, Pan, Gorilla,* and *Macaca* represented the largest single analysis of its kind, and reflected the most extensive test to date of the allopatric and parapatric models of divergence. The strongest indication that divergence between humans and chimpanzees was not consistent with the allopatric model came from estimates of divergence times for the various loci. In particular, there was a spread of 4 million years between the estimate of the most ancient and the most recent divergence times (i.e., ~84–147% of the average divergence time; Patterson *et al.* 2006). Such large discordance among loci is obviously expected under the parapatric, but not the allopatric, model. Patterson *et al.* (2006) derived what they termed a "provocative explanation" for their observation that included initial separation of the lineage leading to humans and chimpanzees, followed by introgressive hybridization between proto-*Pan* and proto-*Homo* before they obtained complete reproductive isolation. This explanation was also used to account for the occurrence of an extinct species (i.e., the Toumaï fossil) whose date preceded the average estimated divergence date calculated from the sequence information. In this case, it was suggested that the hominin characteristics in the Toumaï fossil might reflect the effect of introgression between the lineage eventually resulting in chimpanzees and that from which our own species derived (Patterson *et al.* 2006; but see Ebersberger *et al.* 2007).

Patterson *et al.*'s (2006) explanation is not actually provocative given all the information concerning introgressive hybridization among primates in general and the human/chimpanzee/gorilla web in particular. However, it did elucidate a process that was consistent with various observations. In fact,

Table 3.1 The number of intergenic and coding loci that support a closer phylogenetic relationship between human–chimpanzee ("HC"), chimpanzee–gorilla ("CG"), or human–gorilla ("HG").

	HC	CG	HG
Intergenic loci ($n = 53$)	23 (63.9%)	6 (16.7%)	7 (19.4%)
Coding loci ($n = 76$)	26 (49.1%)	14 (26.4%)	13 (24.5%)

Note: "*n*" = the number of loci examined.

Source: Osada and Wu (2005).

this introgression model not only accounts for the wide range of divergence times among autosomal loci and the observations concerning the dates of some fossil hominins, but it also helps explain a unique pattern of sequence variation detected for the X-chromosome. Unlike the autosomal loci, sequence divergence among human and chimpanzee loci located on the X-chromosome was greatly deflated (Figure 3.4; Patterson *et al.* 2006). The significant reduction in sequence divergence was unique for human–chimpanzee comparisons as indicated by the recovery of expected levels of divergence between X-loci from humans and gorillas (Figure 3.4; Patterson *et al.* 2006). Given that loci on the X-chromosome have, on average, a greater effect on hybrid fitness, selection in hybrids should result in the incorporation of only a single X-chromosome variant (see Patterson *et al.* 2006 for a discussion). This asymmetric introgression would favor the fixation of one set of X-chromosome alleles (in this case from either the proto-human or proto-chimpanzee lineage) and thus result in a pattern of lower levels of sequence divergence times relative to autosomal loci that came from both lineages. This is the pattern observed by Patterson *et al.* (2006).

In general, results from analyses of DNA sequence information for *Pan, Gorilla,* and *Homo* reflect the conclusion that our genome, and those of our closest living relatives, are mosaics (Pääbo 2003; Arnold and Meyer 2006). This mosaicism has been at least partially created through genetic exchange as we—along with gorillas and chimpanzees—diverged from a common ancestor. Now I will review data for taxa within the genera, *Pan* and *Gorilla*, illustrating that genetic exchange did not stop with interactions between proto-humans, proto-gorillas, and proto-chimpanzees.

3.3.2 Evidence of introgression within *Gorilla* and *Pan*

As we have seen, depending on which portion of the genome is chosen, either humans and chimpanzees, or humans and gorillas, or gorillas and chimpanzees resolve as the closest living hominines. Given this observation, it would seem that our null hypothesis should be that when genomic variation is examined within *Pan* and *Gorilla* (and *Homo* in Chapter 4) the traces of introgressive hybridization will be detected as well. Numerous population genetic surveys for both chimpanzees

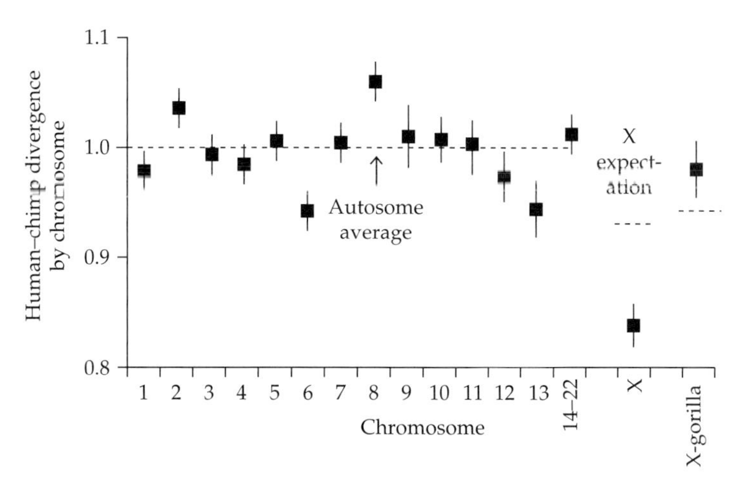

Figure 3.4 A comparison of loci from humans and chimpanzees that occur on the same chromosomes. The dashed line reflects the average divergence among loci located on the autosomal chromosomes. Comparisons between loci on the smaller chromosomes (i.e., 14–22) were pooled because of limited numbers of loci available for analysis. The "X expectation" represents the value predicted for comparisons between human and chimpanzee loci located on the X-chromosome, while the "X-gorilla" value is the observed divergence between loci from humans and gorillas. Though the estimates for the comparisons of humans and chimpanzees are greatly reduced, the observed gorilla–human X-chromosome divergence is not (Patterson *et al.* 2006).

and gorillas have revealed genetic structure in a geographical context and thus allowed tests of this hypothesis. However, these studies have not been without controversy, particularly for the genus *Gorilla*. Before we examine the patterns of genetic variation in chimpanzees, bonobos, and gorillas that are consistent with reticulations within *Gorilla* and *Pan*, it is necessary to discuss the source of this controversy.

In their paper entitled "Unreliable mtDNA data due to nuclear insertions: A cautionary tale from analysis of humans and other great apes", Thalmann *et al.* (2004) reviewed evidence for the translocation of mtDNA sequences into the nuclear genomes of humans and other primates. These nuclear insertions of mtDNA or as they have come to be known, "numts," represent a potential source for artifactual data. Polymerase chain reaction (PCR) amplification of "mtDNA" from different individuals might actually result in the amplification and analysis of (1) nuclear and mtDNA, (2) nuclear, or (3) mitochondrial sequences. Given that nuclear-encoded genes, on average, are known to accumulate mutations at one-tenth the rate of genes contained within the mitochondrial genome (Hartwell *et al.* 2004, p. 534), the occurrence of "1" or "2" could lead to erroneous estimations of evolutionary relatedness, geographic structuring, and reticulate evolution.

Thalmann *et al.* 2004 concluded that gorillas (but apparently not humans, chimpanzees, bonobos, or orangutans) "are notable for having such a variety of numt sequences bearing high similarity to authentic mtDNA that any analysis of mtDNA using standard approaches is rendered impossible." This brought into question inferences made by numerous earlier studies (e.g., Garner and Ryder 1996; Saltonstall *et al.* 1998; Jensen-Seaman and Kidd 2001; Hofreiter *et al.* 2003) concerning the evolutionary processes that have impacted lineages within *Gorilla*—including evidence for reticulations. With this cautionary lesson in mind, more recent analyses of mtDNA variability within primates, and particularly gorillas, have employed methodologies intended to minimize the likelihood of including mtDNA pseudogenes located in the nucleus (e.g., Jensen-Seaman *et al.* 2004). Furthermore, analyses of nuclear-encoded genes have provided data sets that are not affected by the potential artifacts associated with mtDNA studies.

Gorilla

Gorillas were initially placed into a single species, *Gorilla gorilla*, with a separate subspecific designation given to animals from the eastern (*Gorilla gorilla beringei*) and western (*Gorilla gorilla gorilla*) portions of its geographic distribution (Walker *et al.* 1975, p. 477). Recently, the taxonomy of this genus has been changed in that the two subspecies have been given specific status, *Gorilla beringei* and *Gorilla gorilla* (Figure 3.5; see Clifford *et al.* 2004 for a discussion). Some of the evidence for this decision came from the earlier mtDNA studies, and thus one could argue that changing the taxonomy might have been premature (Jensen-Seaman *et al.* 2004). However, studies that included nuclear loci have also revealed differentiation between the eastern and western populations, thus supporting specific recognition (e.g., Jensen-Seaman *et al.* 2001; Figure 3.5). Regardless of the taxonomic status of the eastern and western forms of gorillas, it is apparent that these two lineages have diverged from one another. In the context of this book, it is important to ask whether or not some level of genetic exchange accompanied this divergence. I will highlight four papers that either implicitly or explicitly addressed this question.

Clifford *et al.* (2004) examined mtDNA variation to test for phylogeographic subdivisions in *Gorilla*, and particularly among the western *G. gorilla* subspecies. The major findings from this study included the confirmation of the distinctiveness of the *G. beringei* and *G. gorilla* lineages. Furthermore, genetic variation within *G. gorilla* identified patterns suggesting (1) a subdivision between populations from eastern Nigeria to southeastern Cameroon (i.e., northwestern distribution of this species) relative to all other populations of *G. gorilla* and (2) further substructuring within this latter group of populations (Clifford *et al.* 2004). Clifford *et al.* (2004) also noted two findings that were consistent with past and ongoing introgressive hybridization. First, within *G. gorilla*, there was a zone of admixture between the Nigeria–Cameroon

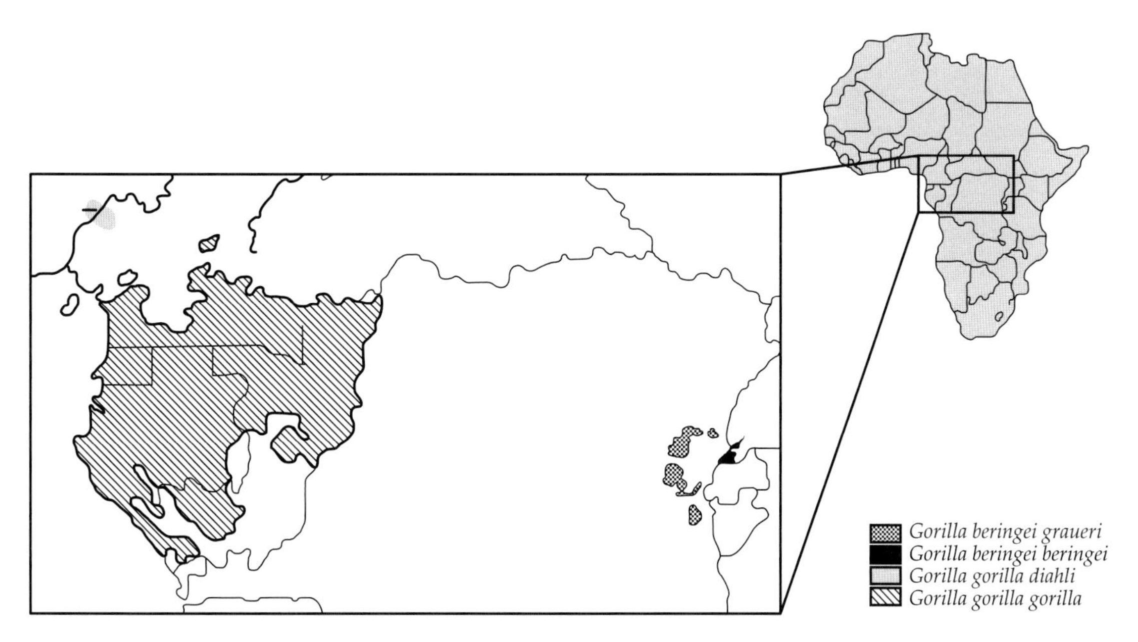

Figure 3.5 Current geographical distributions of two species (with two subspecies each) of *Gorilla* (Thalmann *et al.* 2007).

haplogroup and the haplogroup possessed by the remainder of the western lowland gorilla samples (Figure 3.6). Introgression between these two lineages was suggested for both Gabon and southeastern Cameroon populations by the presence of individuals characterized by the mtDNA sequence variation from both haplogroups (Figure 3.6; Clifford *et al.* 2004). The second observation was based on the identification of presumptive numts. In this case, the authors detected highly similar numt sequences shared by individuals from the two gorilla species (Figure 3.7). Clifford *et al.* (2004) argued that such high similarity of these numts could be explained by their translocation in the common ancestor of *G. gorilla* and *G. beringei*. Alternatively, they argued for recent introgression between the two species, or the differential sorting of ancestral polymorphisms, to explain this finding. Given the evidence for past and ongoing introgression between the divergent forms of *G. gorilla*, it would seem likely that introgression had contributed to the sharing of these numts between *G. gorilla* and *G. beringei*. The detection of mtDNA admixture in these populations has recently been supported by a more extensive phylogeographic analysis by Anthony *et al.* (2007). In this latter analysis, admixture of divergent mtDNA variants was detected in the same geographic regions (i.e., Gabon and Cameroon) identified by Clifford *et al.* (2004).

Kaessmann *et al.* (2001) collected approximately 10,000 base pairs of sequence from the X-chromosome locus, Xq13.3, from 10 *G. gorilla* and 1 *G. beringei* individual, and from individuals of both orangutan subspecies. The goal of their study was to test whether or not the reduced level of genetic variation detected in the human lineage, based on comparisons with chimpanzees, held when *Homo* was compared to *Gorilla* and *Pongo.* Their findings were that humans did indeed possess lower levels of sequence variation relative to gorillas and orangutans (Kaessmann *et al.* 2001). This conclusion was reflected by much greater divergence between samples of the other primates than found among human sequences. However, the greater diversity among the gorilla samples was not accompanied by reciprocal monophyly for samples of the two species (Kaessmann *et al.* 2001). This discordance between the taxonomy of the samples and the molecular phylogeny likely

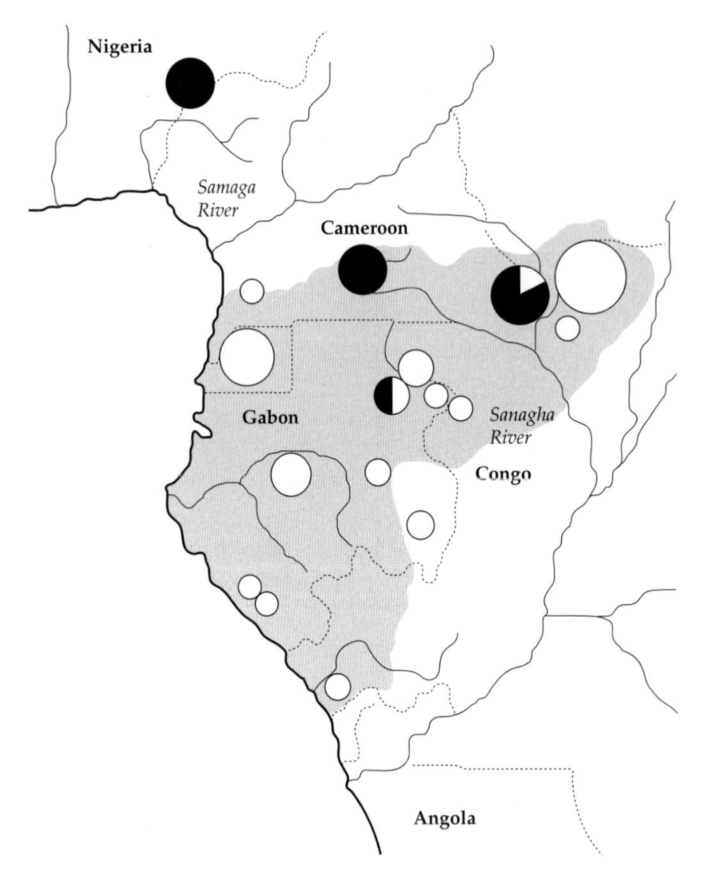

Figure 3.6 The distribution of mtDNA sequence variation among western lowland gorillas (i.e., *Gorilla gorilla*) samples. The filled portion of the circles reflects one haplogroup and the open portion a second haplogroup. The size of the circles reflects the number of samples analyzed per population, and the gray shading across the map indicates the present-day distribution of gorilla populations. Note that admixtures of individuals carrying the two divergent mtDNA forms are found in southeastern Cameroon and Gabon (Clifford *et al.* 2004).

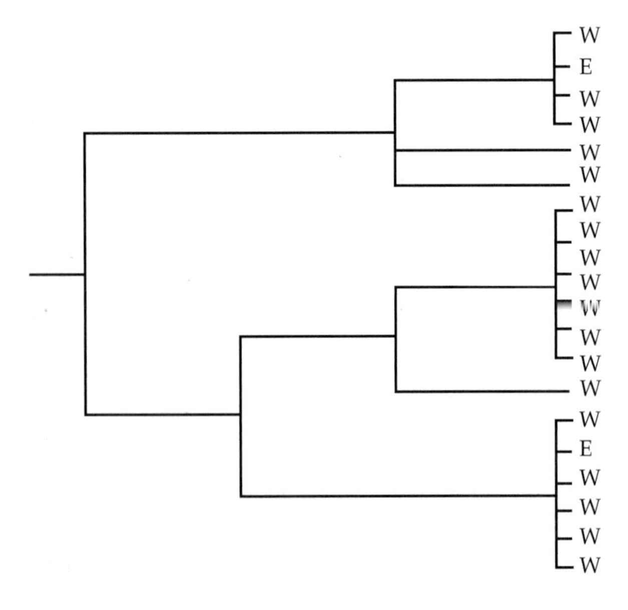

Figure 3.7 Phylogeny of *Gorilla gorilla* (i.e., "W" for western lowland gorillas) and *Gorilla beringei* (i.e., "E" for eastern lowland gorilla) based on presumptive numt sequences. Note that there is little sequence divergence between the two eastern sequences and the western sequences with which they cluster (Clifford *et al.* 2004).

reflects the occurrence of gene flow between these lineages as they diverged.

The final two papers concerning the population genetics of *Gorilla*, which included information that provided tests for reticulation, come from the work by Vigilant and her colleagues. The first was a review article that summarized studies of nuclear DNA variation. From their survey, Vigilant and Bradley (2004) concluded that the limited number of nuclear data sets (1) were unable to resolve *G. gorilla* from *G. beringei* and (2) suggested a recent origin for gorilla lineages, relative to that found for chimpanzees and bonobos. In the second paper, Thalmann *et al.* (2007) collected sequence information from 16, noncoding, autosomal loci from 15 *G. gorilla* (including individuals from both subspecies) and 3 *G. beringei* individuals; they examined approximately 14,000 base pairs of sequence from each animal. To determine whether or not the evolutionary divergence between *G. gorilla* and *G. beringei*, as well as between the two *G. gorilla* subspecies, occurred with introgression, Thalmann *et al.* (2007) employed the analytical methodologies developed by Hey, Wakeley, and their colleagues (e.g., Wakeley and Hey 1997; Hey and Nielsen 2004). These methodologies allow a test of parameter sets that range from no gene flow between the diverging lineages to complete panmixia. They accomplished this by ascertaining the goodness-of-fit between the data and the expectations from the various models. Thalmann *et al.*'s (2007) sequence data did not support either a no-gene-flow model, or a model in which the gorilla ancestral population was in panmixia until very recently. Instead, the pattern of sequence variation at the 16 autosomal loci were consistent with the following scenario: (1) the *G. gorilla* subspecies' divergence was accompanied by repeated admixture caused by climatic oscillations that translated into habitat shifts; (2) divergence of *G. gorilla* and *G. beringei* occurred from 0.9 to 1.6 mya with subsequent introgression until *c.*80,000 years ago; and (3) there was asymmetry in the introgression between the lineages that gave rise to the two species with gene flow occurring largely from *G. beringei* into *G. gorilla* (Thalmann *et al.* 2007).

The studies of *Gorilla* indicate that lineages within this genus form a network rather than a simple bifurcating tree. Once again primates are seen to be an excellent example of evolutionary diversification with genetic exchange and thus of the web-of-life metaphor. Significantly, this is found to be the case for one of the two most closely related taxa relative to our own species. I will now discuss information for our other closely allied sister clade that includes chimpanzees and bonobos.

Pan paniscus and Pan troglodytes

The genus *Pan* is now recognized as containing two species, *Pan troglodytes* (the "common chimpanzee") and *Pan paniscus* (the "pygmy chimpanzee" or "bonobo"; Figure 3.8). Divergence times between these two species, estimated from the level of sequence variation detected at mtDNA, autosomal, X-, and Y-chromosome loci (Pesole *et al.* 1992; Kaessmann *et al.* 1999; Stone *et al.* 2002; Yu *et al.* 2003), have ranged from less than 1 million to nearly 2.5 mya. These estimates indicate a relatively high level of genetic divergence between the two lineages, but only for some loci. For example, analyses of mtDNA variation among *P. paniscus* and *P. troglodytes* individuals have reported monophyly for the two species (Morin *et al.* 1994). In contrast, a number of studies have determined patterns of sequence variation at nuclear loci suggesting the impact of genetic exchange during the divergence of chimpanzees and bonobos.

Deinard and Kidd (1999, 2000) reported mtDNA and nuclear sequence variation between *P. paniscus* and *P. troglodytes* (and between the *P. troglodytes* subspecies, *Pan troglodytes troglodytes, Pan troglodytes verus and Pan troglodytes schweinfurthii*; see Figure 3.8 for distributional map). Consistent with previous studies, mtDNA sequence variation resulted in a monophyletic grouping of chimpanzees relative to bonobos (Deinard and Kidd 2000). However, nuclear sequence variation indicated a significantly different evolutionary inference. Analysis of the HOXB6 locus in chimpanzees and bonobos resolved an admixed network of relationships among the sequences from the two species (Deinard and Kidd 1999). A second study by Deinard and Kidd (2000), which examined sequence variation between *P. paniscus* and *P. troglodytes*, confirmed this finding for the HOXB6

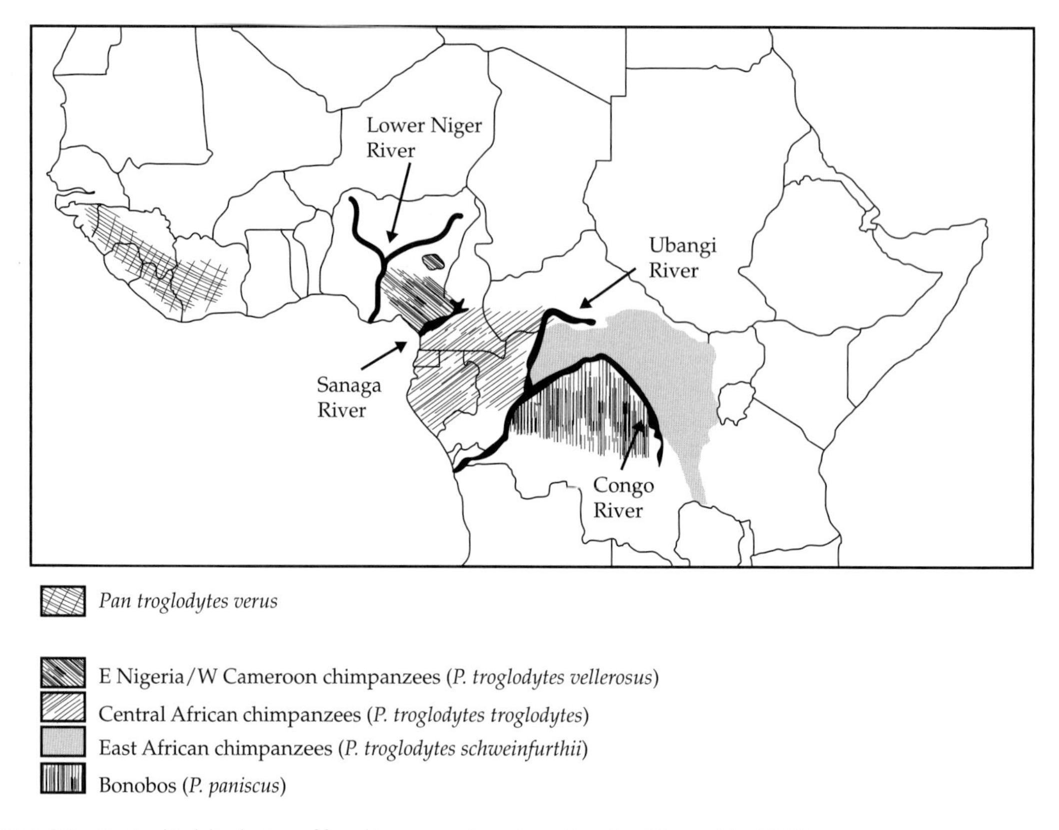

Figure 3.8 Geographical distributions of four chimpanzee subspecies and bonobos (Won and Hey 2005).

locus; some chimpanzees and bonobos were more similar to one another than they were to members of their own species. However, HOXB6 was not the only locus demonstrating closer evolutionary relationships between members of the two species than to conspecific individuals. DNA sequence data for a second autosomal locus, APOB, as well as the X-chromosome locus, PABX defined such patterns as well. For APOB, some *P. paniscus* and *P. troglodytes verus* individuals were more closely related than they were to other members of their own species (Figure 3.9; Deinard and Kidd 2000). Similarly, sequences detected at the PABX locus grouped *P. paniscus* animals as closely to *P. troglodytes verus* and *P. troglodytes schweinfurthii* as the latter animals were to other *P. troglodytes* individuals (Deinard and Kidd 2000).

Kaessmann *et al.* (1999) compared sequence variation at the X-chromosome locus Xq13.3 from chimpanzees and bonobos. Approximately 10,000 base pairs were sequenced from *P. troglodytes troglodytes, P. troglodytes verus, P. troglodytes schweinfurthii,* and from *P. paniscus* individuals. Similar to each of the nuclear loci examined by Deinard and Kidd (1999, 2000), as well as mtDNA loci surveyed previously (e.g., Morin *et al.* 1994), variation at Xq13.3 resolved all *P. paniscus* into a monophyletic group. However, also like previous findings, some of the chimpanzee individuals demonstrated a closer evolutionary relationship with bonobos than they did to other chimpanzees (Figure 3.10; Kaessmann *et al.* 1999). This was reflected by the fact that *P. troglodytes* sequences differed from one another at as many as 22–29 base pair positions, while they differed from *P. paniscus* at only 13–23 base pairs. This led Kaessmann *et al.* (1999) to reflect on the observation that bonobos and chimpanzees can hybridize and that during the divergence of

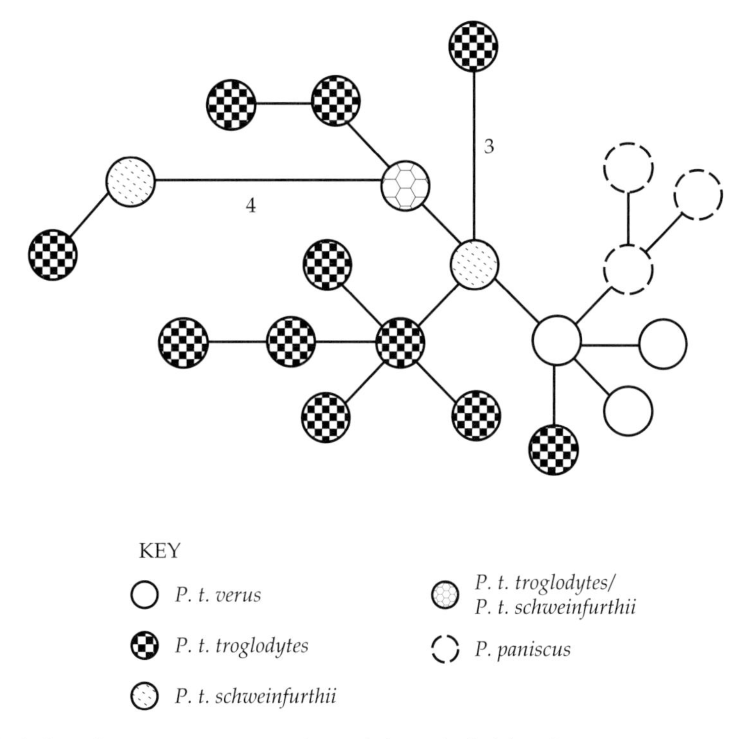

Figure 3.9 Network of relationships among *P. paniscus* and *P. troglodytes* individuals based on sequence variation at the autosomal locus APOB. Each circle reflects a sequence variant at this locus. Closed circles indicate *P. troglodytes* sequences. Dashed circles represent *P. paniscus* genotypes. Numbers along connecting lines indicate instances in which genotypes differed by more than one mutation. The sequence variation indicates that some *P. troglodytes verus* individuals possessed genotypes that were more similar to *P. paniscus* than to other *P. troglodytes* animals (Deinard and Kidd 2000).

these lineages "certain loci, for example, Xq13.3, may have crossed the 'species barrier' much later than other loci"—that is, via introgressive hybridization (but see Won and Hey 2005).

Though the above discussion indicates the close evolutionary relationship between the two species of *Pan*, it should be pointed out that they are easily distinguished on the basis of phenotype (Walker 1975, p. 476). Thus, although estimates of genetic divergence are often low, the degree of morphological differentiation between *P. paniscus* and *P. troglodytes* is significant. As Fischer *et al.* (2004) suggested, the "uncoupling" of loci that determine phenotype versus those that do not, might reflect the action of a variety of evolutionary processes such as genetic drift, natural selection, and so forth. The above findings suggest that reticulation also plays a role in separating the effects of these loci.

Pan troglodytes subspecies

Chimpanzee systematics has a checkered and confusing history, with this group split into as many as 17 species and 34 subspecies. However, many taxonomic treatments have been best represented by the one species, *P. troglodytes* being divided into three subspecies, *P. t. verus* ("west African" chimpanzee), *P. t. troglodytes* ("central African" chimpanzee), and *P. t. schweinfurthii* ("east African" chimpanzee; Morin *et al.* 1994). More recently, genetic variation at mitochondrial loci have also led to the recognition of two west African lineages and the naming of a fourth subspecies, *P. t. vellerosus* (Gonder *et al.* 1997).

Morin *et al.* (1994) discussed the limited understanding of chimpanzee evolutionary relationships based on the nondiagnostic morphological characteristics of the various subspecies. Specifically, they stated, "Vernacular names and

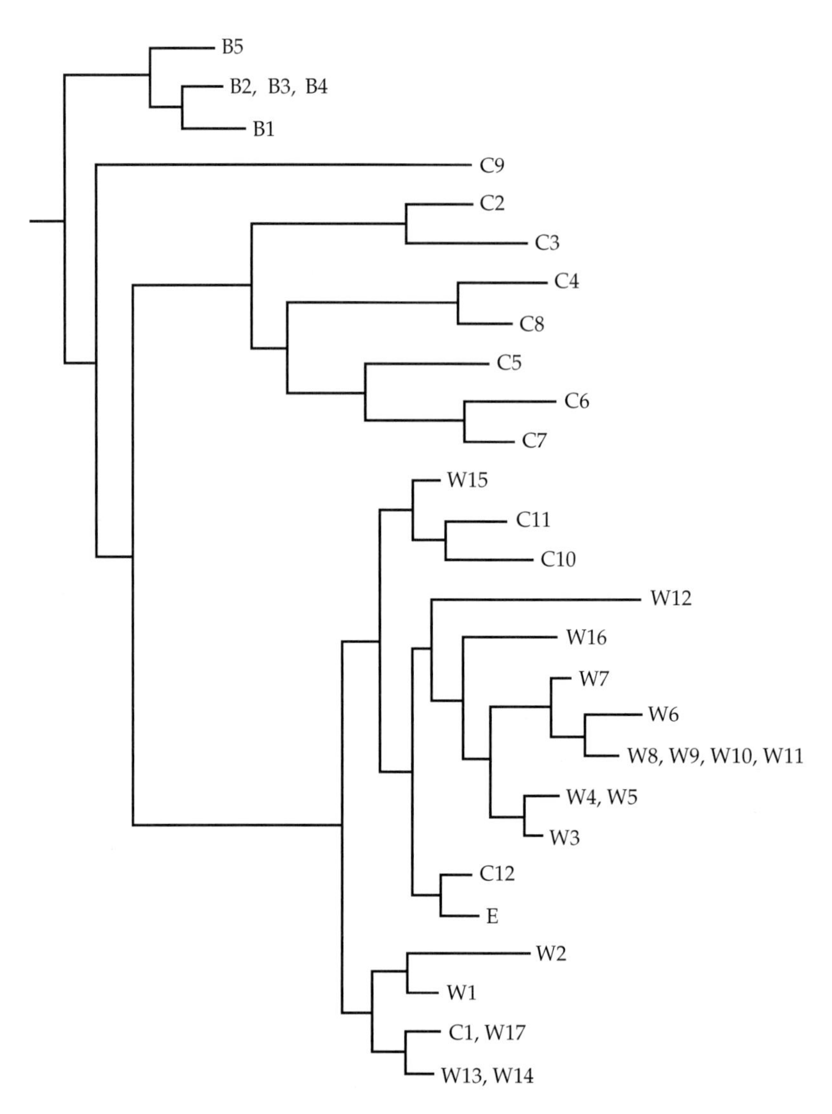

Figure 3.10 Phylogeny of *P. troglodytes* (chimpanzee) and *P. paniscus* (bonobo) samples based on ~10,000 base pairs of noncoding X-chromosome DNA from locus Xq13.3. "C", "E," and "W" indicate sequences from the central, eastern, and western chimpanzee subspecies. "B" indicates samples from bonobos. The western and central chimpanzees are not monophyletic and the single eastern subspecies sample falls within a clade containing individuals of the other two subspecies. Also, some of the chimpanzee individuals are genetically more distant from one another than they are from bonobo individuals (Kaessmann *et al.* 1999).

minor craniometric variation notwithstanding, these *geographically defined* [my emphasis] allopatric subspecies cannot be distinguished morphologically" (Morin *et al.* 1994). Indeed, this high level of morphological similarity, suggesting close evolutionary relatedness, is consistent with a lack of postzygotic reproductive isolation between the different subspecies, as reflected by no increase in levels of mortality among the hybrid progeny (Ely *et al.* 2005). Because the chimpanzee subspecies could not with confidence be separated on the basis of morphological characteristics, captive individuals of uncertain geographic origin have been of questionable value for evolutionary studies aimed at defining genetic variation. To obviate this problem, Morin *et al.* (1994) collected samples from wild chimpanzees across the geographic range of the subspecies to assess mtDNA sequence variation within and between the three forms. Morin *et al.* (1994) found that both the mtDNA cytochrome b gene and the hypervariable control region sequences consistently placed the west African *P. t. verus* samples into a monophyletic clade relative to the other two subspecies. However, the cytochrome b sequences placed *P. t. troglodytes* and *P. t. schweinfurthii* individuals into a common assemblage. In addition, though the control region sequences defined a monophyletic clade containing only *P. t. schweinfurthii* samples, *P. t. troglodytes* individuals did not form a monophyletic grouping (Morin *et al.* 1994). These results are consistent with the evolutionary diversification of *P. troglodytes* being accompanied by periods of gene flow between some of the lineages.

Several studies have indicated the presence of contemporaneous introgressive hybridization between the subspecies, suggesting also that they may not be completely allopatric in their distributions. For example, both Stone *et al.* (2002) and Becquet *et al.* (2007) reported hybrid genotypes in wild chimpanzee populations. The first study involved a survey of variation among chimpanzee subspecies at the Y-chromosome NRY locus. This analysis detected one of 47 "*P. t. verus*" individuals that had an NRY allele typical for this subspecies, but a mtDNA genotype characteristic for *P. t. vellerosus* (Stone *et al.* 2002). In their analysis, Becquet *et al.* (2007) used 310 nuclear loci to genotype both chimpanzee and bonobo individuals. The pattern of genetic variation in *P. troglodytes* suggested the presence of three lineages that corresponded to the western, central, and eastern subspecies (Becquet *et al.* 2007). Furthermore, these authors stated that they had found "little evidence of gene flow between" these lineages. However, 2 of the 51 wild-collected samples possessed hybrid genotypes. One of the samples was inferred to be of central × eastern origin and the other of western × central derivation (Becquet *et al.* 2007). These two individuals reflect 4% of the wild chimpanzees having recombinant genotypes and, given the extremely limited areas of overlap, suggests significant and ongoing impacts from introgression among the subspecies.

The question of the genetic uniqueness of chimpanzee subspecies has continued to generate interest from both an evolutionary and conservation point of view. Fischer *et al.* (2006) examined levels of nuclear, genetic differentiation among western, eastern, and central lineages. Their findings led them to conclude that there was little, if any, support for the recognition of subspecies. They arrived at this inference by sequencing ~22,000 base pairs from 26 autosomal regions from *P. t. verus*, *P. t. troglodytes*, *P. t. schweinfurthii*, and *P. paniscus*. They compared the sequence differentiation between the subspecies and between chimpanzees and bonobos (Figure 3.11). Fischer *et al.* (2006) also examined the sequence variation contained in these regions in two subspecies of orangutans and three human populations. Figure 3.11 illustrates the degree of sequence differentiation detected for each of the taxonomic categories (i.e., between populations [humans], between subspecies [chimpanzees and orangutans] and between species [chimpanzees and bonobos]). No fixed (for alternate alleles) sequence differences were detected between the central and eastern chimpanzee subspecies (or between the orangutan subspecies) and very few fixed differences were detected in comparisons between the western subspecies and either the eastern or central subspecies (Figure 3.11). Fischer *et al.* (2006) argued that the level of differentiation among the three chimpanzee (and two orangutan) subspecies was "comparable to that seen among human populations, calling the validity of the 'subspecies' concept in apes into question." Their conclusion is also consistent with the lack of strong differentiation between chimpanzee subspecies detected by Yu *et al.* (2003) in an analysis of 50, randomly chosen, nuclear loci. Surprisingly, in this latter study the smallest sequence divergence was found between the most geographically separated forms (i.e., the western and eastern subspecies), while the largest differentiation occurred between the central and eastern samples (Yu *et al.* 2003). However, it is important to note that the level of between-subspecies differentiation was nearly the same as that found

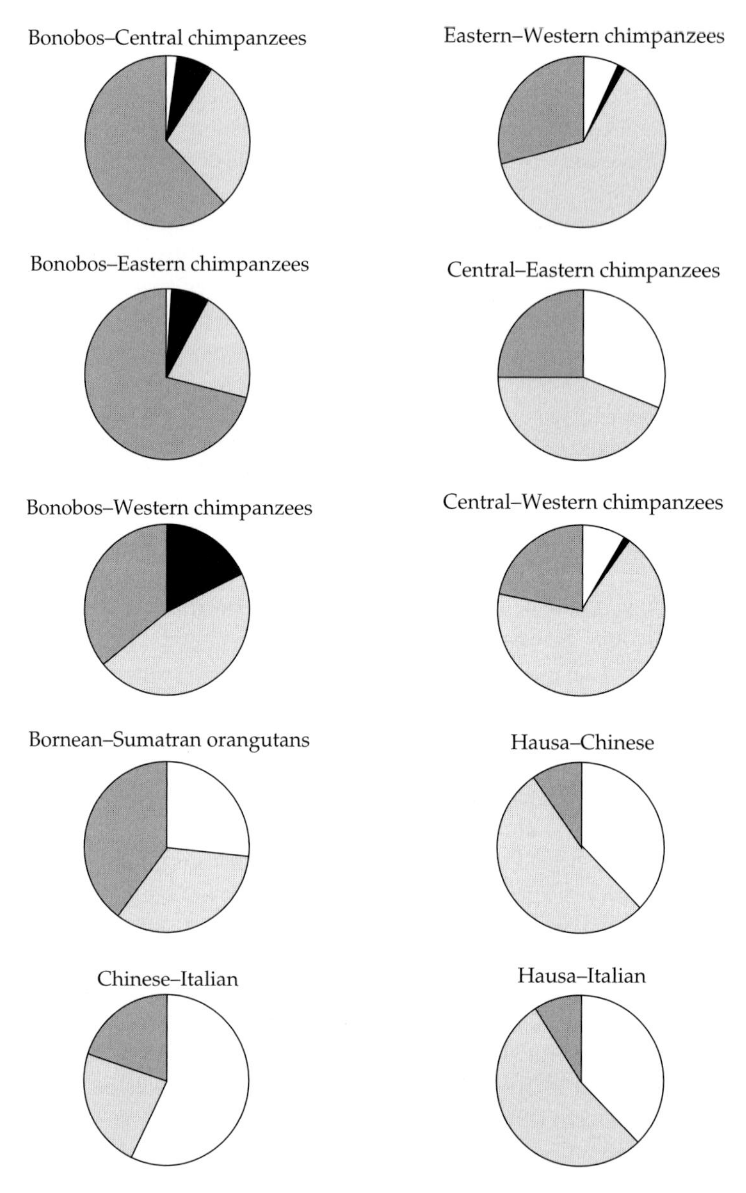

Figure 3.11 Pie charts indicating the proportions of autosomal sequences that are (1) shared (white area of charts), (2) fixed for different base pairs (black area of charts), (3) unique to one group or the other (light or dark gray areas of charts, respectively) for bonobo–chimpanzee, chimpanzee subspecies, orangutan subspecies, and human population comparisons. The chimpanzee and orangutan subspecies show approximately the same levels of sequence divergence as human populations, indicating the low level of differentiation between taxonomic units within either *Pan* or *Pongo* (Fischer *et al.* 2006).

within subspecies, indicating the high genetic similarity between all three lineages. Yu *et al.*'s (2003) and Fischer *et al.*'s (2006) nuclear sequence data support the conclusion of past introgression among the diverging lineages now recognized as different subspecies.

Difficulties in resolving chimpanzee subspecies into monophyletic units have also been encountered

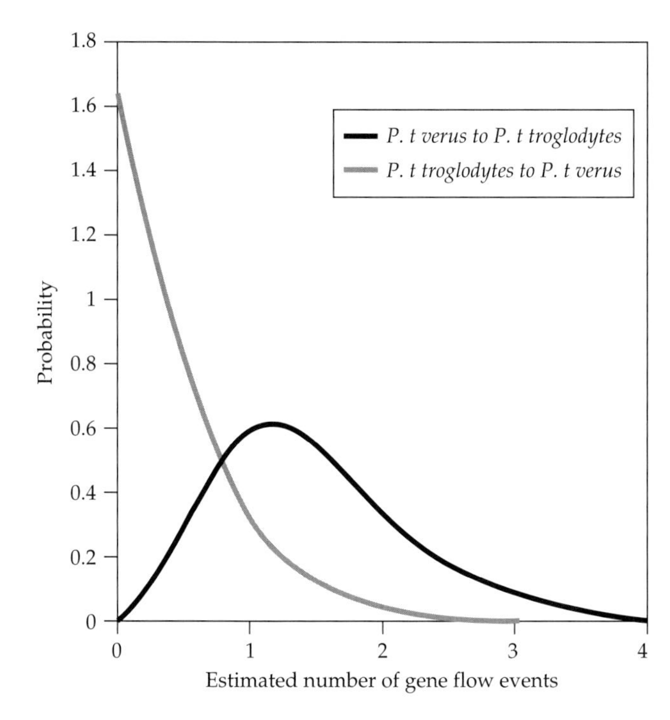

Figure 3.12 Probability distribution of gene flow events between the chimpanzee subspecies, *P. t. verus* and *P. t. troglodytes*. The peak shape and position indicates an estimate of one gene flow event, on average, from *P. t. verus* into *P. t. troglodytes*, but an estimate near zero for introgression in the opposite direction. This indicates that introgression is occurring between these two subspecies, but that it is unidirectional (i.e., occurs from *P. t. verus* into *P. t. troglodytes*, but not vice versa; Won and Hey 2005).

by other workers. As discussed above, Kaessmann *et al.* (1999) and Deinard and Kidd (2000) detected closer relationships between some chimpanzee samples and those of bonobos than to other chimpanzee individuals (Figures 3.9 and 3.10). In addition, both of these studies detected phylogenetic admixture among individuals of the western, eastern, and central subspecies. Figure 3.10 reflects the lack of monophyly for X-chromosome sequences from central and western chimpanzees, and the high similarity between the single eastern chimpanzee sample and individuals belonging to both of the other subspecies (Kaessmann *et al.* 1999). Interestingly, Deinard and Kidd (2000) concluded that the sequence data from nuclear loci could "be used to differentiate common chimpanzee subspecies." However, the genotypic networks constructed from HOXB6, APOB, and PABX show closer relationships between some members of different subspecies than to individuals from the same subspecies (e.g., *P. t. troglodytes* and *P. t. schweinfurthii*; Figure 3.9). These data also support the hypothesis of genetic exchange during the divergence of the chimpanzee lineages.

The last study that I will discuss reflects an explicit test for introgressive hybridization between two of the *P. troglodytes* subspecies: *P. t. verus* and *P. t. troglodytes*. Won and Hey (2005) utilized the data sets collected by Morin *et al.* (1994), Deinard and Kidd (1999), Kaessmann *et al.* (1999), Stone *et al.* (2002), and Yu *et al.* (2003) to estimate gene flow between these subspecies. Their analysis resulted in nonzero migration values in the direction of *P. t. verus* > *P. t. troglodytes* (Figure 3.12; Won and Hey 2005). In contrast, no gene flow was detected from *P. t. troglodytes* into *P. t. verus* (Figure 3.12). Introgression, thus, appears to be unidirectional, but gene flow values indicated that a majority of the loci examined (37 of the 48 loci included) had signatures of gene flow events (Won and Hey 2005). As these authors stated, the finding of introgression between these two subspecies, given their disjunct geographical distributions, was surprising. However, Won and Hey (2005) suggested a scenario involving the recently named subspecies, *P. t. vellerosus* that could account for this finding: (1) divergence from a common ancestor of lineages leading to *P. t.*

verus and *P. t. troglodytes*; (2) divergence from a common ancestor of lineages leading to *P. t. verus* and *P. t. vellerosus*; and (3) present-day geographical proximity of *P. t. vellerosus* and *P. t. troglodytes* results in gene flow from *P. t. vellerosus* into *P. t. troglodytes*. Such introgression would transfer alleles similar to those found in the related *P. t. verus* into *P. t. troglodytes* resulting in Won and Hey's (2005) analysis identifying introgression from the former into the latter, when in reality the introgression involves the related *P. t. vellerosus*. In any case, as with each of the other chimpanzee analyses, these results reflect the affect of introgression during the diversification of this assemblage.

3.4 Summary and conclusions

The descriptions of morphological and genetic variation in hominines reviewed in this chapter support the hypothesis that reticulate evolution has been widespread and evolutionarily salient. From fossil finds indicating the likelihood of introgression and hybrid speciation between australopithecines, to estimates of gene flow between proto-humans, proto-chimpanzees, and proto-gorillas, the network that includes our species reflects exactly the same effects described in more distantly related primates, and indeed, bacteria, plants, and other animals. This really should not be shocking since we and our closest relatives are part of, rather than somehow outside of, the natural world and thus the web of life. However, resistance to our species' inclusion in the category of organisms impacted by genetic exchange during divergence has been strong.

The words of Jolly (2001) reflect well one of the conclusions that can be drawn from the data presented in this and the next chapter concerning the implications of the reticulate evolution within "our" clade: "it is important to stress that if we reduce recognizable 'forms' of *Homo*...to subspecific status because of the possibility of some marginal gene flow between them, this would not imply that they were 'ephemeral' or 'evolutionarily unimportant,' any more than these terms could be applied to, say, anubis baboons." Instead of focusing on whether or not our favorite primate—extinct or extant—is in danger of being reduced in taxonomic standing due to hybridization and the application of a certain species concept, we should recognize the evolutionary diversification that has originated from reticulations. In the next chapter, I will examine the evidence of how reticulate evolution among lineages of *Homo* as well has genetically, behaviorally, and phenotypically enriched our own species.

CHAPTER 4

Reticulate evolution in the genus *Homo*

Hybridization models allow for some gene flow between anatomically modern humans migrating from Africa and Archaic populations outside Africa.

(Tishkoff and Verrelli 2003)

...a wealth of new data has supported a longer chronology of hominin presence outside Africa beginning at ~1.6–1.8 Ma....

(Antón and Swisher 2004)

If these lice indeed codiverged with their hosts ca. 1.18 million years ago, then a recent host shift from an archaic species of *Homo* to modern *H. sapiens*...would require direct physical contact between modern and archaic forms of *Homo*.

(Reed *et al.* 2004)

...the presence of such anomalies in individuals should be considered a potential indicator of hybridization...In this light, it is interesting to consider the possibility that the type skull of *Homo floresiensis* might demonstrate comparable evidence....

(Ackermann *et al.* 2006)

Importantly, haplotypes found in Eurasia suggest interbreeding between then contemporaneous human species.

(Hayakawa *et al.* 2006)

This morphological mosaic indicates admixture between regional Neandertals and early modern humans dispersing into southern Iberia.

(Duarte *et al.* 1999)

4.1 Reticulate evolution within *Homo*

As an interested bystander, I often wonder why it seems to be so important to so many researchers that *Homo* species have not co-occurred, and even more importantly, have not hybridized. Is it because these workers feel that the status of "their" fossil species will somehow be of less importance if it is concluded that it interbred with another form? Is it because a particular species concept (i.e., the "biological species concept" that requires reproductive isolation between different species) has been applied, leading to the circular argument that "We know that lineage X is a separate species, and by definition different species do not interbreed thus lineage X individuals cannot have mated with lineage Y individuals"? Given that these rationales are related, it is likely that resistance to the observations consistent with introgression between "archaic" and "anatomically modern" *Homo* species reflects the influence from both. However, it is unnecessary, and evolutionarily inaccurate, to assign inherent values that elevate the importance of some lineages on the basis of taxonomic category (Darwin 1859). Yet, this taxonomic bias is nowhere more obvious than when the inference of genetic exchange is made (Arnold 1997, 2006). As I have already illustrated, genetic exchange is a component of the evolutionary derivation of organisms. As we have seen from Chapter 3, this includes numerous primates. As we shall see in the following sections, it includes our own species as well.

4.2 Models of *Homo* evolution: Replacement, Multiregional, and Hybridization

Tishkoff and Verrelli (2003) argued that "Comparative [genetic] studies across ethnically diverse human populations and across human and nonhuman primate species is important for reconstructing human evolutionary history." In particular, genetic analyses can be used to test various models constructed to explain the geographic distribution and morphological variation observed in extinct species and contemporaneous *Homo sapiens* populations. Figure 4.1 illustrates three such models. Each model was formulated initially to explain patterns of variation observed in both the fossil record and contemporary humans. For example, recent iterations of the Multiregional model reflect both (1) the regional associations of morphological traits in fossil species (e.g., *Homo erectus*) and modern *H. sapiens* populations and (2) the sharing of many traits between different regions (Tishkoff *et al.* 1996; Wolpoff 1996; Wolpoff *et al.* 2001). The first observation suggests that human populations from different regions evolved separately from one another. The second observation suggests, however, that high levels of gene flow accompanied the evolution of *H. sapiens* from multiple, regional foci (Tishkoff *et al.* 1996; Wolpoff 1996; Wolpoff *et al.* 2001; Tishkoff and Verrelli 2003).

In contrast to multiregional evolution from *H. erectus*, the Replacement model (Figure 4.1) reflects the conclusion that all present-day human populations shared a common *H. sapiens* ancestor that arose in Africa and subsequently spread into every region now occupied, with accompanying extirpation (either passive or active) of, but no hybridization with, any archaic forms (Tishkoff and Verrelli 2003; Fagundes *et al.* 2007). Findings from the fossil record consistent with the hypothesis of replacement without introgression include the following:

1. The earliest (i.e., 90–120,000 years ago), anatomically modern *H. sapiens* samples have been discovered in Africa and the adjacent Middle East (Stringer and Andrews 1988).
2. The earliest indications of modern human behavioral attributes have been detected in eastern and southern African populations with dates ranging from as early as 250,000 to as late as 70,000 years ago (McBrearty and Brooks 2000; Henshilwood *et al.* 2002; Mellars 2004).

In comparison to the other two models, the Hybridization/Introgression model (Figure 4.1; also referred to as the "Assimilation" model) reflects an intermediate amount of hypothesized introgression. With respect to African and "other" population centers, the pattern of introgression was likely bidirectional and temporally and spatially idiosyncratic due to the episodic nature of *Homo* dispersal events (e.g., see Tishkoff and Verrelli 2003; Martinón-Torres *et al.* 2007). Similar to the Multiregional model, the Hybridization/Introgression model is supported by fossils that demonstrate the presence of admixtures of morphological traits described as characteristic of different lineages (Duarte *et al.* 1999; Trinkaus 2007), though these fossils are often assigned to the categories of "intermediate" or "ancestral" (e.g., Stringer and Andrews 1988; Bermúdez de Castro *et al.* 1997).

4.3 Tests of the Multiregional, Replacement, and Hybridization/Introgression Models of human evolution

As with any evolutionary hypotheses, the Multiregional, Replacement, and Hybridization/Introgression models of human evolution depend

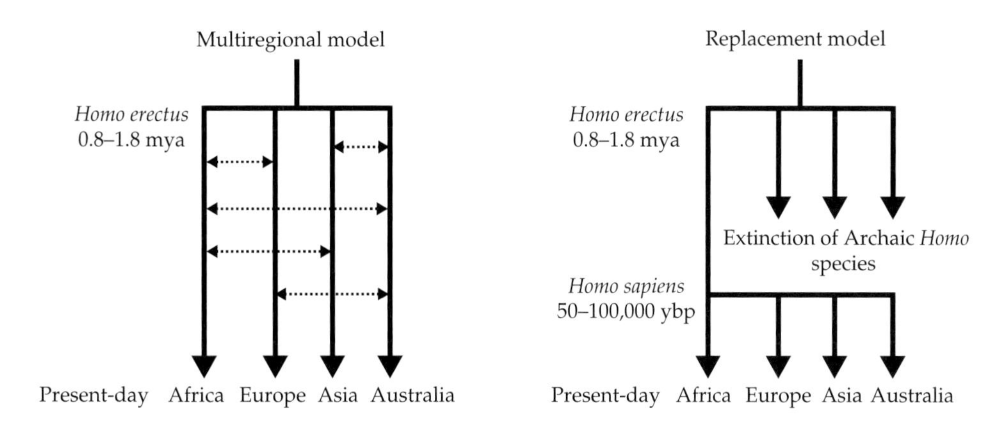

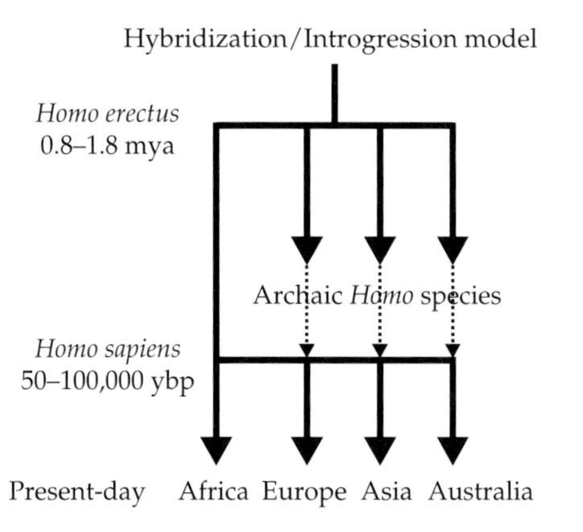

Figure 4.1 Three models formulated to explain the evolution and geographic distribution of present-day *H. sapiens*. The Multiregional model reflects separate origins of present-day human populations located in different geographic regions, but with gene flow/introgression having occurred as the regional populations diverged. The Replacement model (i.e., "Out-of-Africa" model) reflects a single origin (from Africa) for *H. sapiens* with its subsequent spread into regions containing archaic *Homo* species, but without any accompanying introgressive hybridization with these taxa. The Hybridization/Introgression model reflects a single origin of *H. sapiens* with its subsequent spread into regions containing archaic *Homo* species accompanied by introgressive hybridization with these taxa (Wolpoff 1996; Tishkoff and Verrelli 2003).

on a set of assumptions and result in various predictions. Also like many hypotheses, the assumptions and predictions of these three models overlap somewhat. In particular, the Multiregional and Hybridization/Introgression models overlap broadly in both their assumptions and predictions (see Tishkoff and Verrelli 2003 for a discussion). For example, both models predict temporal and spatial overlap between lineages of *Homo*. Furthermore, this overlap is predicted to have resulted in genetic exchange via introgressive hybridization, and the introgression between different forms of *Homo* should be reflected in morphological and genetic admixtures in extinct and extant populations. Because these two models share many of their assumptions and predictions, and because of the focus of this book, I will consider the Multiregional and Hybridization/Introgression models to be synonymous in comparisons to the alternative Replacement model.

In contrast to the other two hypotheses, the Replacement model does not predict spatial and temporal overlap between populations and individuals belonging to different *Homo* taxa, though it does not exclude this possibility. This model does, however, predict no introgressive hybridization, and thus no anatomical, behavioral, genetic, or other evidence of exchange should be present. Furthermore, because the point of origin for all present-day human populations is assumed to be Africa, in its most extreme form the Replacement model predicts a lack of any genetic influence from populations outside of this region on extinct and extant taxa. The following discussion will examine evidence that allows a test of the alternate Hybridization/Introgression and Replacement models. In particular, I will discuss data that test for (1) spatial and temporal overlap between numerous *Homo* species and (2) introgressive hybridization between these lineages.

4.3.1 Testing the models of human evolution: fossil evidence of the co-occurrence of species

For the Hybridization/Introgression model to be relevant for explaining human evolutionary history, the various taxa must have overlapped in time and space. Support for the Replacement model would come from observations of no such overlap. Figure 4.2 illustrates the times of persistence and the geographical overlap of some of the major forms of *Homo*; this includes *H. erectus, Homo heidelbergensis, Homo neanderthalensis, Homo floresiensis,* and *H. sapiens*. Combined temporal and geographical overlap occurred between *H. erectus/H. floresiensis* and *H. neanderthalensis/H. heidelbergensis*. Our own species overlapped in space and time with *H. erectus, H. floresiensis,* and *H. neanderthalensis* (Swisher *et al.* 1996; Antón and Swisher 2004; Morwood *et al.* 2004; Pääbo *et al.* 2004; Futuyma 2005; Finlayson 2005). The data reflected by Figure 4.2 indicate only the confirmation of co-occurrence of different species to broad geographical regions (e.g., the spatial and temporal overlap of *H. neanderthalensis/H. sapiens* in Europe and *H. erectus/H. sapiens* in Southeast Asia—Swisher *et al.* 1996; Mellars 2004). However, this observation is consistent with one prediction of the Hybridization/Introgression model of human evolution. Furthermore, that the dispersal of various lineages often entailed multiple events (Antón and Swisher 2004; Mellars 2004; Martinón-Torres *et al.* 2007) would suggest, as for other organisms (see Arnold 2006 for examples), an increased likelihood of genetic exchange due to repeated opportunities for contact and introgression.

Though the overlaps in space and time reflected in Figure 4.2 are necessary for reticulate evolution, they are not sufficient. The fact that fossils of our own species and those of other hominins occur in

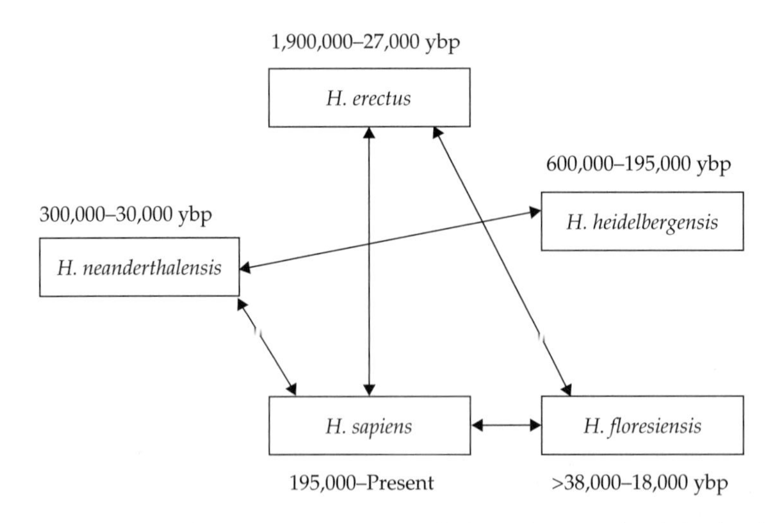

Figure 4.2 Estimated dates of existence of *Homo* species. Boxes connected by lines indicate taxa that inhabited overlapping geographical regions. "ybp" = "years before present." Data from which this figure was constructed were taken from Antón and Swisher (2004), Morwood *et al.* (2004), Pääbo *et al.* (2004), Futuyma (2005), and Finlayson (2005).

the same general, geographic setting, and temporal horizon does not necessarily indicate that the different species occupied overlapping ranges at the same time. Significantly, a number of studies have supported the hypothesis of simultaneous co-occurrence of different *Homo* species. For example, Gravina *et al.* (2005) tested the hypothesis that populations of the latest *H. neanderthalensis* and earliest *H. sapiens* overlapped at the Grotte des Fées de Châtelperron in east-central France. These authors used radiocarbon dating of a fossil sequence that included the Chatelperronian (i.e., late-Neanderthal) and Aurignacian (i.e., early modern human) cultures. The results from their dating of the Chatelperronian and Aurignacian artifacts are illustrated in Figure 4.3. Artifacts associated with *H. neanderthalensis* bracketed, in time horizons, those artifacts typical of *H. sapiens* (Gravina *et al.* 2005). This finding was consistent with the "chronological coexistence—and therefore potential demographic and cultural interactions—between the last Neanderthal and the earliest anatomically and behaviourally modern human populations in western Europe" (Gravina *et al.* 2005). Though Zilhão *et al.* (2006) published a critique of this work suggesting, among other things, that the excavation levels pictured in Figure 4.3 had been mixed due to human disturbance, Mellars *et al.* (2007) verified their earlier conclusions (i.e., Gravina *et al.* 2005) with further analyses of the excavations. At Grotte des Fées de Châtelperron *H. sapiens* and *H. neanderthalensis* overlapped in space and time, thus matching one prediction of the Hybridization/Introgression model of human evolution.

In addition to the evidence of co-occurrence of Neanderthals and anatomically modern humans, Spoor *et al.* (2007) have reported data consistent with spatial and temporal overlap between the most ancient species in this assemblage, *Homo habilis* and *H. erectus*. The analysis by Spoor *et al.* (2007) represented a test of two related hypotheses: (1) *H. habilis* evolved into *H. erectus* via **anagenesis**; and (2) *H. habilis* and *H. erectus* belonged to the same evolutionary lineage. The data collected from fossils discovered in Kenya were inconsistent with both of these hypotheses. Specifically, specimens assignable to *H. habilis* and *H. erectus* were discovered in the Ileret area near Lake Turkana. Significantly, the *H. habilis* and *H. erectus* fossils were assigned ages of 1.44 and 1.55 million years ago (mya), respectively. These ages are the reverse of that expected given the hypothesis that there was an anagenetic transition from *H. habilis* to *H. erectus*. In terms of the second hypothesis (i.e., that these two taxa belonged to the same

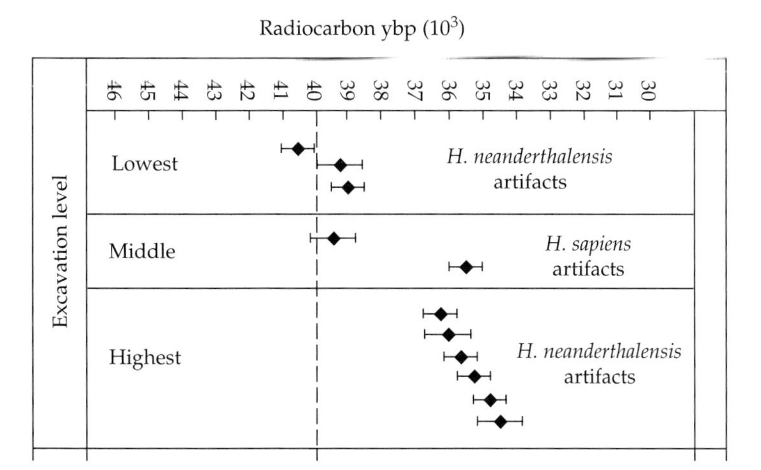

Figure 4.3 Radiocarbon dates for artifacts collected from three horizons (i.e., "Excavation Levels") at the Grotte des Fées de Châtelperron in east-central France. Artifacts dated were typical for either *H. neanderthalensis* or *H. sapiens* cultures (Gravina *et al.* 2005).

evolutionary lineage), Spoor *et al.* (2007) described diagnosable differences between the fossils of each form. For the question of co-occurrence, and thus the potential for introgressive hybridization, it is significant that their data indicate the sympatric distribution of *H. habilis* and *H. erectus* "in the same lake basin for almost half a million years" (Spoor *et al.* 2007).

The final example of likely spatial and temporal overlap involves the newly described *H. floresiensis* and *H. sapiens*. *Homo floresiensis* was described from fossil finds on the Indonesian island of Flores. Though some have suggested that this taxon is merely a developmentally anomalous form of *H. sapiens* (Jacob *et al.* 2006), the majority of studies have defined characteristics consistent with *H. floresiensis'* specific status (e.g., Tocheri *et al.* 2007). This remarkable species was defined by an adult height (1 m) and cranial capacity (380 cm^3) equivalent to that of the smallest australopithecines (Brown *et al.* 2004; Morwood *et al.* 2004). The significance of the recent discovery of this new hominin form was the falsification of the hypothesis that *H. erectus* and *H. sapiens* were the only hominins present in this geographical region (Figure 4.2). Though Brown *et al.* (2004) and Morwood *et al.* (2004) argued for the divergence of the diminutive *H. floresiensis* in genetic isolation from its ancestral *H. erectus*, it was hypothesized that following its origin this species came into contact with the expanding *H. sapiens*. Indeed, Morwood *et al.* (2004) postulated the coexistence of *H. sapiens* and *H. floresiensis* for thousands of years. Thus, like the co-occurrence of *H. neanderthalensis* and *H. sapiens* and that of *H. habilis* and *H. erectus*, the spatial and temporal overlap of *H. sapiens* and *H. floresiensis* would have provided the opportunity for introgression between different lineages within the *Homo* clade.

4.3.2 Testing the models of human evolution: evidence of hybrid fossils

As Darwin noted "That our palæontological collections are very imperfect, is admitted by every one…Only a small portion of the surface of the earth has been geologically explored, and no part with sufficient care" (Darwin 1859, pp. 287–288). Yet, this sentiment is not restricted to the data sets of 150 years ago, as reflected by Doug Futuyma's conclusion (in his textbook *Evolution*) that "On the whole, however, the fossil record is extremely incomplete" (Futuyma 2005, p. 71). It may thus come as a surprise that numerous workers, even those who were not advocates of the Multiregional and/or the Hybridization/Introgression models, have described patterns of morphological variation from the hominin fossil record consistent with gene flow between evolutionary lineages.

Just as with extant populations, several classes of observations are consistent with the hypothesis of introgressive hybridization among archaic and modern forms of *Homo*. First, individuals and populations of one species or another might possess both traits thought to be diagnostic for their own taxon as well as for a taxon with which they are known to have overlapped spatially and temporally. Related to this first observation, a second data set indicative of potential introgression is the assignment of specimens to the categories of "intermediate" or "transitional" between two fossil species. For example, Lordkipanidze *et al.* (2007) defined a fossil hominin from Dmanisi, Georgia that was described as possessing "a surprising mosaic of primitive and derived features." Specifically, this fossil reflected a combination of traits characteristic for either *H. habilis* or anatomically modern humans (Lordkipanidze *et al.* 2007). Third, individuals from a given taxon might demonstrate anomalous developmental patterning, an oft-seen expression of hybridization in contemporary hybrid zones. Though intriguing, I am aware of only one suggestion of this last class of observation. This relates to morphological features found in *H. floresiensis* that were inferred by Ackermann *et al.* (2006) to have possibly resulted from hybridization. It is interesting that Larson *et al.* (2007) also detected an "anomalous" shoulder structure in *H. floresiensis* though they explained this skeletal element by invoking the retention of an ancestral condition shared with some *H. erectus* forms. Because of the lack of any other inferences concerning this third category, I will not include it in the following discussion.

In Chapter 3, I reviewed the evidence for hybridization involving various species of *Australopithecus*.

The same type of evidence exists for the *Homo* network as well. In particular, several studies illustrate the detection of one or more of the above patterns of morphological variation among fossil species of *Homo*. For example, Stringer and Andrews (1988) concluded that most fossil evidence across the range of overlap between *H. sapiens* and *H. neanderthalensis* did not support the Multiregional model. In particular, they argued that; (1) the well-represented fossil record for Europe and Asia lacked transitional fossils between these two forms; (2) there was little evidence of continuity between these two species in other geographical regions as well; and indeed (3) *H. sapiens* predated the occurrence of Neanderthals in the **Levant**. Yet, Stringer and Andrews (1988) did find evidence for genetic/evolutionary continuity between *H. neanderthalensis* and *H. sapiens*, but surprisingly this evidence came from the more incomplete African fossil remains. Specifically, these workers reported the identification of "intermediate" fossils in South Africa, Tanzania, Ethiopia, and Morocco. This observation supported the Multiregional model (Stringer and Andrews 1988), and is obviously consistent with the occurrence of introgressive hybridization between *H. sapiens* and *H. neanderthalensis*.

Hawks and Wolpoff (2001) and Wolpoff *et al.* (2001) have reported analyses of human evolution that also allow conclusions concerning the descriptive and predictive power of the Hybridization/Introgression model. In the first of these studies, the "Accretion model" of Neanderthal evolution was tested. This model explains the evolution of this species as occurring in partial or complete genetic isolation from other lineages, and in particular, that the isolation of the European *H. neanderthalensis* populations resulted in their gradual accumulation of diagnostic morphological traits (see Hawks and Wolpoff 2001 for a further description of this model). In contrast to the expectations of the Accretion model, the morphological variation among the fossil samples indicated genetic continuity rather than isolation between Neanderthals and other members of *Homo*. This conclusion was seen as support for the model of Multiregional evolution (Hawks and Wolpoff 2001) and therefore supports the Hybridization/Introgression model as well.

The second analysis—that of Wolpoff *et al.* (2001)—also found evidence consistent with the hypothesis of gene flow between ancient and recent hominins. Specifically, these workers tested for "dual-ancestry" of fossil populations versus complete replacement of archaic by modern lineages of *Homo* in geographically peripheral regions. Their choice of peripheral population samples was based on the alternative predictions made by the Multiregional (and Hybridization/Introgression) model that assumes gene flow between different species meeting at the peripheries relative to the Replacement model that assumes no such introgression (Wolpoff *et al.* 2001). The findings from this analysis suggested a genetic continuity between earlier and later lineages. For example, fossil finds from Australia were assigned to a morphological category that included characteristics from both anatomically modern humans and *H. erectus* (Wolpoff *et al.* 2001). Similarly, fossil samples from Moravia, Czech Republic possessed not only features diagnostic for *H. sapiens*, but also some recognized as *H. neanderthalensis*-like. From both of these peripheral populations, then, Wolpoff *et al.* (2001) detected the characteristic signature of introgressive hybridization or, in their terminology, dual-ancestry.

"The ubiquitous and variable presence of these morphological features...can only be parsimoniously explained as a product of modest levels of assimilation of Neandertals into early modern human populations." This quote from Trinkaus (2007), like the conclusions from the forgoing analyses, indicates a finding consistent with the genetic admixture of *H. sapiens* and *H. neanderthalensis*. Indeed, Trinkaus (2007) argued that there was no doubting the occurrence of genetic assimilation of Neanderthals by the advancing *H. sapiens* populations. Rather, the question was how much introgression of the former into the latter had occurred. To test for the degree to which modern human populations were impacted by genetic exchange with *H. neanderthalensis*, morphological characteristics of "European early modern humans" or "EEMHs" (Trinkaus 2007) were assessed. The features analyzed included aspects of cranial, dental, mandible, clavicle, scapulae, and metacarpal morphology. From this detailed assessment of morphological variability among the EEMH fossils, Trinkaus (2007) detected the presence of both morphological characteristics diagnostic for the *H. sapiens* lineage that migrated out of Africa, as well as traits that were characteristic for the Neanderthals contacted by populations of this expanding species.

The final study, illustrating findings consistent with the Hybridization/Introgression model, has been the source of much debate (e.g., see Tattersall and Schwartz 1999). In an analysis of the remains of an ~4-year-old child from an Upper Paleolithic burial site in Portugal, Duarte *et al.* (1999) reported morphological characteristics suggestive of genetic contributions from both *H. sapiens* and *H. neanderthalensis*. Specifically, Duarte *et al.* (1999) examined the cranium, postcrania, mandible, and dentition from the fossil specimen. Features that were Neanderthal-like included both body proportion and some of the mandible structures. In contrast, other mandible characteristics as well as dental and skeletal structures aligned the fossil with EEMHs. Finally, the temporal bone demonstrated intermediacy between the two *Homo* species. Duarte *et al.* (1999) referred to the gestalt of the fossil as being a mosaic of the diagnostic characters found in one species or the other. They echoed the findings of each of the above studies by concluding that their results falsified the Replacement hypothesis and, instead, supported models that included the component of introgressive hybridization (Duarte *et al.* 1999).

4.3.3 Testing the models of human evolution: molecular evidence

As discussed above, tests of the Replacement versus Hybridization/Introgression and Multiregional models have been a major catalyst for studies of human evolution using fossil material. Similarly, analyses of DNA sequence variation in extant and extinct populations of *Homo* have often been motivated by a desire to test hypotheses concerning the (1) geographic point of origin for *H. sapiens* and (2) presence or absence of admixture between different lineages (e.g., Hammer 1995; Pääbo 1999; Wells 2002; Shimada *et al.* 2007; Hawks *et al.* 2008). These two hypotheses are related and reflect aspects of the models that alternately assume introgression or no introgression.

The history of molecular tests of the alternative hypotheses is often seen as beginning with the work of A.C. Wilson's group, specifically with the publication of Cann *et al.* (1987). This report, in the journal *Nature*, described an analysis of mtDNA sequence variation as assayed by restriction enzyme cleavage. By analyzing a total of 147 samples of extant humans, from sites across the geographic range of this species, Cann *et al.* (1987) inferred a time since divergence from a common ancestor of *c*.200,000 ybp and a point of origin for the most ancestral population as being Africa (Figure 4.4). The effect of this publication was earth shaking, and has contributed to the development of numerous research programs built around the clarification of when, where, and how anatomically modern humans arose (see Vigilant *et al.* 1991; Serre *et al.* 2004 for references to analyses based on the findings of this paper). However, the analysis of molecular data to test for processes that have impacted human evolution (as shown in Chapter 3) have not been restricted to the most recent timescales. In particular, since Cann *et al.*'s

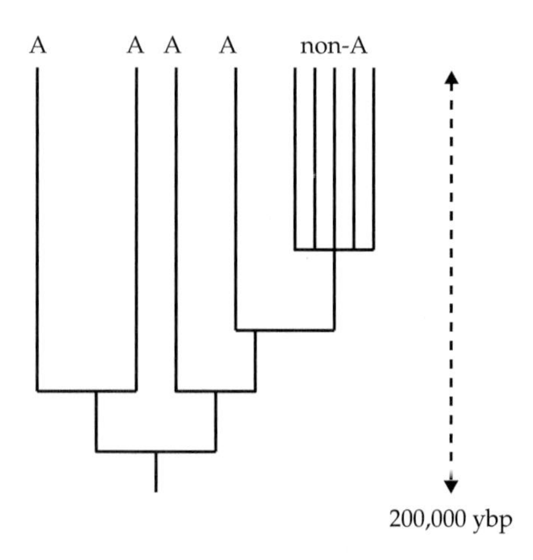

Figure 4.4 Evolutionary relationships and the timing of origin of modern human populations derived from mtDNA sequence variation. The observations that African ("A") sequences fall on either side of the root of the phylogeny and that non-African ("non-A") sequences are resolved as more recent additions to the phylogeny are consistent with Africa as the point of origin for modern *H. sapiens*. *Sources*: Cann *et al.* 1987; Vigilant *et al.* 1991.

(1987) report, the role of genetic exchange between species of *Homo* has been examined repeatedly. Thus, the following discussion will be divided into evidence consistent with introgression between (1) *H. sapiens* and *H. neanderthalensis* and (2) *H. sapiens* and either *H. neanderthalensis* or more anciently derived *Homo* spp. (e.g., *H. erectus*).

In fairness to the literature on human molecular evolution, many workers have concluded that there is little or no evidence to support the hypothesis of introgression between *H. sapiens* and archaic *Homo* spp. (e.g., Krings *et al.* 1997; Ovchinnikov *et al.* 2000; Caramelli *et al.* 2003; Pääbo *et al.* 2004; Serre *et al.* 2004). These conclusions have nearly always been based solely on mtDNA variation (but see Kaessmann and Pääbo 2002). In particular, mtDNA sequence variation has been collected from Neanderthal fossil material and compared with those sequences from *H. sapiens*. These analyses have consistently found larger sequence variation between *H. neanderthalensis* and *H. sapiens* than within either species. Figure 4.5 illustrates the results of such an analysis in which extant and ancient mtDNA sequences from *H. sapiens* samples were compared to those of Neanderthals.

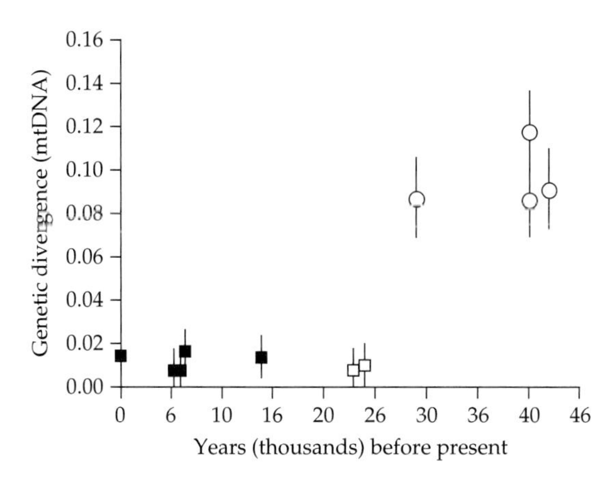

Figure 4.5 Comparison of mtDNA genetic distances between both extant and ancient *H. sapiens* (filled and empty squares) and *H. neanderthalensis* (empty circles). The filled square at the "0" time point reflects 2566 samples of extant Europeans. Note that the two open squares are anatomically modern samples that are temporally most closely associated with *H. neanderthalensis*, but which are genetically closely associated with all other *H. sapiens* samples (Caramelli *et al.* 2003).

Caramelli *et al.* (2003) not only detected greater levels of between species than within species sequence variation, but also demonstrated that samples of *H. sapiens* that were temporally closer to some *H. neanderthalensis* individuals were still genetically more similar to other anatomically modern human samples (Figure 4.5).

Notwithstanding the many excellent studies of both ancient and modern mtDNA variation in *Homo*, because mtDNA is largely a nonrecombining molecule passed down from mother to child, it acts as a single locus and thus, compared to biparentally inherited nuclear genes, is of limited utility for detecting gene exchange. Takahata *et al.* (2001) voiced this conclusion in stating "our inability to detect ancient admixture results largely from single-locus information, and many independent regions are needed to improve the power." Indeed, by gathering sequence data from X-, Y-, and autosomal loci, Takahata *et al.* (2001) detected 1 of 10 genomic markers that demonstrated ancestry not in Africa, but in Asia. This result—the presence of both African and Asian ancestral sequences in contemporary *H. sapiens* populations—supports the Hybridization/Introgression and Multiregional models. Similarly, some sequence variation at the mtDNA locus supports these models as well, with (1) variants identified as African present in clades of non-African samples (Ingman *et al.* 2000) and (2) estimates of the number of mtDNA introgression events being low, but not zero (Currat and Excoffier 2004).

Before proceeding to a discussion of the specific findings from analyses of mtDNA, Y-chromosome, X-chromosome, and autosomal loci, it is important to emphasize that the first two types of loci (i.e., mtDNA and Y-chromosome) provide less resolution for tests of introgression between modern and archaic lineages of *Homo*. I have mentioned one reason already, mtDNA (and Y-chromosome loci as well) are inherited as a unit from one parent. This limits the amount of information that can be gleaned from these loci. However, related to this is the fact that since Y-chromosome and mtDNA loci are inherited from only one parent, the effective population size (i.e., the number of individuals that contribute to sexual reproduction) consists of only those individuals of one sex (males for the Y-chromosome loci and females for the mtDNA loci). Taken together—inheritance as a single locus and through only one sex—these factors cause the Y-chromosome and mtDNA loci to be useful only for tests of introgression that have occurred in the most recent portion of human evolutionary history (Templeton 2002). This limited informational content has resulted in conclusions apparently biased toward support of the Replacement model of human evolution when in fact the data do not allow a rigorous test of any of the models (Templeton 2007a).

H. sapiens × *H. neanderthalensis*

Alan Templeton has played a major role in furthering the debate concerning the predictive and explanatory power of the various models of human evolution (e.g., see Templeton 2002, 2005). The following quote from one of his most recent papers reflects the conclusions drawn from his analyses of genetic variation in extant human populations:

> Genetic data have indeed played a critical role in studies on human evolution…but not in the manner thought two decades ago nor in the popular science literature of today. Far from supporting the out-of-Africa replacement hypothesis, the genetic data are definitive and unambiguous in rejecting replacement. (Templeton 2007a)

Thus, according to Templeton, the Hybridization/Introgression and Multiregional models explain best the patterns seen.

Figure 4.6 illustrates a model of the expansion events that resulted in genetic exchange between *Homo* lineages (Templeton 2002, 2005, 2007a, b). The patterns of genetic variation at autosomal, X-chromosome, Y-chromosome, and mtDNA loci provide views of different time horizons beginning with the migration of *H. erectus* from Africa through historical times (Figure 4.6). The overall pattern is one of repeated contact and introgression between the various lineages and taxa of the genus *Homo* (Templeton 2007b). Specifically, three migration events were detected by Templeton's (2007b) analysis. The first event, dated at 1.9 mya, involved the movement of *H. erectus* from its African origin. Within approximately 500,000 years, recurrent genetic exchange was proceeding between the African and Eurasian populations (Templeton 2007b). A second movement from an African source population occurred from 0.39 to 0.97 mya (Figure 4.6). This latter event involved not only the expansion of peoples from Africa, but their interbreeding with Eurasian populations as well (Templeton 2002, 2005, 2007b). The third major migration event occurred between 96 and 169,000 ybp (Figure 4.6) and is coincidental with the spread of anatomically modern humans from Africa (Stringer and Andrews 1988). Like the previous incursions from Africa, introgression occurred between divergent lineages (Templeton 2002, 2005, Templeton 2007b). Numerous analyses of sequence variation illustrate well the data that have informed discussions concerning genetic exchange between *H. sapiens* and *H. neanderthalensis*, thus contributing to tests of the Replacement versus Hybridization/Introgression models, including Templeton's model illustrated in Figure 4.6.

Zietkiewicz *et al.* (2003) tested the Replacement and Hybridization/Introgression models by collecting and analyzing sequence variation at the X-chromosome, dystrophin locus. When these data were examined phylogenetically and geographically, several conclusions were reached (Zietkiewicz *et al.* 2003). First, two lineages of African origin were identified; one lineage remained in Africa, while one participated in a migration event from Africa. Second, the lineage that migrated from Africa is shared among present-day populations distributed across different continents. Third, the African-specific variants demonstrate the effects of recombination between different lineages within Africa. Fourth, though there were two African-associated lineages that contributed to the present-day distribution of dystrophin sequences, allelic variation indicated also the role of both non-African sequence evolution and introgression between divergent *Homo* lineages, thus falsifying the Replacement model of human evolution. For example, Figures 4.7 and 4.8 reflect the phylogenetic and geographic distribution of the dystrophin "B006" sequence variants (Zietkiewicz *et al.* 2003). As illustrated in Figure 4.7, the almost exclusively non-African B006 alleles are most closely associated with the

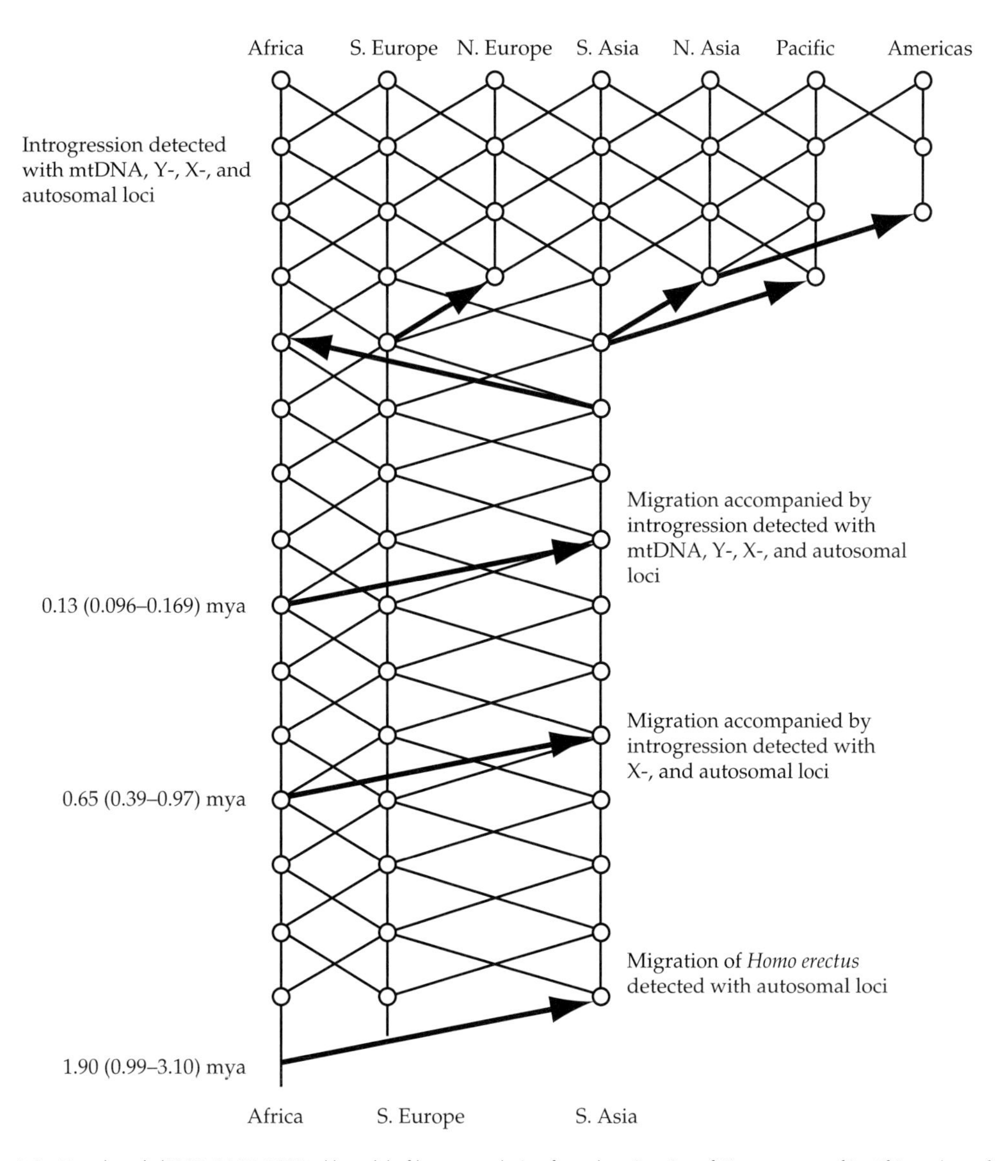

Figure 4.6 Templeton's (2002, 2005, 2007a, b) model of human evolution from the migration of *H. erectus* out of its African place of origin to the present. Three migration events, accompanied by coincidental or subsequent introgression are dated at 1,900,000, 650,000, and 130,000 ybp. Each of these events, as well as more recent migrations and introgression episodes, are marked by the indicated loci (Templeton 2002, 2005, 2007b).

root of the phylogeny. Furthermore, the restriction of the B006 lineage to an almost exclusively non-African geographic distribution (Figure 4.8) identifies this as a variant that likely preceded the origin of the more recent African lineage that migrated into other continents *c.*130,000 ybp (Figure 4.6). These observations, along with estimates of time since common ancestor, suggest that the populations carrying these alleles left Africa as early as 160,000 ybp and then admixed with those lineages involved in the most recent out-of-Africa migration event (Zietkiewicz *et al.* 2003).

Two related data sets that have also facilitated tests of the various human evolution models come from the Neanderthal genome sequencing project (Lambert and Millar 2006). Both analyses compared the sequence variation of *H. neanderthalensis* with that previously identified for the genomic regions of *H. sapiens*. Noonan *et al.* (2006) examined ~54,000 base pairs representing >1100 loci from *H. neanderthalensis*. In regard to tests for introgression from Neanderthals into anatomically modern humans, maximum likelihood estimates of such admixture were zero. However, as pointed out by Noonan *et al.* (2006), the 95% confidence interval for this estimate ranged from 0% to 20% thus suggesting the need for additional *H. neanderthalensis* sequence data. Yet, this range of values also indicates the likelihood of introgression.

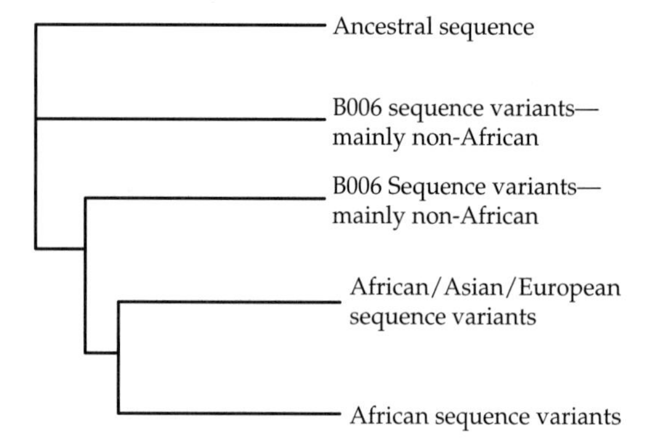

Figure 4.7 Phylogenetic distribution of the "B006" sequence variants of the X-chromosome, *dystrophin* gene. Note that the sequences nearest the root of the tree (i.e., inferred to be the most ancient) are not African, but instead are from other geographic regions. *Source*: Zietkiewicz *et al.* 2003.

The second analysis of Neanderthal sequence variation, carried out by Green *et al.* (2006), involved the assaying of 1,000,000 base pairs from the Neanderthal genome. The sequences analyzed originated from both the mtDNA and nuclear genome, and were used to calculate time since divergence of *H. sapiens* and *H. neanderthalensis* and the ancestral population sizes for both species. Most significantly for the present discussion, Green *et al.* (2006) also tested for patterns indicative of introgressive hybridization. Though they did not discuss testing for the occurrence of introgression

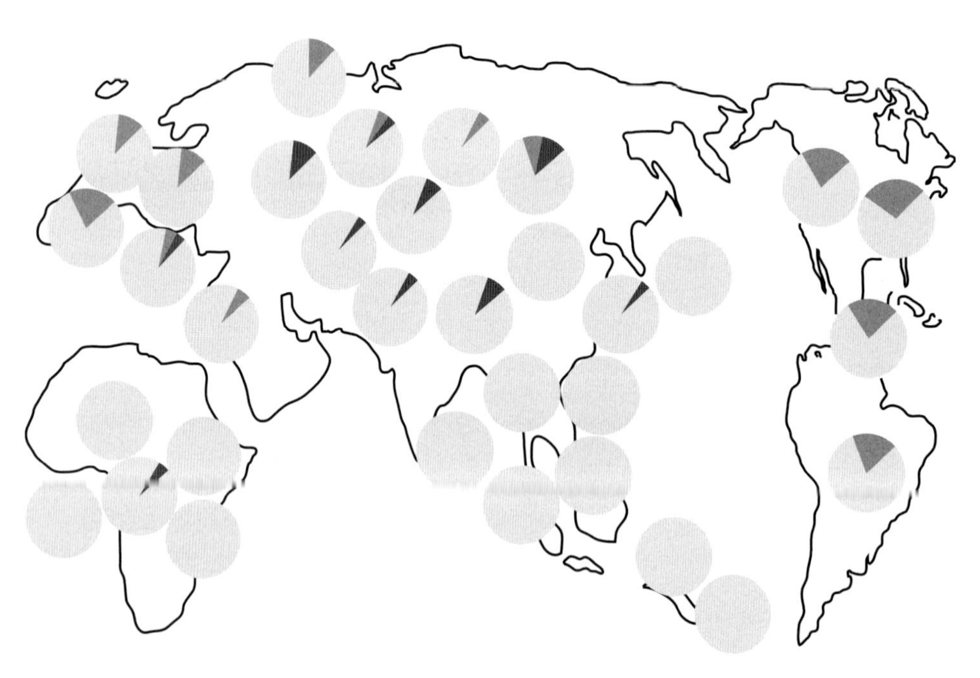

Figure 4.8 Geographic distribution of the *dystrophin* gene sequence lineage, "B006." This lineage was found to be almost exclusively non-African and demonstrates a pattern consistent with introgression between the recently migrating African lineages and more ancient lineages occupying Eurasia (Zietkiewicz *et al.* 2003).

from *H. neanderthalensis* into *H. sapiens*, they did argue for the reverse genetic exchange event. Specifically, they detected an approximately 30% frequency of derived alleles (relative to those in chimpanzee) at the *H. neanderthalensis* loci. This led to the following conclusions: (1) the high level of derived alleles was compatible with introgression between these two species; and (2) because the Neanderthal X-chromosome loci demonstrated higher divergence relative to the autosomes, the introgression may have mainly involved matings between *H. sapiens* males and *H. neanderthalensis* females (Green *et al.* 2006). Results from both Noonan *et al.* (2006) and Green *et al.* (2006) reflect the probability that the genomes of Neanderthals and our own species are mosaics due to bidirectional introgressive hybridization (but see Wall and Kim 2007 for concerns over the data quality from the Green *et al.* 2006 study).

The final example I will discuss, which falsifies the Replacement model and supports the admixture-accompanying-human-evolution models, involves a sequence analysis of the autosomal gene, *microcephalin*. Evans *et al.* (2006) tested not only for admixture between archaic and modern lineages of *Homo*, but also for the possible effect that such introgression might have had on the development of adaptations in *H. sapiens*. They addressed both of these goals by examining the worldwide distribution of genetic variation at a locus associated with a putative *H. sapiens* adaptation, that of increased brain size. The *microcephalin* locus is a regulator of cerebral cortex volume, with a loss of function of this gene expressed in reduced volumes comparable to early hominids (Jackson *et al.* 2002). Surprisingly, the only major neurological expression from this loss of function is a lessening of cognitive abilities (Jackson *et al.* 2002). The recent demonstration (Dediu and Ladd 2007) of a strong association of linguistic tone and derived alleles at this and another locus (i.e., the *ASPM* locus) that affect brain size, further support the hypothesis of an adaptive advantage for some *microcephalin* mutations.

The evolutionary trajectory of the *microcephalin* locus has included an acceleration of mutation accumulation since the simian ancestors, with the largest increase in evolutionary rates seen along the *Homo* lineage (Evans *et al.* 2004). The high levels of variation and the increased rate of change, particularly within *H. sapiens*, have resulted in the inference of a role for strong positive selection in the evolutionary history of this locus (Wang and Su 2004; Evans *et al.* 2005). The adaptive potential and evolutionary patterns for the *microcephalin* gene led Evans *et al.* (2006) to determine the age and potential source of the alleles present in contemporary human populations. Figure 4.9 illustrates a model constructed to explain the following findings (from Evans *et al.* 2004, 2005, 2006): (1) the non-D and D alleles diverged from a common ancestor c. 1.7 mya; (2) the non-D and D alleles characterized the *H. sapiens* and *H. neanderthalensis* ancestral lineages, respectively; (3) approximately 37,000 ybp there was an introgression event of the D allele from *H. neanderthalensis* into *H. sapiens*; (4) strong, directional selection favored the D allele leading to its current worldwide distribution and 70% frequency. As with any adaptive trait introgression event (e.g., Martin *et al.* 2005, 2006), the transfer of the *microcephalin* D allele is hypothesized to have provided an advantage to the recipient lineage. Evans *et al.* (2006) posit that as *H. sapiens* spread from its origin it would have encountered *H. neanderthalensis* populations that were well adapted to the local environments. Thus, "It is perhaps not surprising then that modern humans, although likely superior in their own way, could in theory benefit from adopting some adaptive alleles from the populations they replaced" (Evans *et al.* 2006).

H. sapiens × Homo spp.?

Separating the examples given in the previous section from the following studies is artificial for two reasons. First, it may be that the gene exchange events described below, like those above, took place between *H. sapiens* and *H. neanderthalensis*; in some cases I have been unable to determine this from the authors' descriptions. Second, whether or not the "archaic" taxon that contributed to the introgressive hybridization was at the stage in their evolutionary trajectory where they would have been recognized as one species or another is problematic. So, why did I make this division between the former and the present sections? This also has two rationales. First, one of my emphases

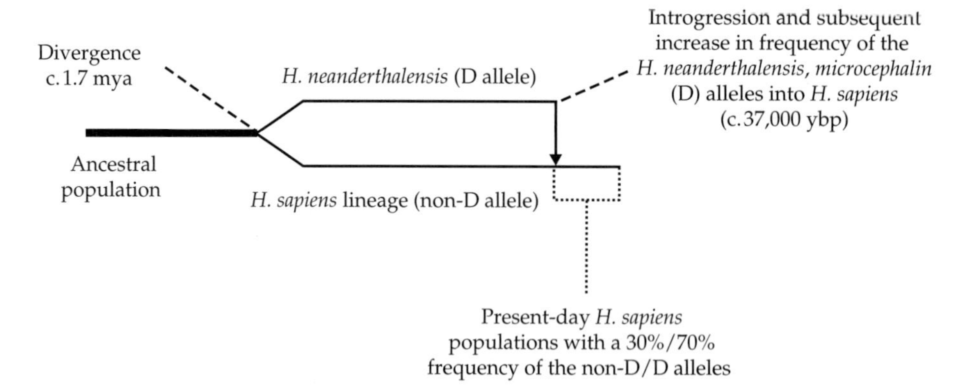

Figure 4.9 Model to explain the high frequency occurrence of the ancient, *microcephalin* D allele in extant *H. sapiens*. The model includes the ancient derivation of both the non-D and D alleles from a common ancestor, the fixation of the alternate alleles in the *H. neanderthalensis* and *H. sapiens* lineages, and the introgression of the D allele from Neanderthals into anatomically modern humans c. 37,000 ybp (Evans *et al.* 2006).

in this book and elsewhere (e.g., Arnold 1997, 2006) is that taxonomic status is relatively unimportant in predicting the role played by genetic exchange in the evolution of assemblages. Instead, the most important consideration is whether the interacting lineages have diverged to the point at which their hybrid products have the potential of displaying novelty (ecological, behavioral, etc.) relative to their parents (Arnold 1997, 2006). The second reason for the sectional division, related to the first rationale, is that I also have an agenda to demonstrate the continuous evolutionary impact of reticulate evolution in the groups discussed throughout this text. Thus, the fact that the taxa interacting with *H. sapiens* have been named, for example, *H. erectus* or *H. neanderthalensis* reflects a continual role for introgression between the lineage leading to modern humans and their sister taxa. In Chapter 3, I reviewed all of the data of which I am aware, indicating the continuity of introgression and hybrid speciation during the evolutionary history of many primate clades including that containing our own genus. I want to emphasize the same point for the species belonging to *Homo*.

Plagnol and Wall (2006) described an analysis of associations between > 100 autosomal loci assayed from extant humans. The model with the best fit to their data included a 5% rate of admixture between some archaic species and *H. sapiens*. Significantly, the admixture apparently occurred in both European and West African populations of *H. sapiens*. As Plagnol and Wall (2006) indicate, with the date of admixture set at *c.*50,000 ybp, the archaic source population for the European introgression is likely *H. neanderthalensis*. In contrast, they argued that Neanderthals could not have been the source for the West African introgressive hybridization. Instead, the donor of the archaic genes transferred into *H. sapiens* would likely have been a form recognized as transitional between more anciently derived lineages and our own species (e.g., see Stringer and Andrews 1988 for a discussion). Thus, the data from Plagnol and Wall (2006) indicate introgression involving our own species with not only *H. neanderthalensis*, but with other archaic *Homo* taxa as well.

Garrigan *et al.* (2005) provided further support for introgression between anatomically modern humans and another species of *Homo*. These workers examined the sequence variation at the X-chromosome pseudogene, *RRM2P4*. The phylogenetic and geographic distribution of *RRM2P4* alleles supported an Asian point of origin for the variants present in *H. sapiens*. Furthermore, the estimated date of origin for this lineage was ~2 mya (Garrigan *et al.* 2005). Finally, the archaic *RRM2P4* lineage not only appears to have arisen in Asia, but also is rarely found in Africa; the frequency of this lineage is 53% and 0.6% in present-day Asian and African populations, respectively. Given (1) the antiquity of the *RRM2P4* lineage (corresponding to the fossil date of origin of the genus *Homo*) and (2) its Asian association, Garrigan *et al.* (2006) inferred an admixture model. Specifically, they argued for introgression between out-of-Africa *H. sapiens* migrants with indigenous *H. erectus*. This interaction would account for the high frequency of this ancient lineage in Asia and Europe, relative to extant African populations. A more recent analysis of sequence variation at this locus, in an additional 131 African and 122 non-African individuals, also tested for admixture between archaic lineages and anatomically modern humans (Cox *et al.* 2008). This latter analysis detected geographic patterning that was not consistent with the Replacement model of human evolution, but rather supported introgressive hybridization between the dispersing, modern human populations, and the archaic ancestors of humans (Cox *et al.* 2008).

Hayakawa *et al.* (2006), like Garrigan *et al.* (2005), examined genetic variation at a pseudogene locus in extant human populations. This autosomal locus contains the deactivated human *CMPNacetylneuraminic acid hydroxylase* gene (*CMAH*). It has been estimated that the molecular process that created this pseudogene occurred *c.*3.2 mya (i.e., in members of the genus *Australopithecus*) and was transposable element-mediated (Hayakawa *et al.* 2001). Hayakawa *et al.* (2006) collected sequence data for *CMAH* from human populations located in Africa, Europe, Asia, and the Americas. From these sequences they detected two lineages that diverged from a common ancestor soon after the inactivation event (*c.*2.9 mya). These two lineages—the "P" and "non-P" forms—gave rise to all of the worldwide distributed, extant genotypes; the P lineage is restricted to Africa, while the non-P clade is found both in Africa and the rest of the world (Hayakawa *et al.* 2006). One conclusion from this latter observation is that some African lineages were relatively isolated, while others contributed to the migration of *Homo* into other parts of the world. Furthermore, in terms of introgression, Hayakawa *et al.* (2006) detected geographical and genetic patterns consistent with genetic exchange between the expanding anatomically modern humans and the indigenous, archaic (i.e., *H. erectus* or *H. neanderthalensis*) species. Thus, as with the results from the studies by Garrigan *et al.* (2005) and Plagnol and Wall (2006), those of Hayakawa *et al.* (2006) indicate that the out-of-Africa migrations—first by *H. erectus* and then by *H. sapiens*—brought archaic and modern species into contact that led to introgressive hybridization.

4.3.4 Testing the models of human evolution: additional evidence

In this section, I will discuss data from one final study. This study is unique relative to all other analyses discussed in this chapter (and elsewhere in this book) in that Reed *et al.* (2004) used the genetic variation of a human-associated organism (i.e., the head and body louse species, *Pediculus humanus*) to test the hypothesis of physical contact between archaic and modern species. Though unique in terms of tests for reticulation, this study is not unique in regard to analyses of the coincidental evolutionary effects that *H. sapiens* can have on associated organisms (e.g., transport of the Pacific rat with Polynesian migrations [Matisoo-Smith and Robins 2004] and the migration of the gut bacteria, *Helicobacter pylori* with the out-of-Africa dispersal by *H. sapiens* [Linz *et al.* 2007]).

Figure 4.10 illustrates the various conclusions that can be drawn from Reed *et al.*'s (2004) data. In particular, they concluded the following concerning the evolutionary history of both *P. humanus* as well as *H. sapiens* and *H. erectus*: (1) *P. humanus* consists of two anciently derived (*c.*1.18 mya) lineages; (2) one lineage is distributed worldwide and went through a population bottleneck with its *H. sapiens* host *c.*100,000 ybp; (3) the other *P. humanus* lineage occurs only in the New World; (4) the lice divergence is tightly correlated in time with the speciation of *H. sapiens* and *H. erectus*, and cospeciation analyses indicate a codivergence between the two lice lineages and these two *Homo* species; (5) a recent (*c.*150,000 ybp) host shift brought the two lice lineages together leading to introgressive hybridization; (6) direct physical contact between the archaic (*H. erectus*) and modern (*H. sapiens*) hosts would have been necessary for the transfer

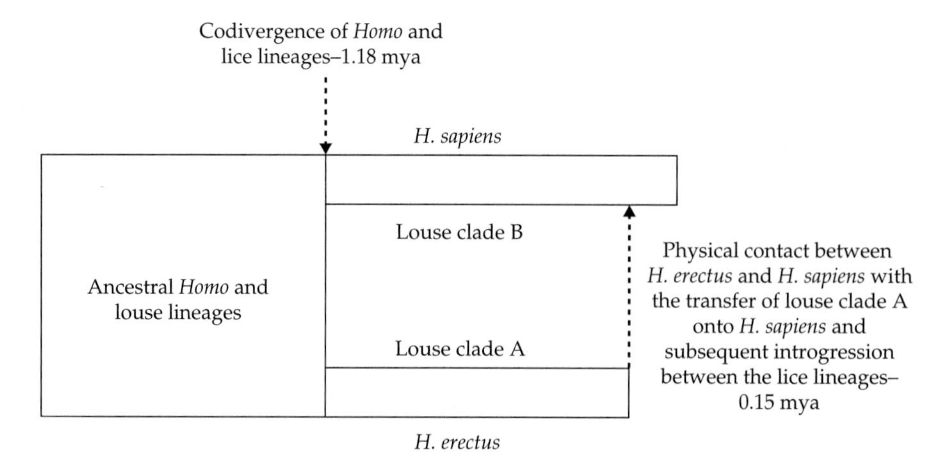

Figure 4.10 Model of the coevolutionary events associated with the primate head and body louse, *Pediculus humanus*, and its hosts *H. erectus* and *H. sapiens*. The divergence of the two lice clades is proposed to have occurred due to the speciation of *H. erectus* and *H. sapiens* from a common ancestor c.1.18 mya. Following a 1-million-year isolation on their respective primate hosts, the two lice lineages were brought back into contact by direct physical interaction—likely involving sexual reproduction—between *H. erectus* and *H. sapiens*. The contact between the two forms of *P. humanus* resulted in introgressive hybridization (figure constructed from information found in Reed *et al.* 2004).

of the lice lineages. Reed *et al.* (2004) indicated that "direct physical contact" between the two primate host species could have taken the form of fighting, sharing of clothing, and so on. However, given the wealth of evidence supporting introgression between various species in this assemblage (see above), it would seem more likely that the lice lineages were exchanged during reproductive interactions.

4.4 Summary and conclusions

This chapter was constructed around the theme of reviewing data sets that allow tests of the Replacement model and all other models of human evolution that assume some level of introgression between *Homo* taxa. When the data are examined critically, even from studies concluding otherwise, signatures of some level of genetic exchange between *H. sapiens* and archaic species such as *H. erectus* and *H. neanderthalensis* are apparent. This is true whether or not whole genomes (or large portions of them) are utilized or sequence variation is assayed at only one or a few loci. One of the significant facts that should be gleaned from this information is that the genus *Homo* is not an unusual primate in regard to introgressive hybridization and hybrid speciation. Thus, our lineage and those of our closest relatives have, like all other clades of primates (see Chapter 3), been impacted by gene exchange between sister taxa. As the dates of most of the references in this chapter indicate, there is a new and burgeoning literature. Thus, evidence for repeated introgressive hybridization will become more evident as additional analyses of genetic variation, in both our own species and those of archaic species (e.g., *H. neanderthalensis*), are reported.

CHAPTER 5

Reticulate evolution and beneficial organisms—part I

Sympatric populations of white-tailed deer and mule deer... share a common mitochondrial DNA restriction map genotype.

(Carr *et al.* 1986)

The distribution of chloroplast genotypes in this group indicates that sympatric species of oak in the eastern United States do not represent fully isolated gene pools, but are actively exchanging genes.

(Whittemore and Schaal 1991)

A preponderance of interspecific allelic interactions involved one locus each in the two different subgenomes of (allotetraploid) *Gossypium*, thus supporting several other lines of evidence suggesting that intersubgenomic interactions contribute to unique features that distinguish tetraploid cotton from its diploid ancestors.

(Jiang *et al.* 2000)

This finding confirms that inter-specific hybridization between wolves and dogs can occur in natural wolf populations.

(Vilà *et al.* 2003)

...swamp buffaloes have an unraveled mitochondrial history, which can be explained by introgression of wild buffalo mtDNA into domestic stocks.

(Kierstein *et al.* 2004)

Grapevine was selected because of its important place in the cultural heritage of humanity beginning during the Neolithic period... This analysis reveals the contribution of three ancestral genomes to the grapevine haploid content.

(Jaillon *et al.* 2007)

5.1 Reticulate evolution and the formation of organisms utilized by humans

In this chapter, and in Chapters 6 and 7, I will turn my attention to describing the role that genetic exchange—introgressive hybridization, hybrid lineage formation, and horizontal transfer—has played in the evolution of organisms that provide *Homo sapiens* with nourishment, protection, and stimulation. The evolutionary origin and trajectory of countless organisms (from which we derive everything from the food that we eat to drugs that we ingest or inhale) have been affected by reticulate events. For example, Vilà *et al.* (2005) demonstrated the importance of introgressive hybridization in the evolution of domestic mammals such as dogs, pigs, goats, sheep, and cattle. By assaying the genetic variation for the major histocompatibility complex loci ("MHC") in these species, Vilà *et al.* (2005) detected 39–66 alleles. As these authors reflected, the two avenues for allelic variation in domestic animals are new mutations and introgression from wild ancestors. Since the mutation rate, over the relatively short time period since the initial domestication events, could not account for the high MHC allelic variation, it was concluded that the alleles had been inherited from the ancestral species (Vilà *et al.* 2005). Furthermore, the number of founders necessary to explain such a rich pool of alleles greatly exceeded (possibly by orders of magnitude) the expected, initial population sizes (Vilà *et al.* 2005). Each of these observations led to the conclusion that repeated introgression events had occurred during the evolution of the domesticated taxa. Specifically, Vilà *et al.* (2005) stated that "contrary to common assumption, domestic and wild lineages might not have been clearly separated throughout their history."

The diversity of species utilized by humans is enormous. Likewise, the breadth of classes of utilization is very great. As emphasized earlier, the role of reticulate evolution in the formation of these taxa is pervasive. An additional example of the tremendous diversity, in terms of both the domains of life from which the species are drawn as well as the types of human-associated functions for which they are put to use, is reflected by the grass genus *Elymus*. This group is employed to add a naturalized gestalt to developed environments (http://www.bluestem.ca/ornamental-grass.htm), thus contributing to *H. sapiens'* heightened enjoyment of residences and cityscapes.

Like all of the earlier examples of human-associated species, ornamental grasses belonging to the genus, *Elymus*, also reflect the role of reticulate evolution. In particular, this complex reflects hybrid speciation, and specifically, allopolyploid species formation. Helfgott and Mason-Gamer (2004) used sequence information from the *phosphoenolpyruvate carboxylase* gene (i.e., *PepC*) to test the hypothesis that the North American representatives of this genus were the products of allopolyploid speciation from crosses between members of *Pseudoroegneria* and *Hordeum*. A previous study by Mason-Gamer (2001), involving an analysis of sequence information for the nuclear *waxy* gene, had supported this hypothesis. However, the tribe, Triticeae, is known for frequent discordances between phylogenies that are based on different genomic regions (e.g., see Kellogg *et al.* 1996). This made it necessary to confirm the identification of the progenitors of this putative allopolyploid network (Helfgott and Mason-Gamer 2004). Figure 5.1 illustrates the phylogenetic placement of the *Elymus* taxa relative to other members of the Triticeae. Significantly, the *PepC* sequences place the *Elymus* species into two clades, one that includes *Pseudoroegneria* and another that contains *Hordeum*. This paraphyletic distribution of *Elymus* is consistent with the presence of two genomes, and specifically, the genomes (and thus *PepC* alleles) from *Pseudoroegneria* and *Hordeum* (Helfgott and Mason-Gamer 2004). This is the expected result if these latter two genera gave rise to the North American polyploid *Elymus* lineages through hybridization.

As the earlier examples illustrate, it is possible to identify a wide array of organisms on which humans depend for sustenance and chemical and visual stimulation whose evolution reflects the web-of-life metaphor (Arnold and Larson 2004; Arnold 2006). In the following discussion, I will greatly broaden this observation to provide a sense of the wealth of examples of such organisms. Most of the data sets I will discuss come from the burgeoning genomics literature. In this chapter and Chapters 6 and 7, I will use the genome sequence information for organisms as diverse as bacteria, dogs, rice, cattle, cotton, and wheat to test for patterns consistent with reticulate evolution. Yet, the same data also motivate and inform experiments designed to understand gene function and gene evolution (e.g., Bowers *et al.* 2005), and may also facilitate an increase in the production potential of organisms on which humans depend (Paterson *et al.* 2005). The categories to be addressed in the present chapter include organisms with functions as varied as companionship, drugs, building material, and clothing. In Chapters 6 and 7, I will discuss the affect of genetic exchange on lineages from which *H. sapiens* derives animal- and plant-based foodstuffs, respectively.

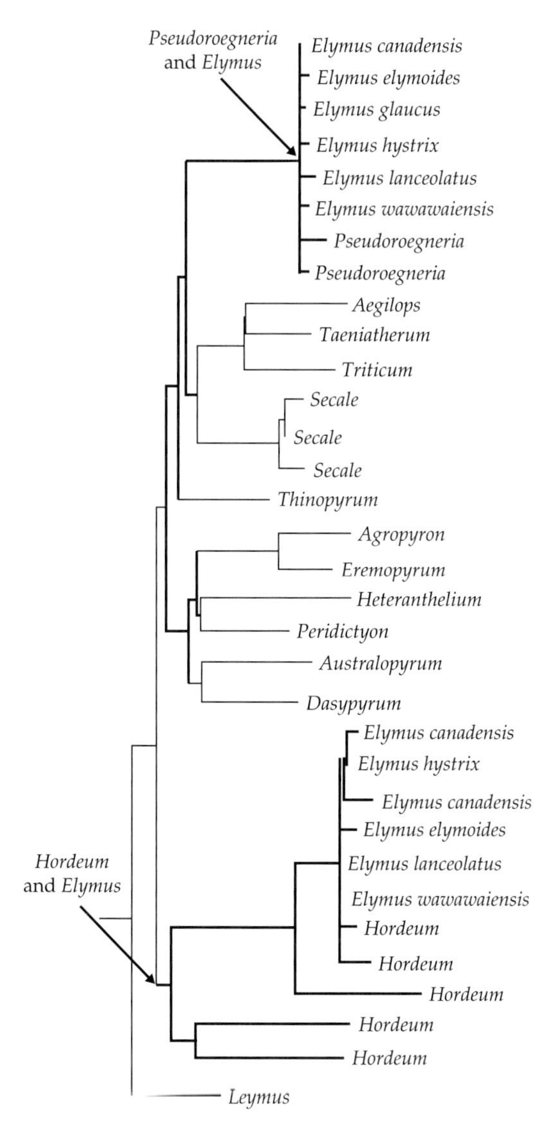

Figure 5.1 Phylogenetic relationships among various members of the tribe Triticeae. In particular, this phylogeny supports the hypothesis of an allopolyploid derivation of the ornamental grass genus, *Elymus*. The grouping of the alternate *PepC* alleles with either of the two putative ancestral taxa—that is, *Pseudoroegneria* or *Hordeum*—is consistent with the allopolyploid origin hypothesis for *Elymus* (from Helfgott and Mason-Gamer 2004).

5.2 Reticulate evolution and human companions

The domestication of animals, plants, and microorganisms that provide us with some measure of protection, relaxation, and companionship has grown, at least for Western cultures, into a major lifestyle and economic factor. For example, in 2003—in the United States alone—62% of households owned animals, pet owners spent ~30 billion dollars on their charges, with nearly 14 billion dollars being spent on pet food (Barnes 2004). Pet food costs in the United States were expected to rise to nearly 17 billion dollars by 2008 (Barnes 2004).

Discussing the following three animal clades (i.e., dogs, cats, and genets) in the category of "companions," and indeed even having such a category, demonstrates the author's ethnocentricity. In other words, for the vast majority of the world's peoples, these animals are not primarily companions, but instead are used mainly as a protein source. Old and New World human populations have utilized *Canis familiaris* (i.e., domestic dogs) as a beast of burden and an important portion of their diet (e.g., see http://coombs.anu.edu.au/~vern/wild-trade/eats/eats.html). For example, domestic dogs formed a substantial part of the menu during some great gatherings of Native Americans. At the "Laramie Council" of 1851, "Father De Smet, the famous Catholic missionary, who was there, declared that 'no epoch in Indian annals probably shows a greater massacre of the canine race'" (Ambrose 1996, p. 54). It is thus with apologies to the viewpoint of the majority of the cultures of *H. sapiens* that I discuss the role of genetic exchange in the evolution of canids, felids, and genets as that of the evolution of our companions. Yet, it is also true that this function likely has been as pivotal in human cultural and economic development (at least in the West; Barnes 2004) as the domestication of sources of food.

5.2.1 Reticulate evolution and canids

Reviewing analogous results from closely and distantly related canids can bolster support for the influence of reticulate processes in the evolutionary history of domestic dogs. Figure 5.2 reflects the phylogenetic relationships of the canid assemblage that Lindblad-Toh *et al.* (2005) termed "wolf-like." This clade contains the domestic dogs as well. Within this clade, the gray wolf (*Canis lupus*)

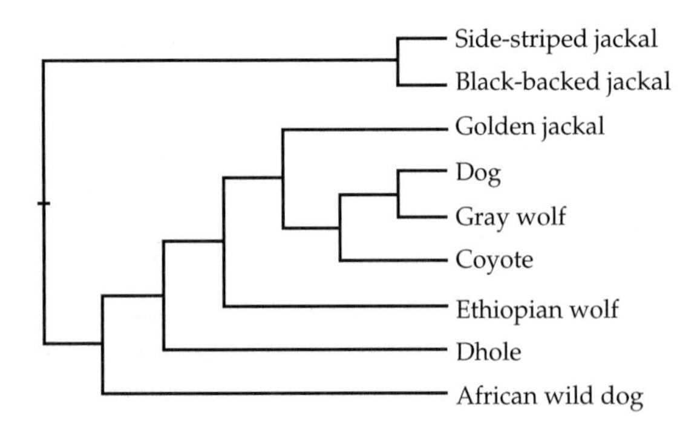

Figure 5.2 Phylogenetic relationships among "wolflike" canids (Lindblad-Toh *et al.* 2005) based on ~15,000 base pairs of coding and noncoding sequences (Lindblad-Toh *et al.* 2005).

and domestic dog (*C. familiaris*) are most closely related demonstrating only 0.04% and 0.21% divergence in coding and noncoding sequences, respectively (Lindblad-Toh *et al.* 2005). These two species form an assemblage with three other species with which *C. familiaris* is known to hybridize in the wild—*Canis latrans* (coyote), *Canis simensis* (Ethiopian wolf), and *Canis aureus* (golden jackal; Lindblad-Toh *et al.* 2005). The phylogenetic arrangement of the wolflike web indicated in Figure 5.2 reflects not only the close evolutionary/genetic relationships of many of the associated species, but also the propensity for introgression among these taxa. For example, a test for hybridization between domestic dogs and coyotes in the southeastern United States found that 12 "coyotes" (i.e., 11% of the sample) possessed mtDNA genotypes characteristic of *C. familiaris* (Adams *et al.* 2003). The fact that this introgression was apparent in coyote populations from as far apart as the states of West Virginia and Florida suggested an ancient event, before the colonization of the southeastern United States by *C. latrans* (Adams *et al.* 2003).

Swift and kit foxes

Mercure *et al.* (1993) tested the level of genetic divergence as an estimate of the evolutionary distinctiveness of the swift and kit foxes (*Vulpes velox* and *Vulpes macrotis*, respectively). Their analysis of mtDNA from sequence variation resulted also in a test of the amount of genetic exchange between these taxa. The geographic pattern of mtDNA variation was consistent with introgression between *V. velox* and *V. macrotis*. In particular, where the two overlapped in southeastern New Mexico, 41% and 59% of the animals had swift fox and kit fox genotypes, respectively. Furthermore, there was also an admixture of mtDNA lineages in southeastern Arizona as well. These findings led Mercure *et al.* (1993) to assign a zone of contemporary hybridization to an area that was a few hundred kilometers in width. In addition to the evidence for present-day hybridization, these authors also reported the occurrence of swift fox genotypes in the range of kit fox (in western Nevada). This suggested past dispersal by *V. velox* into *V. macrotis* populations, with subsequent introgressive hybridization (Mercure *et al.* 1993).

Red wolves

Lindblad-Toh *et al.* (2005) did not include in their phylogeny (Figure 5.2) another lineage closely related to the gray wolf, domestic dog, and coyote whose origin and evolution illustrates the potential role of introgressive hybridization; this species is the red wolf (*Canis rufus*). The history of investigations into this unusual, North American canid is replete with controversies. Most of the controversies are related to the role of reticulate evolution (e.g., see Dowling *et al.* 1992). Recently, as conservation of this taxon has begun to involve reintroductions into North Carolina, it has become apparent that the canid predilection for interspecific matings is present in *C. rufus* as well. Indeed, Adams *et al.* (2007) stated: "Hybridization with coyotes (*Canis latrans*) continues to threaten the recovery of endangered red wolves (*Canis rufus*) in North Carolina and requires the development of new strategies to detect and remove coyotes and hybrids." Using 18 microsatellite loci on DNA isolated from 89 fecal samples, they were able to match 73 of the 89 samples to 23 individuals. From this cohort, they identified six individuals (~25%) that demonstrated admixtures of coyote and red wolf mtDNA and nuclear alleles. This high frequency of introgressed individuals substantiates Adams *et al.*'s (2007) concern. In fact, a previous study using simulations to test the effects of various factors on the degree to which introgression would impact red wolves resulted in the conclusion that contact between red wolves and coyotes would need to be prevented if the reintroduction program was to be successful (Fredrickson and Hedrick 2006).

Like the recent reports of introgressive hybridization plaguing attempts to reintroduce *C. rufus*, there have been numerous investigations into whether or not introgression has played a significant role in the origin and evolutionary trajectory of this species as well. The red wolf had been endemic to the southeast United States, but was extirpated from most of its range due to the combined effects of habitat loss, culling, and introgressive hybridization with coyotes (Wayne and Jenks 1991). Introgression from coyotes into the remaining red wolves likely occurred due to the greatly reduced numbers of conspecific mates (see Arnold 1997, 2006 for reviews of similar cases). Indeed, the establishment of a captive-breeding program preceded the extinction of this species from nature by only ~1 year (Wayne and Jenks 1991).

Wayne and his colleagues published a series of papers in which they provided data, supporting the hypothesis of a hybrid origin for *C. rufus* from matings between coyotes and gray wolves (e.g., Wayne and Jenks 1991; Roy *et al.* 1994, 1996; Wayne and Gittleman 1995). These studies involved the analysis of mtDNA and nuclear loci in an attempt to define the origin and evolutionary trajectory of this species. For example, Wayne and Jenks (1991) examined mtDNA restriction fragment and sequence variation present in captive red wolves, 77 other canids, and from red wolf skins collected before the period in which introgression with coyotes was thought to have occurred. The detection of both coyote and gray wolf mtDNA genotypes, in the red wolf samples, indicated the hybrid nature of the red wolves. The alternate hypotheses to explain this observation were that (1) the red wolf was formed originally through hybridization and introgression between *C. lupus* and *C. latrans* or (2) following the origin of the red wolf, there was significant introgression throughout its range involving both gray wolves and coyotes (Wayne and Jenks 1991). The application of mtDNA and nuclear sequence analyses gave further definition and provided more rigorous tests of these alternative hypotheses. In particular, the nuclear genotypes detected in *C. rufus* were intermediate relative to *C. latrans* and *C. lupus* (Roy *et al.* 1996). In addition to further defining the hybrid status of this taxon, these additional studies indicated that the red wolf was most likely a hybrid derivative of recent origin (Wayne and Gittleman 1995; Roy *et al.* 1996). Furthermore, as noted from recent studies of reintroduced red wolf populations (see above), *C. rufus* continues to be impacted by introgression with coyotes (Roy *et al.* 1996). Thus, the red wolf reflects well the widespread and long-term impact of reticulate evolution in the canid clade.

Gray wolves

The inference that introgressive hybridization between gray wolves and coyotes has given rise to the red wolf lineage also suggests the role that introgressive hybridization may have played in the evolution of *C. latrans* and *C. lupus* (Lehman *et al.* 1991; Wayne *et al.* 1992). Numerous data sets reflect the signature of genetic exchange between these latter two species. Once again, the earliest studies of genetic variation involved mtDNA sequence analyses. Lehman *et al.* (1991) carried out a geographically widespread survey of mtDNA sequence variation for North American coyote and wolf populations. The distribution of mtDNA genotypes suggested that 7 of the 13 variants in the wolf samples were derived through introgression from coyotes. Figure 5.3 reflects the phylogenetic relationships of these variants that are consistent with this conclusion. As is clear from this evolutionary tree, some of the "wolf" genotypes were found to be more closely related to coyote sequences than they were to other wolf genotypes (Figure 5.3). Furthermore, in a region encompassing portions of the state of Minnesota and the Canadian provinces of Ontario and Quebec, the frequency of the coyote mtDNA variants within the gray wolf population was found to be >50% (Lehman *et al.* 1991).

In contrast to the extremely high-frequency introgression of coyote mtDNA into wolf populations, there was no detectable introgression of mtDNA in the opposite direction. The basis for this latter inference is clear from the observation of no *C. latrans* samples falling within the separate

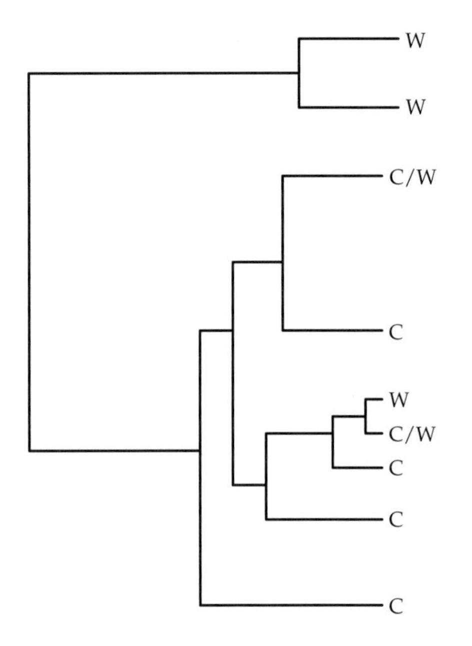

Figure 5.3 Evolutionary tree constructed from mtDNA sequence variants detected in natural populations of either gray wolf ("W") or coyote ("C"). Note that several of the "wolf" lineages display closer relationships to coyote sequences than to other wolf genotypes (data from Lehman *et al.* 1991).

C. lupus clade in the top portion of Figure 5.3 (Lehman *et al.* 1991). This finding was consistent with the following hypothesis. The disturbance of the wolf habitat (i.e., due to the transformation of forested regions into farmlands) had resulted in a smaller population of *C. lupus*. This perturbation and reduction in wolf numbers could facilitate the interaction between male wolves and female coyotes along the ecotones between the open habitat occupied by *C. latrans* and the remaining forested regions containing *C. lupus* (Lehman *et al.* 1991).

To further assess the evolutionary role of introgression between coyotes and wolves, Wayne and his colleagues also examined variation in the nuclear genome. Specifically, Roy *et al.* (1994) utilized 10 microsatellite loci to assay variation in *C. lupus* and *C. latrans* (as well as *C. rufus*; see earlier discussion). Like the mtDNA data, the nuclear loci indicated (1) that the introgression had occurred between coyotes and wolves, but (2) the introgression was asymmetric. Roy *et al.* (1994) thus detected a significant effect on the allele frequencies within *C. lupus* due to introgression from coyotes. However, there was little effect detected from introgression of wolf alleles into populations of coyotes. Roy *et al.* (1994) inferred a similar pattern (to that suggested by Lehman *et al.* 1991) of male wolf hybridization with female coyotes, and then backcrossing by the F_1 hybrids with *C. lupus* rather than *C. latrans* individuals.

C. latrans is not the only canid species that has impacted the evolutionary trajectory of wolf populations. In fact, conservation concerns for this and other canids are most often associated with the role that introgression from domestic dogs plays in the genetic assimilation of native species. For example, Gottelli *et al.* (1994)—through an analysis of both mtDNA and microsatellite sequence variation—detected hybridization between females of the rare and endangered Ethiopian wolf (*C. simensis*) and male domestic dogs. Due to their extremely limited numbers (reflected by their status as the "most endangered canid"; Gottelli *et al.* 1994), the prevention of introgressive hybridization with *C. familiaris* was viewed as critically important. Similarly, Vilà *et al.* (2003) tested the hypothesis that a hybridization event between domestic dogs and the rare Scandinavian wolves had occurred. A combined analysis of maternally, paternally, and biparentally inherited markers (mtDNA, Y-chromosome, and microsatellite loci, respectively) confirmed that one individual was most likely an F_1 hybrid between a female Scandinavian wolf and a male domestic dog. Due to the level of concern over *C. familiaris* genes entering the wolf population through backcrossing by such F_1 individuals, officials decided to identify and kill all of the littermates of this individual (which was found dead; Vilà *et al.* 2002). Following this action, one of the littermates remained unaccounted for.

Four additional reports can also be used as exemplars of tests for genetic exchange between wolves and domestic dogs. Each study reflected the introgression of genes from dogs into some native wolf populations, but with some of the authors concluding that the impact from introgression between gray wolves and domestic dogs was relatively slight. First, Vilà and Wayne (1999) reviewed results from various studies, and produced genetic data sets of their own, to test for introgression between populations of gray wolves and domestic dogs. They concluded that introgression of *C. familiaris* alleles into *C. lupus* was not significant and was not the major conservation issue that it was for rarer forms, such as the Ethiopian wolf (Vilà and Wayne 1999). Randi *et al.* (2000), through an analysis of mtDNA, came to a similar conclusion for Italian populations of *C. lupus*. Specifically, they did not detect introgression of mtDNA from domestic dogs into the Italian populations. Likewise, no hybrid individuals were detected among 130 wolves sampled from the Alps (Fabbri *et al.* 2007).

In contrast to the earlier studies, wolves sampled from Eastern Europe have been shown to possess genes from *C. familiaris* (Randi *et al.* 2000; Andersone *et al.* 2002). Furthermore, in two subsequent analyses of Italian wolves, Randi and Lucchini (2002) and Verardi *et al.* (2006) detected introgression of dog genes at frequencies of 0.9% and 5.0%, respectively. Randi and Lucchini's (2002) study incorporated both wild and captive-bred wolves and the use of 18 microsatellite loci, thus allowing a more rigorous test for admixture. Verardi *et al.*'s (2006) study examined twice as many wild wolves, using linked and unlinked microsatellite markers. This latter study detected not only a much greater proportion of hybridization than found previously, but also that this introgression was not relatively recent (Verardi *et al.* 2006). Andersone *et al.* (2002) also utilized a combination of mtDNA and microsatellites to test for admixture in Latvian wolf populations. They too detected introgression. Of 31 samples examined, six pups and six adult wolves were inferred to be of hybrid origin. However, this frequency is not necessarily reflective of the actual degree of hybridization, since these samples appeared to have been selected due to phenotypic expressions of hybridity (Andersone *et al.* 2002). Notwithstanding the preselection, the detection of introgressed forms once again indicates the interaction between *C. lupus* and *C. familiaris*. Therefore, though introgression from dogs into wolves (of all species) may be relatively rare, it does occur and is likely producing evolutionary novelty (Arnold 1997, 2006). It is also of potential conservation importance for certain isolated, and population-size-limited, canid lineages.

Domestic dogs

Understanding the role of genetic exchange in the origin and evolution of *C. familiaris*, given its importance as a key human-associated species, is the goal of this section, with the preceding examples acting as analogies for domestic dogs. Like domesticated animals (and plants) in general (e.g., Pedrosa *et al.* 2005; Ríos *et al.* 2007), there has been much interest demonstrated in deciphering the genetic variability of present-day lineages and the geographic origin of *C. familiaris*. Numerous authors have demonstrated that current breeds of dogs possess high levels of genetic variability, with some breeds demonstrating evidence of genetic diversification due to their isolation from other breeds (Kim *et al.* 2001; Leonard *et al.* 2002; Irion *et al.* 2003; Parker *et al.* 2004). Furthermore, genetic analyses have also allowed tests for the geographic point of origin for various *C. familiaris* lineages. For example, Savolainen *et al.* (2002) detected numerous mtDNA genotypes, indicating the derivation of dogs from numerous, maternal wolf lineages. However, the presence of much more genetic variability in East Asian lineages pointed to the likelihood of this region as the source for the origin of domestic dogs, with the timing of the event estimated to be *c.*15,000 ybp (Savolainen *et al.* 2002). Another analysis by Leonard *et al.* (2002) tested the hypothesis that New World *C. familiaris* had originated *in situ* rather than from Old World lineages. Mitochondrial DNA analyses falsified this hypothesis due to the demonstration of close genetic associations between ancient New World dog remains and current New and Old World lineages (Leonard *et al.* 2002).

The great genetic diversity among present-day breeds (as well as ancient remains) of *C. familiaris* has been used to infer a significant role for reticulate evolution in the origin and trajectory of this species. Various studies (Vilà *et al.* 1997, 1999; Sundqvist *et al.* 2006; reviewed by Wayne and Ostrander 2007) reflect findings that support this conclusion. Like the admixture of gray wolf and coyote lineages (Figure 5.3)—a result that is a signature of introgressive hybridization between these species (Lehman *et al.* 1991)—Figure 5.4 illustrates the same pattern for lineages of *C. familiaris* and *C. lupus*. The mtDNA genotypes detected in many of the current dog breeds are most closely related to wolf genotypes. The large degree to which this is the case (Figure 5.4) reflects the large affect that introgression between dog and gray wolf lineages has had on the genetic structure and presumably the evolutionary development of *C. familiaris* (Vilà *et al.* 1997). Indeed, Vilà *et al.* (1997) suggested that the genetic enrichment afforded by multiple introgression events from wolves into the domestic dog lineage might have yielded the material necessary for the diverse phenotypes reflected by the plethora of breeds (Vilà *et al.* 1999).

As with the findings of Vilà *et al.* (1997, 1999), Sundqvist *et al.* (2006) also detected patterns of genetic variation consistent with genetic exchange between the ancestral *C. lupus* and the derived *C. familiaris* lineage. Specifically, these authors collected sequence information from a sample of several hundred male dogs and > 100 male gray wolves for mtDNA, Y-chromosome, and/or microsatellite loci. These analyses allowed conclusions concerning the origin of various contemporary breeds of *C. familiaris*. Of more importance to the present discussion was Sundqvist *et al.*'s (2006) detection of high levels of variability at the mtDNA and Y-chromosome loci; such extreme heterogeneity was consistent with the hypothesis of extensive introgression between wolves and dogs (Vilà *et al.* 2005; Sundqvist *et al.* 2006). Thus, as with canids in general, domestic dogs have apparently diverged from other lineages while at the same time contributing to and benefiting from introgressive hybridization. This indicates that one of our closest companions (and for many *H. sapiens* populations, one of our most important food sources) reflects web-of-life processes.

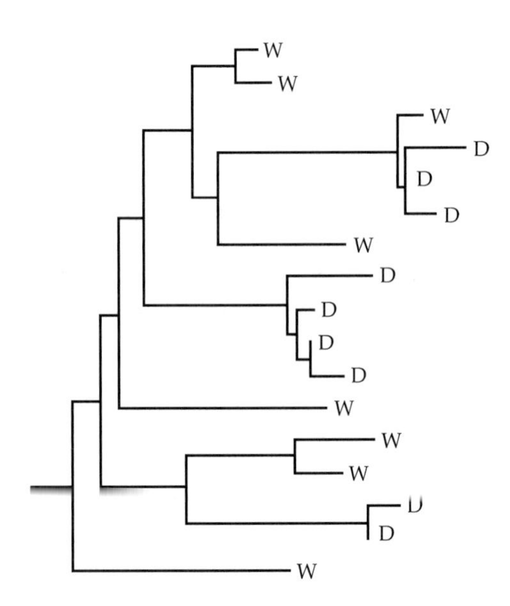

Figure 5.4 Associations of mtDNA genotypes present in gray wolves ("W") and domestic dogs ("D"). The close relationship between the wolf and dog sequences reflects multiple introgression events that transferred genetic material from gray wolves into the domestic dog lineage (see Vilà *et al.* 1997 for the data from which this phylogeny was constructed).

5.2.2 Reticulate evolution and the domestic cat/wildcat assemblage

Driscoll *et al.* (2007) have referred to domestication of wild lineages "as one of the more successful 'biological experiments' ever undertaken." Their analysis of a sample of 979 cats, including domestic cats (*Felis silvestris catus*), European wildcats (*Felis silvestris silvestris*), Near Eastern wildcats (*Felis silvestris libyca*), central Asian wildcats (*Felis silvestris ornata*), southern African wildcats (*Felis silvestris cafra*), and Chinese desert cats (*Felis silvestris bieti*), revealed the geographic partitioning of genetic variability consistent with a Near Eastern origin for the domestic cat. Furthermore, the mtDNA sequence data collected by Driscoll *et al.* (2007) indicated a sharing of genotypes among the subspecies that had diverged *c.*230,000 ybp. This estimate is much more than an order of magnitude greater than the age of domestication estimated from archaeological finds (*c.*9,500 ybp; see Vigne *et al.* 2004 for a discussion).

The genotypic diversity detected by Driscoll *et al.* (2007), like that of domestic dogs, reflects the derivation of *F. s. catus* from multiple maternal lineages. Also like the origin of *C. familiaris*, the domestication of *F. s. catus* has apparently been accompanied by past and ongoing introgressive hybridization. Figure 5.5a and b reflect findings from an analysis of nuclear (rRNA) and mtDNA (NADH-5) gene sequences. Johnson and O'Brien

(1997) used these data to examine relationships in the family Felidae as a whole. Figure 5.5a and b indicates the relationships between the domestic cat and its closest sister lineages and thus allow a test for introgressive hybridization. A phylogeny constructed from the combined nuclear and mtDNA data sets placed the domestic cat and the European and African wildcats into a clade that also included the Sand, Jungle, and Black-footed cat (Figure 5.5a; Johnson and O'Brien 1997). By definition (because only one individual of each subspecies was included in this first phylogeny), *F. s. catus, F. s. silvestris*, and *F. s. cafra* fell onto separate lineages. However, when sequences from multiple individuals of these taxa were examined an admixture of lineages, reflective of introgression, was resolved (Figure 5.5b; Johnson and O'Brien 1997). In particular, the domestic cat was found to cluster with both *F. s. silvestris* and *F. s. cafra* (Figure 5.5b).

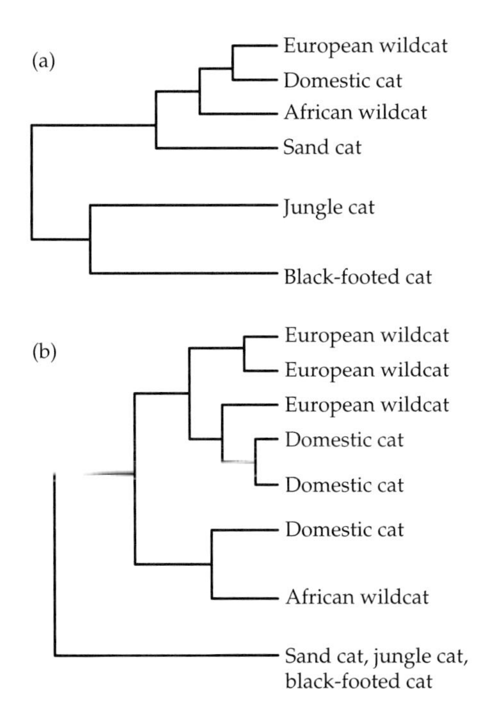

Figure 5.5 Phylogenies for the assemblage that includes the domestic cat and its closest extant sister taxa derived from a combination of nuclear rRNA and mtDNA *NADH-5* gene sequences (results from Johnson and O'Brien 1997). (a) Phylogeny that utilizes only one individual each from the various lineages. (b) Phylogeny that includes multiple individuals of domestic cats and wildcats. The intermixing of the domestic cat genotypes with the two wildcat lineages suggests the action of introgressive hybridization.

A number of analyses, subsequent to that of Johnson and O'Brien (1997), allowed a greater resolution concerning introgressive hybridization and the evolution of *F. s. catus* and its wildcat allies. Beaumont *et al.* (2001) sampled nuclear microsatellite variation from 230 "wild-living" cats and 74 cats from households across Scotland and England. The genetic data along with morphological characteristics indicated the following: (1) there were two distinct subdivisions within the wild cats, but with intermediates between these two groups; (2) there was introgression inferred from domestic cats into the wild-living populations; and (3) gene flow from the wild-living cats into the house cats was estimated to be relatively low (Beaumont *et al.* 2001). Randi *et al.* (2001) utilized both microsatellite and mtDNA data to estimate reticulate evolution involving wildcats and domestic cats. Using these loci, they assayed genetic variation for 50 *F. s. catus*, 48 *F. s. silvestris* (from Italy), 23 *F. s. libyca* (from Italy and Africa), and 7 captive-bred hybrids. Though Randi *et al.* (2001) concluded that the wildcat taxa were reproductively isolated from domestic cats, and that introgression was "very limited," 8% and 4%, respectively, of the *F. s. silvestris* and *F. s. libyca* individuals were inferred to be of hybrid origin. Similarly, Lecis *et al.* (2006) also detected an 8% admixture frequency for European wildcats from Italy. This latter analysis also estimated as much as 31% of the Hungarian wildcat population to be of hybrid derivation (see Pierpaoli *et al.* 2003 as well). Finally, Oliveira *et al.* (2007) examined the genetic variation present in both domestic cats and European wildcats from Portugal. As with each of the previous studies, this analysis detected significant introgression between these two gene pools. A statistical analysis using the microsatellite variation assigned 6 of 34 "wildcats" to the "domestic cat" class (Oliveira *et al.* 2007). This led to the conclusion common for most of the earlier analyses that "hybridization is of major concern for the appropriate implementation of wildcat conservation strategies" (Oliveira *et al.* 2007).

The conservation concerns associated with genetic assimilation of native wildcats through introgression with the domestic cat are of significance. In this regard, Driscoll *et al.* (2007) concluded that because of introgression between wildcats and domestic cats, mtDNA from the latter was commonly found in the native wildcat populations. Notwithstanding the important conservation implications of domestic/native hybridization, in the context of the current discussion, it is more significant to address the hypothesis that introgression from native cat lineages has, and possibly still is, enriching the genetic structure of the human-associated *F. s. catus*. Each of the earlier studies of cat populations suggests the role that wild > domestic reticulate events may have played in the evolution of domestic cat populations. Many of the "wild-living" cat populations are not only typified by morphological traits associated with the various wildcat subspecies, but also demonstrate "feral cat" phenotypes (e.g., Beaumont *et al.* 2001; Lecis *et al.* 2006; Driscoll *et al.* 2007). These feral cats could thus act as a genetic bridge for not only introgression into wildcat populations, but into domestic lineages as well. Indeed, Driscoll *et al.*'s (2007) extensive analysis of nearly 1000 domestic cats reflects admixtures consistent with introgression from native populations into domestic house cats. The evidence for this conclusion came from discordance between nuclear and mtDNA markers. For example, 1.5% of the "domestic cats" possessed a mtDNA genotype characteristic of wildcat lineages (Driscoll *et al.* 2007). The presence of (1) feral, domestic cats, (2) wildcats with feral cat morphological traits, and (3) domestic cats with genetic material typical of wildcats reflects past and ongoing bidirectional introgression between *F. s. catus* and its wildcat progenitors. Though this introgression appears to be asymmetric—with present-day gene flow largely preceding in the direction of domestic > wild populations—the domestic lineages are nonetheless being impacted by wildcat genes and thus reflect not only an origin, but also a continued evolutionary trajectory affected by reticulation.

5.2.3 Reticulate evolution and genets

Genets (genus *Genetta*) are carnivorous taxa that belong to the family Viverridae, and reflect a mainly sub-Saharan Africa distribution (Gaubert *et al.* 2004a). Several *Genetta* species have been domesticated, with some of their characteristics described as strongly bonding with their owners, playfulness, curiosity, and the ability to use their retractable claws to climb bare legs (http://www.juliesjungle.com/genet.php). Gaubert and his colleagues have published numerous papers in which they have discussed the range distributions, ecological settings, phylogenetic relationships, and species status for members of the genus *Genetta* and sister genera such as *Osbornictis, Poiana*, and *Prionodon* (Gaubert *et al.* 2002, 2004a, b, 2005, 2006; Gaubert and Begg 2007). Several findings from these analyses reflect the role of introgressive hybridization in the evolutionary development of *Genetta*. As is usual, phylogenetic treatments have not only resolved problematic evolutionary patterns (e.g., the relationships of the purported sister genera of *Genetta*; Gaubert *et al.* 2004b), but also provided a means of testing for reticulation. One finding made by Gaubert *et al.* (2002) that could be considered consistent with the occurrence of reticulate evolution was the phylogenetic placement of *Genetta johnstoni*—using morphological traits—into a separate clade relative to all other *Genetta* species. However, this finding was later determined to more likely reflect the role of convergent morphological evolution that resulted in *G. johnstoni* sharing traits with members of other genera, rather than with other species of *Genetta* (Gaubert *et al.* 2004b).

Notwithstanding the falsification of a reticulate origin hypothesis for *G. johnstoni*, introgressive hybridization between genet lineages has been inferred from other analyses. For example, Gaubert *et al.* (2004a) typified the "large-spotted genet" assemblage as reflecting numerous cases of introgressive hybridization. Indeed, three of the six species within this clade (*Genetta maculata*, *Genetta tigrina*, and *Genetta pardina*) were identified as likely participating in reticulation (Gaubert *et al.* 2004a). An analysis mainly designed to test the utility of natural history collections for defining the ecological ranges of genets and evolutionary pattern/process among the species also resulted in support for genetic exchange. In particular, Gaubert *et al.* (2006) collected pelage and cranial traits for individuals of *Genetta cristata* and *Genetta servalina*, both outside and within an area of sympatry. Though morphological admixtures were absent outside of the zone of overlap, they were present in individuals collected within the sympatric region, a pattern consistent with the occurrence of introgressive hybridization between these two species (Gaubert *et al.* 2006).

Introgressive hybridization between *G. cristata* and *G. servalina* was also indicated by a subsequent analysis of mtDNA and Y-chromosome loci sequence variation. Gaubert and Begg (2007) thus detected mtDNA sequences characteristic of *G. servalina* in an individual possessing *G. cristata* morphological traits. Likewise, this analysis identified introgression involving other genet taxa as well. For example, asymmetric gene flow was discovered between *G. tigrina* and *Genetta felina* (i.e., introgression initiated by hybridization between *G. tigrina* males and *G. felina* females; Gaubert and Begg 2007). Furthermore, Gaubert and Begg (2007) reviewed evidence that suggested a hybrid origin for the lineage that gave rise to *Genetta genetta, Genetta angolensis, Genetta poensis, Genetta schoutedeni, Genetta bourloni, G. pardina, G. maculata, G. tigrina*, and *G. felina*.

One additional study by Gaubert *et al.* (2005) reflects well the degree to which reticulate evolution has affected the patterns of morphological and genetic variability in genets. It also exemplifies the facilitating effect that environmental factors may have on genetic exchange. In their study, Gaubert *et al.* (2005) collected data for nine discrete morphological traits along with sequence variation at the mtDNA cytochrome *b* locus. Though morphological traits alone were unable to define all hybrid individuals, the combination of these data with the mtDNA sequence information detected numerous examples of "cryptic" introgression (Gaubert *et al.* 2005). In particular, this study resulted in the inference of introgression between (1) *G. tigrina* and *G. genetta* and (2) *G. tigrina* and *G. felina*. Significantly, Gaubert *et al.* (2005) characterized the degree to which the latter two species introgressed with the phrase, "massive hybridization." Finally, these authors concluded that the type specimen of the species, *Genetta mossambica* was a hybrid between *G. maculata* and *G. angolensis*. In considering environmental factors that might contribute to occurrences of zones of sympatry and thus introgression, Gaubert *et al.* (2005) identified seasonal rainfall patterns and periods of frost as two possible key components. However, regardless of the causal factors, the extensive analyses by these workers have identified the degree to which genetic exchange has impacted this assemblage of African carnivores that have been incorporated into the pool of human pet species.

5.3 Reticulate evolution and human burden bearers

The category of "human burden bearers" may seem somewhat odd, or tangential, to the present discussion, but like all other categories of human-benefiting organisms these species have been instrumental in the development of *H. sapiens* societies. The two groups of organisms that I have chosen to place under this category—water buffalo and donkeys—have, like most species used as exemplars of other categories, been utilized by humans for multiple, beneficial purposes. For example, a food product from water buffalo (*Bubalus bubalis*) has become a reflection of all-things-Italian:

> Italy and mozzarella are synonymous...The buffalo [from which the milk for some Italian mozzarella cheese comes] was introduced into Italy in the seventh century...Mozzarella became widespread throughout south Italy from the second half of the eighteenth century...mozzarella di bufala cheese is not a recent arrival. (www.mozzarelladibufala.org)

However, reflecting also the appropriateness of the usage of buffalo as an example of reticulate evolution impacting a beast of burden, "In ancient times [in Italy], the buffalo was a familiar sight in the countryside...used as a draught animal in plowing compact and watery terrains." Thus, beasts of burden one day may be food the next, or may even provide both benefits simultaneously.

5.3.1 Reticulate evolution and the donkey

Like many of the examples of reticulate evolution of human-associated species, donkeys (*Equus asinus*) demonstrate phylogenetic patterns of nonmonophyly of related lineages (Figure 5.6). Specifically, Beja-Pereira *et al.* (2004) found that there were two clades into which *E. asinus* lineages were placed; both clades included the donkey and a wild African ass taxon (Beja-Pereira *et al.* 2004). The first clade consisted of *E. asinus* paired with the Nubian wild ass, *Equus africanus africanus*. The second assemblage included the donkey and the Somali wild ass, *Equus africanus somaliensis*. Beja-Pereira *et al.* (2004) concluded that the degree to which the sequences found in the two donkey lineages differed indicated a divergence time of 0.3–0.9 mya. This greatly exceeded the estimated 10,000 ybp for the domestication event, and thus led to Beja-Pereira *et al.*'s (2004) conclusion that the two donkey lineages reflected separate origins from maternal wild ass lineages (i.e., Nubian and Somali) characterized by anciently diverged mtDNA.

Though the results from Beja-Pereira *et al.* (2004) strongly support an African (rather than an Asian) point of origin for *E. asinus*, the phylogenetic relationships reflected in Figure 5.6 can alternatively be explained by a single origin for the donkey from either the Nubian or Somali wild ass lineage. However, such a scenario would require an introgression event subsequent to its origin. Thus, like all the other examples given in this book, introgression from the alternate lineage would have resulted in the phylogenetic signal detected by Beja-Pereira *et al.* (2004; Figure 5.6). Like a two-origin model, this hypothesis requires two events and thus is equally parsimonious, but instead of separate origins with no gene flow explaining the current sequence divergence (i.e., Beja-Pereira *et al.*'s conclusion), there would have been a single origin plus an introgression event. Deciding between these two hypotheses would be facilitated by detailed population-level analyses in which gene flow from domestic into wild populations was tested for.

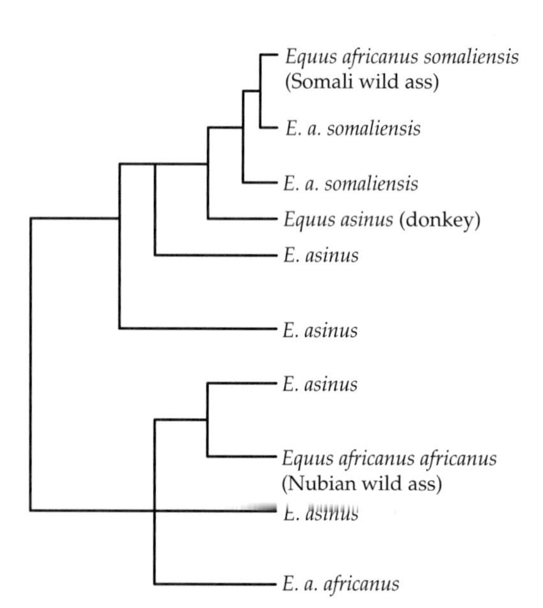

Figure 5.6 Weblike relationships among wild asses (Nubian and Somali) and the domestic donkey defined using mtDNA sequence variation (Beja-Pereira *et al.* 2004). The distribution of *E. asinus* into two clades containing either *E. a. africanus* or *E. a. somaliensis* is consistent with a single origin of the donkey in Africa followed by mtDNA introgression.

5.3.2 Reticulate evolution and the water buffalo

The wild and domesticated water buffalo, *B. bubalis*, network illustrates the role of introgressive hybridization between ancestors and their domesticated derivatives. For many domesticated forms, there are multiple origins in different geographical regions. This is reflected by a polyphyletic assortment of the domesticated taxa, that is domesticated forms founded from multiple, independent wild lineages. Such is the case for pigs, cattle, and sheep (see Chapter 6). This conclusion has also been reached for domesticated water buffalo in which there are two morphologically, and genetically, recognizable types—the river and swamp forms (Sena *et al.* 2003; Vallinoto *et al.* 2004). Lau *et al.* (1998) designated the swamp variant to be the ancestral water buffalo phenotype. Furthermore, they posited that this form arose in Asia and spread to the Indian subcontinent in which the river buffalo subsequently evolved. Following their divergence, two domesticated lineages arose from the two foci of water buffalo evolution. The domesticated, swamp buffalo then extended its range through mainland Southeast Asia with the concomitant result of "interbreeding with wild buffalo" populations (Lau *et al.* 1998).

In contrast to the model of water buffalo evolution and domestication proposed earlier, Kierstein *et al.* (2004) detected a geographic distribution of genetic variability that suggested a different center of origin. These workers collected mtDNA D-loop sequence variation for 19 swamp buffalo and 61 river buffalo individuals. A comparison of the 36 mtDNA genotypes detected from this analysis with previously published sequences from Southeast Asia and Australia led to an inference of domestication of both water and river lineages on the Indian subcontinent approximately 5,000 ybp (Kierstein *et al.* 2004). An additional center of domestication was also postulated for China. Following the domestication events, contact between the different lineages of *B. bubalis* resulted in widespread introgression from wild into domesticated animals (Kierstein *et al.* 2004). Although there are different hypotheses concerning the points of origin for the various forms, each model reflects the role of reticulation in the origin and evolution of domestic water buffalo (Lau *et al.* 1998; Kierstein *et al.* 2004).

5.4 Reticulate evolution and human clothing

The evolution of human clothing material has involved the interactions between humans and numerous plant and animal species. Once again, the examples to be discussed under this topic—cotton, whitetail and mule deer, and caribou/reindeer—are neither exhaustive nor reflective of the only category into which these organisms fit in terms of their utility for humans. Thus, I am not discussing the reticulate evolution of cattle in this section even though the development of leather products is of fundamental importance to the manufacture of human clothing. Instead, I have chosen to discuss cattle breed evolution under the heading of the evolution of animal food products in Chapter 6. Likewise, the human use of deer and caribou has not been limited to their skins, but includes decorations, food, and hunting implements as well. However, cotton fiber and deer and caribou skins provide not only examples of the reticulate evolution of plant and animal products used for clothing, but also the divergent, reticulate processes of allopolyploidy (cotton) and introgressive hybridization (deer and caribou/reindeer).

5.4.1 Reticulate evolution and cotton

The world's cotton fiber is produced from four species, *G. [Gossypium] arboretum* L. (n = 13, A genome), *G. herbaceum* L. (n = 13, A genome), *G. barbadense* L. (n = 26, AD genome), and *G. hirsutum* L. (n = 26, AD genome). The two tetraploid species dominate world cotton production. (Zhao *et al.* 1998)

This quote reflects the fact that the evolution of one of the major clothing materials worldwide was, at its inception, reticulate. Indeed, phylogenetic analyses of the entire genus have demonstrated widespread hybrid (i.e., allopolyploid) speciation events (Seelanan *et al.* 1997). Likewise, the tetraploid cultivars, defined as containing two genomes, are allopolyploid derivates from hybridization between diploid lineages that were either "A" or "D" genome taxa.

Hypotheses concerning the geographic point of origin of cotton have had to take into account the indigenous occurrence of diploids and polyploid taxa in Africa, Central America, South America, Asia, Australia, and the islands of Hawaii and Galapagos (Zhao *et al.* 1998). The tetraploid species, *G. hirsutum*, exemplifies the uncertainty of the geographic origin of species from which humans derive much of their clothing material. Furthermore, Brubaker and Wendel's (1994) description of *G. hirsutum* as "the species of cotton that now dominates world cotton commerce" indicates its exemplar status for this discussion. In particular, as much as a decade ago, this species along with *G. barbadense*, already accounted for much of the estimated US$20 billion cotton market (Jiang *et al.* 1998). Given *G. hirsutum*'s and *G. barbadense*'s New World distributions, the biological basis of nearly the entire cotton industry likely evolved somewhere in this region. Phylogenetic analyses of wild and domesticated populations of *G. hirsutum* confirmed this conclusion. Specifically, native Yucatan peninsula populations were inferred to be the source for (1) the Guatemalan and southern Mexico cultigens and thus (2) the modern "Upland" cultivars derived from Mexican highland populations that were founded from the southern Mexico and Guatemalan lineages (Brubaker and Wendel 1994).

As already mentioned, most significantly for the present discussion is the observation that the species utilized for cotton fiber production are derived from the reticulate process of allopolyploid speciation. Though the combination of A and D lineages through allopolyploidy is recognized to have occurred in the New World, surprisingly the A (maternal) parent appears to have been an Old World species, while the D genome (paternal) parent was of New World derivation (Wendel 1989). In addition, the bringing together of divergent, diploid, genomes provided the genetic basis for the fiber quality necessary for cotton to be usable as a clothing material.

As with the unexpected contribution of the Old World lineage, Jiang *et al.* (1998) found that loci from the D genome progenitor contributed most to fiber quality and yield in the cultivar. The surprising nature of this finding results from the observation that the D genome progenitor "does not produce spinnable fibers" (Jiang *et al.* 1998). The genome of origin for the fiber genes was further supported by analyses of multiple genetic mapping populations under a variety of environmental regimes (Rong *et al.* 2007). Indeed, this latter study provided a great deal of resolution concerning the evolution of both *G. hirsutum* and *G. barbadense* and resulted in the following conclusions: (1) the A and D genome loci contributing to fiber quality and quantity are mostly different, indicating the interactions of many, "corresponding" genomic regions; (2) there are numerous, undetected loci that contribute to these traits; and (3) fiber characteristics are apparently affected by a "complex network of interacting genes." In summary then, the evolution of this major source of human clothing has resulted from the "merger of two genomes with different evolutionary histories in a common nucleus" (Jiang *et al.* 1998).

5.4.2 Reticulate evolution and deerskins

The use of deerskins for clothing by North American native peoples was a widespread phenomenon. Dances that commemorated the killing of the animals reflected this use (Hibben 1992). Two species of North American deer, *Odocoileus virginianus* (the whitetailed deer) and *Odocoileus hemionus* (mule deer) were commonly utilized in western-most North America. Transfer of this clothing technology to European-extraction explorers was recorded in the writings of these later travelers. For example, the journals from the Lewis and Clark expedition (May 13, 1805) reported that

Captain Clark who was on shore the greater part of the day killed a mule and a common [i.e., white-tailed] deer, the party killed several deer and some elk principally for the benefit of their skins which are necessary to them for clothing. (http://www.lewisclarkeandbeyond.com/journals/)

Apparently, beginning with their earliest migrations, through their travels in the nineteenth century, one of the key ingredients for the survival of *H. sapiens* in North America was the protection provided by the deerskin clothing made from white-tailed and mule deer.

The significant morphological differences between *O. virginianus* and *O. hemionus* (Roosevelt 1996, pp. 127–128) were also reflected by behavioral differences leading to different habitat preferences (Avey *et al.* 2003). Theodore Roosevelt in his book, *Hunting Trips of a Ranchman* (1996, p. 128) described the microhabitat associations of white-tailed and mule deer in the following manner:

One of the noticeable things in western plains hunting is the different zones or bands of territory inhabited by different kinds of game. Along the alluvial land of the rivers and large creeks is found the whitetail. Back of these alluvial lands generally comes a broad tract of broken, hilly country, scantily clad with brush in some places; this is the abode of the black-tailed [i.e. mule] deer.

Recent analyses do, however, belie the significant phenotypic and behavioral differences between *O. virginianus* and *O. hemionus*. For example, nonconcordant phylogenetic patterns based on sequence variation at mtDNA and nuclear loci have been detected (Carr *et al.* 1986; Cathey *et al.* 1998; Bradley *et al.* 2003). Figure 5.7

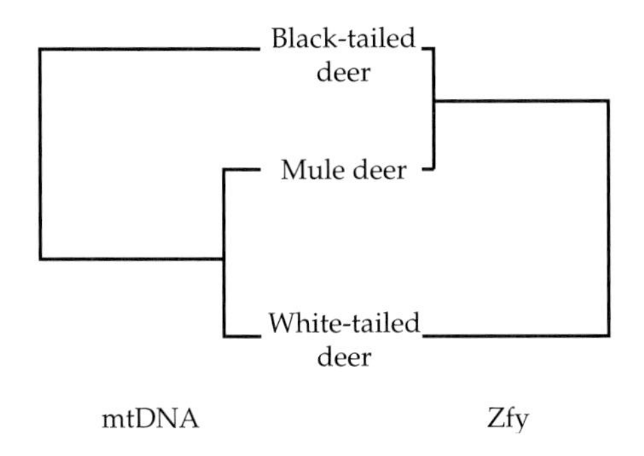

Figure 5.7 Phylogenetic placement of black-tailed, white-tailed, and mule deer based on the maternally inherited mtDNA and paternally inherited Y-chromosome (i.e., "*Zfy*") loci. The *Zfy* phylogeny agrees with behavioral and morphological data that place blacktailed and mule deer into the same species, *O. hemionus*, and separate from the white-tailed deer, *O. virginianus*. The mtDNA phylogeny indicates that introgressive hybridization has likely occurred from the white-tail into the mule deer lineage (Cathey *et al.* 1998).

reflects this discordant pattern for maternally inherited mtDNA versus paternally inherited Y-chromosome loci (the zinc finger gene, *Zfy*; Cathey *et al.* 1998). The *Zfy* sequence data resolved a phylogenetic tree in agreement with the morphological and behavioral data sets. The conspecific black-tailed and mule deer individuals were resolved as sister lineages relative to the white-tailed deer samples (Figure 5.7; Cathey *et al.* 1998). In contrast, mtDNA sequences resulted in the pairing of mule deer and white-tailed deer in a clade separate from the black-tailed deer lineage (Figure 5.7). The closely associated habitats of the two species (Roosevelt 1996; Avey *et al.* 2003) and the detection of present-day hybrid zones (Carr *et al.* 1986; Bradley *et al.* 2003) supported a hypothesis of mtDNA introgression leading to morphologically defined mule deer individuals carrying white-tailed deer mtDNA haplotypes. Over the western North American distribution of these species, introgressed animals (given that all mule deer possess mtDNA from the white-tailed deer lineage) would have thus provided the clothing material for a large proportion of the indigenous peoples and the later migrations by European-derived explorers.

5.4.3 Reticulate evolution and caribou/reindeer skins

One of the prime functions of the [Hudson's Bay Company] Baker Lake post was to buy caribou skins locally and export them to posts where they were badly needed...Without caribouskin clothing, many of the coastal Eskimos would refuse to trap during the bitterly cold months of winter...Caribou clothing was the only type warm enough to allow them to spend days and nights out on the land. (Pryde 1971, p. 22)

Duncan Pryde's remembrances (derived from his 10 years spent living in the Arctic), reflects the key role that caribou skins played in the existence of ancient and contemporary populations of Native North American *H. sapiens*. Yet again, this species reflects multiple uses, particularly as a source of warm clothing and protein. As such, the presence of caribou/reindeer has acted as a facilitator for the survival of humans in extreme environments.

Given its circumpolar distribution the role that the caribou or reindeer, *Rangifer tarandus*, has played in the spread and occupation of harsh environments has not been limited to North America. Furthermore, reticulate evolution has evidently affected the trajectories of both native taxa, as well as the domesticated lineages (in North America referred to as "reindeer"; Cronin *et al.* 2003). For example, a web of genetic interactions was detected for two lineages of tundra caribou in southwestern Greenland. The native *Rangifer tarandus groenlandicus* and the semidomesticated, *Rangifer tarandus tarandus*, were brought into contact through the introduction of the latter by humans (Jepsen *et al.* 2002). Fjords and glaciers have acted as partial barriers to introgression between the native and introduced subspecies. However, Jepsen *et al.* (2002) found evidence for genetic exchange between the wild and domesticated lineages; microsatellite markers characteristic for *R. t. groenlandicus* and *R. t. tarandus* were found to have been transferred between the sympatric caribou and reindeer populations (Jepsen *et al.* 2002).

The finding of significant introgression between native caribou and introduced reindeer in Greenland contrasts with a report concerning native and introduced forms in Alaska. Specifically, Cronin *et al.* (2003) concluded that "there may have been a low level of introgression, but the two forms have maintained different allele frequencies over the 110 years since the introduction of reindeer to Alaska." This conclusion was reiterated in an analysis that also included microsatellite loci (Cronin *et al.* 2006). Yet, these authors did report a high level of genetic similarity between arctic Canadian caribou from Baffin Island and Scandinavian reindeer populations (Cronin *et al.* 2003). This latter finding is consistent with a hypothesis of significant levels of introgressive hybridization between caribou of the Canadian archipelago and those from northern Europe. Likewise, introgressive hybridization could also explain findings from a study of mtDNA variation among *Rangifer tarandus pearyi* (from the Canadian Archipelago), *Rangifer tarandus platyrhynchus* (from Svalbard—an archipelago located approximately midway between Norway and the North Pole) and *Rangifer tarandus eogroenlandicus* (an East Greenland form that has been extinct since 1900). Though each of these taxa belongs to the so-called "small bodied, high-arctic reindeer", Gravlund *et al.* (1998) found that these lineages clustered with different "large-bodied reindeer" taxa and thus did not reflect a monophyletic assemblage. Gravlund *et al.* (1998) concluded that the data were consistent with separate origins for the small-bodied lineages from different large-bodied reindeer. Their results are also consistent with an alternative model of a single origin of the small-bodied forms followed by secondary contact and mtDNA introgression from large-bodied lineages.

A third study by Cronin and his colleagues (Cronin *et al.* 2005) involving an analysis of 18 microsatellite loci and the mitochondrial cytochrome *b* gene locus of *Rangifer tarandus granti* (Alaskan barren ground caribou), *R. t. groenlandicus* (Canadian barren ground caribou), and *Rangifer tarandus caribou* (woodland caribou) also supported the occurrence of reticulate evolution. Cronin *et al.*'s (2005) findings led to the following conclusions: (1) mtDNA introgression has occurred between the domestic reindeer populations introduced into Alaska with the native *R. t. granti* (see also Flagstad and Røed 2003); (2) mtDNA and microsatellite locus introgression has led to low levels of differentiation between the two barren ground subspecies (*R. t. groenlandicus* and *R. t. granti*); and (3) hybridization of the Labrador herd of the woodland caribou (*R. t. caribou*) with *R. t. groenlandicus* has resulted in both nuclear and mtDNA introgression between these morphologically differentiated subspecies. Taken together, the studies of genetic variation among the numerous subspecies of caribou/reindeer lend support to a significant impact from reticulate evolution via introgressive hybridization.

5.5 Reticulate evolution of fuel, building materials, corks, and containers

Though the title of this section might seem overly detailed, it was chosen intentionally to emphasize again the diversity of human-utilized, biological products that reflect reticulate evolution. In addition, though some of these categories could be illustrated by several examples (e.g., the "fuel" subsection could have included many tree species complexes that reflect extensive introgressive hybridization), I have chosen to limit the discussion of each to a single representative case. In this way, I am able to include a number of uses that might be less evident, but yet still emblematic of *H. sapiens*' day-to-day dependence on reticulate assemblages for the cultivation of crops, protection from the elements, and the transportation and storage of foodstuffs and drugs.

5.5.1 Reticulate evolution of wood products—fuel

The genus *Quercus* (oak) is a paradigm for the role that genetic exchange can play in the evolution of lineages (e.g., Stebbins *et al.* 1947; Grant 1981; Howard *et al.* 1997; Williams *et al.* 2001; Dodd and Afzal-Rafii 2004; Lexer *et al.* 2006; Valbuena-Carabaña *et al.* 2007). Furthermore, humans have utilized various members of this assemblage for a wide array of products, including wooden ships, mine timbers, flooring, furniture, and coffins (Brown *et al.* 1949; pp. 539–548). *Quercus* species, particularly those from the "white oak" complex, have also provided firewood due to their excellent fuel value (Brown *et al.* 1949; p. 548).

Whittemore and Schaal (1991) illustrated the role that introgressive hybridization had played in the white oak species complex by analyzing both cytoplasmic (cpDNA) and nuclear (rDNA) genetic variation in five species that demonstrated sympatric distributions in the eastern United States. *Quercus alba, Quercus macrocarpa, Quercus michauxii, Quercus stellata,* and *Quercus virginiana* var. *fusiformis* demonstrate distinct morphological characteristics and adaptations to different, but overlapping, microhabitats (Hardin 1975, 1979; Solomon 1983; Whittemore and Schaal 1991). These species are interfertile, but natural hybrids, as detected by morphological characteristics, have been infrequently identified (e.g., Hardin 1975). For example, none of the trees sampled by Whittemore and Schaal (1991) possessed admixtures of the morphological traits diagnostic for the five species. In contrast, for the six instances in which two species co-occurred, the most common cpDNA genotype was shared by both taxa (Whittemore and Schaal 1991). Introgressive hybridization among the various species had thus affected the genetic structure of these sympatric populations. This was the case even for a hybrid zone involving the evergreen species, *Q. virginiana* and the deciduous species, *Q. stellata*. The occurrence of introgression between these two species was particularly surprising given their extreme morphological differences and the previous lack of evidence for extensive hybridization between them (Whittemore and Schaal 1991).

White oaks, utilized as they have been for such a variety of key products that benefit humans, reflect *H. sapiens*' dependence on natural hybrids for survival and cultural development. Furthermore, these species are indicators of the degree to which the web of life can connect, both genetically and evolutionarily, even distantly related members of a species complex.

5.5.2 Reticulate evolution of wood products—building materials

"*Pinus lambertiana* is one of the *c.*20 species of subsection *Strobus*...This clade is known by the common name 'white pines' and is distributed discontinuously throughout the Northern Hemisphere" (Liston *et al.* 2007). The white pines, like species of *Quercus*, have played an essential role in the development of human populations. In particular, the wood from *P. lambertiana* (commonly known as sugar pine) has provided the raw material for crates, millwork (e.g., interior and exterior trim on houses), roofs, signs, piano keys, and organ pipes (Brown *et al.* 1949, p. 445). Though reticulate evolution is a well-known phenomenon in the genus as a whole (Liston *et al.* 2007), the monophyletic origin of the white pines has been repeatedly demonstrated by a variety of molecular markers (including chloroplast and nuclear loci; Liston *et al.* 1999; Gernandt *et al.* 2005; Eckert and Hall 2006; Syring *et al.* 2007). Furthermore, in discussing *P. lambertiana*, Critchfield and Kinloch (1986) stated that this species was, compared to all other members of subsection *Strobus*, extremely limited in its ability to hybridize with other lineages.

The observation of Critchfield and Kinloch (1986)—coming as they did from extensive, experimental hybridization studies—makes the results of a phylogeographic study by Liston *et al.* (2007) surprising. These latter workers examined the pattern of cpDNA variation in *P. lambertiana* populations stretching from Oregon to Baja California. This survey detected two cpDNA haplotypes, designated "north" and "south," and a narrow hybrid zone between these two lineages located in northeastern California. The northern haplotype was found to be phylogenetically aligned with whitebark (*Pinus albicaulis*) and East Asian, pine lineages, while the *P. lambertiana* haplotype from the southern portion of the sampling area clustered with other North American and Central American taxa (Liston *et al.* 2007). Given the extremely low differentiation between the sugar pine and whitebark pine cpDNA sequences, Liston *et al.* (2007) suggested that during the Pleistocene the latter had likely expanded from Asia into North America, making contact with the former. The secondary contact between these two species would have provided the opportunity for cpDNA transfer from *P. albicaulis* into *P. lambertiana* (Liston *et al.* 2007).

5.5.3 Reticulate evolution of wood products—cork

Costa *et al.* (2004) referred to the cork oak, *Quercus suber*, as the "dominant species" of the southwestern Iberian Peninsula, "montado," agroforestry industry. This conclusion is supported by (1) the cultivation of cork oak in Portugal and Spain covering ~726,000 and 510,000 ha, respectively (Costa *et al.* 2004) and (2) the annual production of ~135,000 tons of cork from the Portugal cultivars alone, which reflects >50% of worldwide cork production (Costa and Oliveira 2001). The total hectares utilized in cork oak production has been estimated at ~2.7 million and includes, in addition to Spain and Portugal, farms in Algeria, Morocco, Italy, Tunisia, and France (Figure 5.8). Cork oak habitats form a key ecological element of the regions in which they occur. As a World Wildlife Report (2006) stated, "These landscapes are one of the best examples in the Mediterranean for balancing conservation and development for the benefit of people and nature. They sustain rich biodiversity and traditional livelihoods." Indeed, the ecological setting of the cork oak forests can contain up to 135 species per square meter, many of which are the best known of the endangered animal taxa (e.g., Iberian lynx, Iberian Imperial Eagle, Black Stork; WWF 2006). Unfortunately, *Q. suber* habitats have come under increasing danger of extinction largely due to the changeover in the wine industry from cork

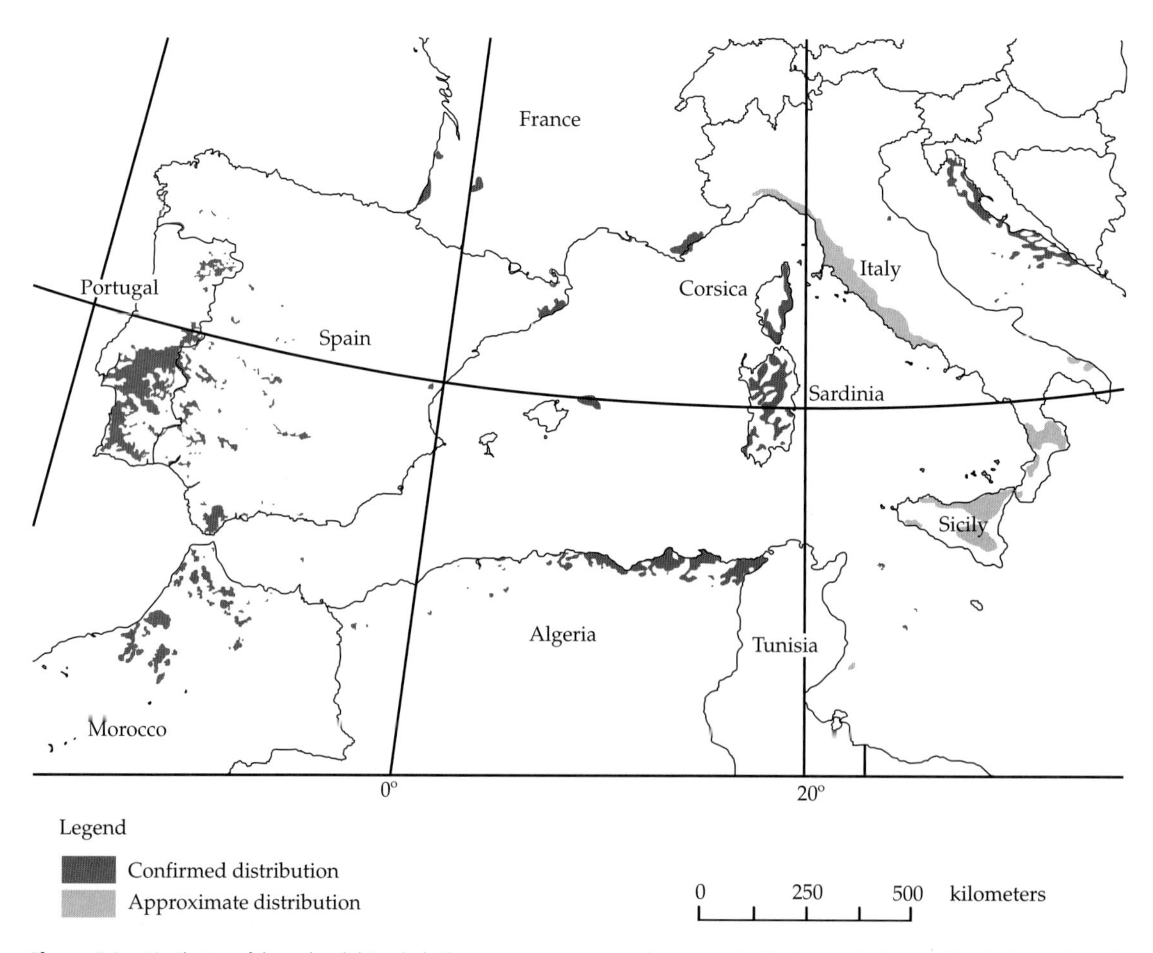

Figure 5.8 Distribution of the cork oak (*Q. suber*). The approximate amount of area occupied by this species is 2.7 million ha (map adapted from WWF 2006).

stoppers to those made of synthetic materials or metal (WWF 2006).

The evolutionary importance of introgressive hybridization involving *Q. suber* with other *Quercus* lineages has been detected by numerous studies. For example, Toumi and Lumaret (1998) used allozyme markers to test for patterns of genetic variation in cork oak. In addition to phylogeographic discontinuities, these authors found apparent introgression of allozyme alleles from holm oak (*Quercus ilex*) into cork oak individuals. Similarly, Coelho *et al.* (2006) described a pattern of shared nuclear alleles (i.e., amplified fragment length polymorphisms [AFLP]) between these two species that was consistent with introgression from *Q. ilex* into *Q. suber*. Several analyses of cpDNA variation have also led to the inference of introgressive hybridization between cork oak and holm oak populations. In particular, Jiménez *et al.* (2004), Belahbib *et al.* (2001), and Lumaret *et al.* (2005) reported cpDNA variation in these two species resulting in the following conclusions: (1) cpDNA introgression had occurred between *Q. suber* and *Q. ilex* in numerous regions of overlap; (2) there was asymmetry in the introgression event with cpDNA genotypes from *Q. ilex* having introgressed into *Q. suber* at a much higher frequency than from *Q. suber* into *Q. ilex*; (3) following introgression of *Q. ilex* cpDNA into the cork oak lineage, mutations have occurred leading to multiple cpDNA types in cork oak (i.e., derived from both the introgressed and native cpDNA); and (4) the presence of *Q. ilex* cpDNA in individuals that were *Q. suber* on the basis of nuclear loci may reflect the "migration" of cork oak through pollen-mediated introgression. The inference of asymmetric introgression into cork oak by holm oak has also been supported by a recent and extensive phylogeographic analysis of cpDNA variation by Magri *et al.* (2007). Indeed, these latter authors reported data supporting the hypothesis of introgression between *Q. suber* and *Q. ilex*, but also between *Q. suber* and the deciduous oak species, *Quercus cerris* (Turkey oak; see later).

Another class of observation also indicates the importance of reticulate evolutionary processes involving *Q. suber*. However, in this case, rather than being the recipient of genetic material through introgression, the cork oak lineage has acted as one of the progenitors for hybrid species. Specifically, both *Quercus afares* and *Quercus crenata* have been recognized as hybrid species. *Q. afares* is a North African endemic that is sympatric with both *Q. suber* and its other putative parent, *Quercus canariensis* (Mir *et al.* 2006). Surprisingly (because cork oak is evergreen, while *Q. canariensis* is a semideciduous species), *Q. afares* reflects an admixture of morphological, physiological, and ecological traits characteristic of the two hypothesized progenitors (see discussion in Mir *et al.* 2006). Consistent with a hybrid origin for *Q. afares* was the finding of combinations of nuclear and cpDNA loci from *Q. suber* and *Q. canariensis* in *Q. afares* individuals/populations (Mir *et al.* 2006). Likewise, a hybrid derivation of *Q. crenata* from crosses involving *Q. suber*, but in this case with *Q. cerris*, has also been inferred (Conte *et al.* 2007). Nuclear genotypes in *Q. crenata* populations indicated its intermediate position between cork and Turkey oak genotypes (Conte *et al.* 2007). The intermediacy of the genotypic scores for *Q. crenata* is consistent with an admixture of alleles and thus a hybrid derivation from *Q. suber* and *Q. cerris*.

Both the introgression of genomic material between cork oak and other species of Mediterranean oaks, as well as *Q. suber*'s role in the origin of hybrid species, again reflects the evolutionary significance of reticulate evolution in general. Furthermore, reticulate evolution has impacted a species that has a keystone position in terms of its ecological, cultural, and economic value.

5.5.4 Reticulate evolution of containers—bottle gourds

"The bottle gourd is predominantly grown for its fruit, which when dry, form a woody rind (exocarp) that is used mostly for the manufacture of containers (for water and food)" (Clarke *et al.* 2006). *Lagenaria siceraria* (the bottle gourd) is known mainly from cultivated populations. However, Decker-Walters *et al.* (2004) described a wild population from a remote region of southeastern Zimbabwe, a finding that was consistent with an African point of origin. Though genetically similar to cultivated *L. siceraria*, this wild population demonstrated genetic distinctiveness (for both nuclear and cpDNA loci) relative to the cultivar (Decker-Walters *et al.* 2004). Furthermore, the wild samples possessed nuclear markers indicative of introgression with individuals of *Lagenaria sphaerica* (Decker-Walters *et al.* 2004). This latter observation suggests that the evolutionary history of a lineage related to the progenitor of the cultivated bottle gourd was reticulate.

The African origin of the bottle gourd was followed by its migration (likely due to humans) to other Old World regions and subsequently, ~10,000 ybp by its transport to the New World (Piperno *et al.* 2000; Erickson *et al.* 2005). It is also hypothesized that *L. siceraria* was either (1) carried by humans from Asia or the New World to Polynesia by 1200 AD or (2) that the Polynesian cultivar was a hybrid derivative from introgression between Asian and New World lineages (see Clarke *et al.* 2006 for a discussion). The sequencing of 36 *L. siceraria* cultivars from Asia, the Americas, and Polynesia, for two cpDNA and five nuclear loci, tested these alternate hypotheses. Clarke *et al.*'s (2006) results were consistent with a biparental contribution from Asia and the New World in the formation of the Polynesian lineage. Specifically, the cpDNA loci of Polynesian *L. siceraria* indicated an Asian contribution, while the nuclear loci reflected introgression from American lineages (Clarke *et al.* 2006). Though whether or not the hybrid origin of the Polynesian bottle gourd was due to human-mediated movement of this species could not be determined, it is clear that this major, cultivated product is the result of reticulate evolution involving lineages from different continents. Thus, introgressive hybridization appears to have been foundational in the origin and evolution of at least the wild African population (Decker-Walters *et al.* 2004) and the Polynesian cultivar (Clarke *et al.* 2006) of *L. siceraria*.

5.6 Reticulate evolution of drugs

H. sapiens' intake of various biological products that can be termed "drugs" is likely to have begun as soon as our species recognized the effects possible from their ingestion. However, regardless of when in our evolutionary history this consumption began, the diversity of compounds that we now utilize is legion. This diversity also reflects the widespread utilization of organisms that have evolved by processes that underlie the web of life—horizontal transfer, introgression, and hybrid speciation. To illustrate the reticulate nature of the evolution of these organisms—and to reflect a portion of the wide array of compounds—I will discuss examples of drugs that are used to treat illnesses (e.g., atropine and essential oils) as well as drugs that we rely upon for relieving tension and providing pleasure (e.g., chocolate, tobacco, and coffee). Notwithstanding their intended function, the application of each of these compounds provides a noticeable response (at least in many humans) that is a paradigm of a drug-induced, physiological effect. This is the case whether or not humans would prefer to categorize a specific product as "drug," "intoxicant," or "food."

5.6.1 Reticulate evolution and atropine

The plant genus *Atropa* has been the source of a taxonomic debate since the late 1800s (see Olmstead and Sweere 1994; Yuan *et al.* 2006 for discussions). In particular, there have been conflicting hypotheses concerning the lineages that are its sister taxa. Yet, in regard to the context of this section, the most important evolutionary questions center on the origin of the medicinally important species, *Atropa belladonna*. As reflected in the following quote, *A. belladonna*'s ability to produce atropine has been of longstanding, medical importance:

> Hardly a practitioner of anesthesia begins administration of an anesthetic agent without the ready availability of atropine... Throughout this long history of pain relief... the principle of safeguarding the thread of life has remained the core of anesthetic practice...

and this "reminds us that we practice the art and science of anesthesiology against a backdrop of legendary proportion" (Holzman 1998). In addition to anesthetic properties, atropine produces numerous physiological effects resulting in its application to (1) relieve spasms of the gastrointestinal tract, (2) reduce excessive secretions by organs, (3) alleviate symptoms of Parkinson's disease, and (4) maintain proper heart function during surgery (http://www.drugs.com/mtm/atropine.html).

As mentioned earlier, the evolutionary derivation of one source of atropine (*A. belladonna*) was debated for over a century. However, the collection of DNA sequence data from both chloroplast and nuclear loci has resolved not only the sister-species relationships of this lineage, but also its hybrid derivation. Chloroplast sequences provided evidence of *A. belladonna*'s placement within the Hyoscyameae clade (Olmstead and Sweere 1994). An analysis of the presence/absence of nuclear, transposable element loci (Yuan *et al.* 2006) supported this phylogenetic alignment; it also indicated that *A. belladonna* possessed genomes from multiple progenitor lineages. Yuan *et al.* (2006) inferred the allopolyploid derivation of this species, resulting in its possessing three genomes (i.e., *A. belladonna* is a hexaploid derivative). Specifically, two of the three copies (Figure 5.9) of the granule-bound *starch synthase I* gene in *A. belladonna* possessed an insertion of a transposable element, while the third copy did not contain the inserted element (Yuan *et al.* 2006). The inference drawn from the pattern of sequence variation in this lineage was that a tetraploid species (possessing the insertion in both of its genomes) belonging to the Hyoscyameae assemblage had hybridized with a diploid taxon (lacking the insertion) from a lineage outside this clade, resulting in the hexaploid *A. belladonna* (Figure 5.9; Yuan *et al.* 2006). This hybridization event gave rise to an organism that has been used for millennia to relieve suffering in humans (Holzman 1998).

5.6.2 Reticulate evolution and essential oils

"The industrial mint crops are cultivated in several countries for their essential oils. The oil, menthol, carvone, lemoline, dementholated oil and terpene fractions from the latter are variously used in the cosmetics, pharmaceuticals, food, confectionery and liquor industries" (Khanuja *et al.* 2000). The range of medicinal applications of some mints also reflects the diversity of human-utilized products from various members of the genus *Mentha*. For example, extractions (containing phenols, flavonoids, menthol, and menthone) of the oils from peppermint (*Mentha piperita*) have been shown to have the potential for antimicrobial, antiviral, antioxidant, antitumor, and antiallergenic properties (McKay and Blumberg 2006). In addition, these extracts have the added benefits of (1) producing relaxation of gastrointestinal tissue, (2) analgesic and anesthetic responses in the central and peripheral nervous systems, and (3) modulating effects on the immune system of model animals (McKay and Blumberg 2006).

Given the medicinal potential for the extracts from not only peppermint, but other mints as well

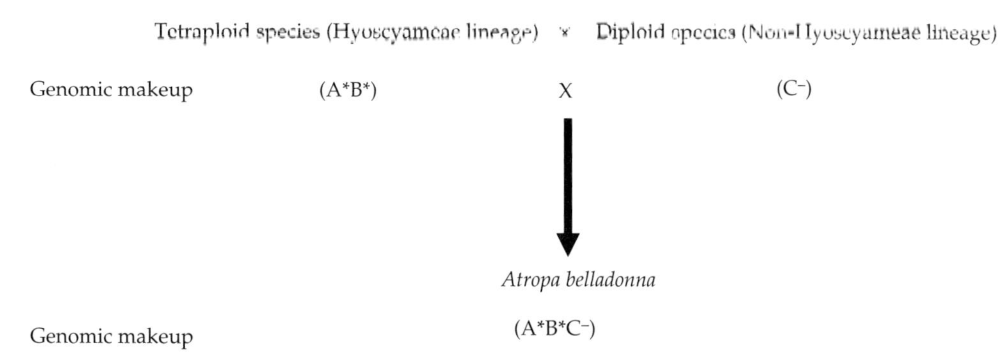

Figure 5.9 Model describing the derivation of *A. belladonna*. A tetraploid species, belonging to the tribe Hyoscyameae, that possessed the insertion of a transposable element in both copies of the granule-bound *starch synthase I* gene (indicated by the asterisks next to the "A" and "B" genome designations) hybridized with a diploid, non-Hyoscyameae species that lacked the insertion (indicated by the minus superscript). This hybridization event resulted in hexaploid *A. belladonna* with two of its three copies of the granule-bound *starch synthase I* gene possessing transposable element insertions (Yuan *et al.* 2006).

(e.g., carvone from the spearmints, *Mentha gracilis* and *Mentha spicata*; Tucker *et al.* 1991; Telci *et al.* 2004), it is significant that hybridization, and specifically hybrid speciation, is widespread in this genus. Thus, the lineages that provide the basis for the essential oils, and so on have a reticulate history (e.g., see Tucker and Fairbrothers 1990; Khanuja *et al.* 2000; Gobert *et al.* 2002). Figure 5.10 (Gobert *et al.* 2002) reflects a portion of the hybrid speciation events recognized for the genus, in this case for three of the major lineages utilized for mint extracts—*M.* × *gracilis* (or *M. gracilis*), *M. spicata*, and *M.* × *piperita* (or *M. piperita*). As indicated in Figure 5.10, the hybrid speciation events produced forms that are recognized as either spearmint or peppermint. However, there is an added complexity to these reticulations because there were three and two separate evolutionary origins for lineages that produce compounds associated with the category of "spearmint" and "peppermint," respectively (Gobert *et al.* 2002). This is reflected in the designation of both *M.* × *gracilis* and *M. spicata* as spearmints, and in the presence of multiple chromosomal forms in *M. spicata* and *M.* × *piperita* (Figure 5.10). These observations indicated the repeated, allopolyploid origin of the present-day *Mentha* taxa. Thus, hybrid speciation underlies the small-scale and industrial-scale isolation of the medicinal products derived from mints.

5.6.3 Reticulate evolution and chocolate

Dillinger *et al.* (2000) have provided an overview of the medicinal uses of chocolate. From its original application by the New World Maya, Olmec and Aztec civilizations through the early twentieth century, there were a number of ailments commonly treated through the ingestion of the product from the beans of *Theobroma cacao* (i.e., cacao or chocolate). These medicinal applications included a mechanism to generate weight gain in emaciated patients; nervous system stimulation for apathetic or exhausted individuals; and the improvement of digestion, kidney, and bowel functions (Dillinger *et al.* 2000). Though these same benefits could be derived from some of the current-day cacao products as well, the current use of these products is as a confection rather than a medicine (Ariefdjohan and Savaiano 2005).

The genus *Theobroma* occurs naturally in tropical America and is made up of 22 species, with *T. cacao* placed into a section separate from all other species (Kennedy 1995). Like *Coffea arabica*, the plant from which humans isolate coffee (see later), cacao production is a major source of revenue in the wet tropics; *T. cacao* is of significant agricultural importance in some areas of West Africa, Brazil, and Malaysia (Kennedy 1995; Motamayor *et al.* 2003). Also, like coffee, the major consumption of

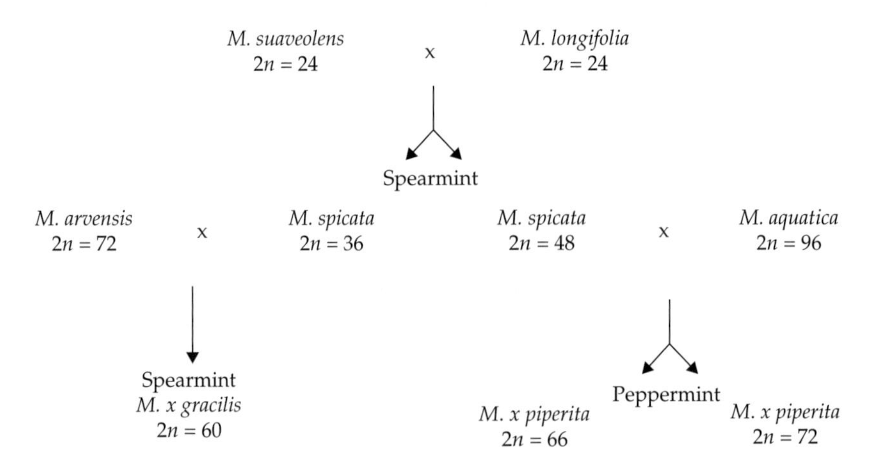

Figure 5.10 The hypothesized relationships and derivations of hybrid lineages within *Mentha*. "$2n$" indicates the chromosome number for each taxon. The chromosomal data indicate the allopolyploid hybrid origin for the lineages that belong to the categories of "spearmint" and "peppermint" (Gobert *et al.* 2002).

chocolate products occurs in Europe and North America (Kennedy 1995; Wrigley 1995). However, unlike the biological source of coffee (which is polyploid), *T. cacao* cultivars are diploid lineages ($2n = 20$) derived from native species. Three significant cultivars of *T. cacao* have been utilized for the production of chocolate; these have been termed Criollo, Forastero, and Trinitario (Kennedy 1995). Though each of these three has formed a foundation for chocolate production, much of the present-day cacao derives from the Trinitario cultivar (Motamayor *et al.* 2003).

The three cacao cultivars have been categorized on the basis of properties associated with their beans (Kennedy 1995). Furthermore, experimental crossing analyses involving various members of the genus detected extremely limited capacities for hybrid formation (Addison and Tavares 1952). Notwithstanding the distinctiveness of the cultivars and the reproductive isolation among *Theobroma* species, an analysis of nuclear loci in the three cultivar lineages found them to be not reciprocally monophyletic (Motamayor *et al.* 2002). This is consistent with the discovery of numerous hybrids within this cultivar assemblage (Saunders *et al.* 2004). Also, the position of Trinitario as the major cacao cultivar relates to its apparent hybrid origin from crosses between the Criollo and Forastero lineages. The events that resulted in the reticulate origin of Trinitario included (1) the introduction of Forastero trees from the Lower Amazon into the native Trinidadian habitat of the Criollo form and (2) subsequent introgressive hybridization between these two lineages (see Motamayor *et al.* 2003 for a discussion).

It is important to note that the asymmetric cultivation of Trinitario is a recent phenomenon; Trinitario did not begin to replace other forms in Central and South America until the 1800s (Motamayor *et al.* 2003). However, the reasons for the replacement of the other cacao-producing forms by Trinitario reflect the potential for genetic exchange to result in important evolutionary novelties. In particular, two main attributes of the Trinitario cultivar have been recognized as the impetus for its replacement of lineages that had been cultivated by the indigenous peoples for >1500 years (Motamayor *et al.* 2002). The characteristics seen as critical for Trinitario's success were its elevated disease resistance and its combining the higher productivity of the Forastero cultivar and the desirable flavor characteristics of the Criollo variant (Trinitario is referred to as "fine cocoa"; Kennedy 1995; Motamayor *et al.* 2003). Both of these attributes likely resulted from the introgressive hybridization that took place between the introduced Forastero and the native Criollo plants on the island of Trinidad. This reticulate event thus provided the genetic basis for the evolution of this medicinal and confectionary product utilized extensively by Old and New World *H. sapiens*.

5.6.4 Reticulate evolution and tobacco

The utility of *Nicotiana tabacum* as a model biological system for scientific studies is extremely broadly based. For example, this species has provided the biological material for testing hypotheses concerning molecular and genomic evolution, gene function, the timing/form of speciation, and the processes leading to the origin of cultivars (Volkov *et al.* 1999; Lim *et al.* 2000, 2004, 2007; Kitamura *et al.* 2001; Ren and Timko 2001; Fulnecek *et al.* 2002; Matyásek *et al.* 2002; Skalická *et al.* 2003, 2005; Clarkson *et al.* 2004, 2005; Dadejová *et al.* 2007; Petit *et al.* 2007). However, the obvious notoriety of this plant centers on its use as one of *H. sapiens'* longstanding drugs of choice. The health risks of cigarette smoking, in particular, have been highlighted by the Centers for Disease Control and Prevention in the following way:

> Smoking harms nearly every organ of the body; causing many diseases and reducing the health of smokers in general. The adverse health effects from cigarette smoking account for an estimated 438,000 deaths, or nearly 1 of every 5 deaths, each year in the United States. More deaths are caused each year by tobacco use than by all deaths from human immunodeficiency virus (HIV), illegal drug use, alcohol use, motor vehicle injuries, suicides, and murders combined. (http://www.cdc.gov/tobacco/data_statistics/Factsheets/health_effects.htm)

Notwithstanding such dire statistics, ~15% of male and 9% of female citizens of the United States consumed >25 cigarettes per day in 2004 (http://www.cdc.gov/tobacco/data_statistics/tables/adult/table_4.htm).

I have pointed out previously that there is no comparison between the individual artisan's skill needed to construct high-end, hand-rolled, long-leaf cigars, and the mechanization with which a packet of cigarettes or a can of snuff is produced (Arnold 2006, p. 176). However, the common elements used to produce each of these diverse products are the leaves from *N. tabacum*. Furthermore, the origin of this widely applied drug involved reticulation. To understand the origin and evolutionary trajectory of the tobacco-producing lineage, it is necessary to review the history of use of *Nicotiana* species as sources for nicotine. An examination across geographic settings and cultures reveals that *N. tabacum* has not been the sole source for nicotine. *N. tabacum* was indeed utilized by Central and South American indigenous groups. In contrast, inhabitants of western North America consumed *Nicotiana bigelovii*, *Nicotiana attenuata*, and *Nicotiana trigonophylla*, while eastern North American, northern Mexican, and West Indian peoples used *Nicotiana rustica*. Finally, *Nicotiana benthamiana* was the source of the tobacco ingested by native Australians (Gerstel and Sisson 1995).

The origin of the tobacco industry in the New World (specifically, in the state of Virginia) involved the species used by the local Native Americans, *N. rustica* (Gerstel and Sisson 1995). However, this species was quickly supplanted by the better-flavored *N. tabacum* (Gerstel and Sisson 1995). Significantly, the origin of *N. tabacum* involved allopolyploid speciation. Indeed, DNA sequence information has identified the maternal and paternal progenitors of the tobacco cultivar to be the diploid species, *Nicotiana sylvestris* and *Nicotiana tomentosiformis*, respectively (Lim *et al.* 2000; Kitamura *et al.* 2001; Ren and Timko 2001; Fulnecek *et al.* 2002). The timing of *N. tabacum*'s origin has been estimated at *c.*5–6 mya (see discussions by Fulncek *et al.* 2002; Clarkson *et al.* 2005). As with many examples of genetic exchange, the combination of the divergent, *N. sylvestris* and *N. tomentosiformis* parental genomes into a single nucleus and cytoplasm gave rise to a derivative lineage with novel adaptations. In this case, the adaptations were reflected in its utility as a source of the drug, nicotine. Gerstel and Sisson (1995) highlighted the novel set of adaptations by stating that "the wild parents of *N. tabacum*...possess a dominant 'converter' gene which...demethylates nicotine into undesirable nornicotine in the leaves. For this reason these species may have been useless." Like all the examples used in this section, reticulate processes (in this case occurring millions of years before the origin of the genus *Homo*) resulted in the biological starting material for a drug-producing, agricultural industry.

5.6.5 Reticulate evolution and alcohol

I have reviewed elsewhere evidence of the affect of genetic exchange not only on the origin of yeast lineages used in the production of various alcoholic beverages, but also on their adaptive evolution (Arnold 2006, pp. 120–121). Briefly, several lines of evidence support a significant role for both hybrid speciation and introgressive hybridization in the formation of the *Saccharomyces* species on which the production of beer, wine, ale, sake, and so on depends. In terms of hybrid lineage formation, the *Saccharomyces cerevisiae* genome is well-documented as having originated through a whole-genome duplication event that may or may not have involved allopolyploid speciation (Scannell *et al.* 2007). In addition, other species of yeast have been determined to be allopolyploid and thus hybrid species (Masneuf *et al.* 1998; de Barros Lopes *et al.* 2002). Finally, introgressive hybridization and horizontal transfer events have been inferred repeatedly for various *Saccharomyces* taxa (e.g., Groth *et al.* 1999; Liti *et al.* 2006; Hall and Dietrich 2007).

As with the evolution of yeast lineages, the origin and evolutionary trajectory of *Vitis vinifera*, the grapevine, has to some extent been closely associated with human cultures. Though used for table grape and raisin production as well, this species is best known as the basis of the wine industry (This *et al.* 2006). The origin of winemaking, as an intentional human activity, is thought to have occurred sometime during the Neolithic (*c.*6,000–10,500 ybp) in the Near-East (McGovern and Hartung 1997; This *et al.* 2006). However, *V. vinifera* is one of ~60 species in the genus, with its origin timed at *c.*65 mya (Aradhya *et al.* 2003; This *et al.* 2006). The two subspecies recognized as the wild (*silvestris* or *sylvestris*) and cultivated (*vinifera* or *sativa*) forms of this species coexist in Eurasia and North Africa, with the wild progenitor being rare (Aradhya *et al.* 2003; Arroyo-García *et al.* 2006; This *et al.* 2006). The morphological/reproductive differences between the cultivated and wild lineages are significant and reflect human-mediated selection. The divergent characteristics include: (1) leaf shape; (2) seed shape and size; (3) hermaphroditic (male and female components present in the same flower) flowers in the cultivar, but sexes on separate flowers in the wild lineage; and (4) large, well-developed clusters of seeds in the cultivar that are not present in the wild *silvestris* plants (This *et al.* 2006).

Both ancient polyploidy as well as contemporaneous introgressive hybridization has shaped the evolution of the grapevine genome. Jaillon *et al.* (2007) compared the genome sequence of a Pinot Noir-derivative genotype of *V. vinifera* to those of *Arabidopsis*, poplar and rice. This analysis resulted in the inference that three genomes had contributed to the grapevine cultivar—*V. vinifera* is a hexaploid (Figure 5.11). The polyploidy event(s) that resulted in the three genomes being brought together to form the hexaploid *V. vinifera* was determined to be ancient rather than recent (Figure 5.11; Jaillon *et al.* 2007). Whether or not the formation of the hexaploid occurred by a single step, or multiple steps, is not yet known. Regardless of the number of events, the origin of this polyploid derivative likely involved hybridization between divergent lineages. This inference derives from the observation that, for plant lineages, allopolyploid (hybrid) speciation occurs much more frequently than does *autopolyploid speciation* (see Stebbins 1947; Soltis *et al.* 2003).

In addition to the ancient hybrid formation of the lineage that led to the grapevine taxon, the origin of various cultivars has been impacted by spontaneous introgressive hybridization between different cultivars and between *V. v. vinifera* and *V. v. silvestris*. Bowers *et al.* (1999), Aradhya *et al.* (2003), and Arroyo-García *et al.* (2006) described this common conclusion. For example, Aradhya

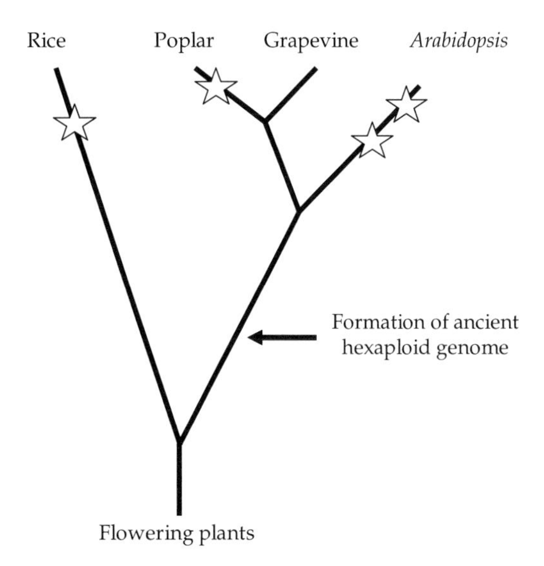

Figure 5.11 Polyploidy events associated with the evolution of the grapevine, *V. vinifera*. Genome sequencing of the grapevine and comparisons of its genome with those of rice, poplar, and *Arabidopsis* indicated an ancient hexaploid formation in the lineage leading to the three dicot taxa (i.e., poplar, grapevine, and *Arabidopsis*). The genome analyses also indicated a series of additional whole-genome duplication events along the lineages leading to rice, poplar, and *Arabidopsis* (each duplication event is indicated by a star; Jaillon *et al.* 2007).

et al. (2003) reviewed data indicating that contact between the wild and cultivated grapevines had led to the establishment of "complex introgressive hybrid swarms in transitions zones"—zones located between the agricultural fields and natural settings. Indeed, this type of spontaneous introgression (i.e., not from controlled crosses by humans) was seen as a key factor in introducing the genetic variation that provided the material for the development of European *V. vinifera* cultivars (Bowers *et al.* 1999; Aradhya *et al.* 2003) and thus the basis of much of the present-day, worldwide, wine industry.

5.6.6 Reticulate evolution and coffee

There are two *Coffea* lineages, *C. arabica* and *Coffea canephora*, used for the vast majority of coffee cultivation in both the Old and New Worlds. Raina *et al.* (1998) concluded that (1) coffee plants were cultivated in >50 countries and (2) this industry resulted in a trade of ~US$18 billion. More recently, Vega *et al.* (2003) quoted a value of US$70 billion for the retail value of coffee sales. Of this trade, *C. arabica* accounted for ~70% and *C. canephora* ~30% of the coffee products (Moncada and McCouch 2004). The greater proportion of the market captured by *C. arabica* reflected the higher quality beverage (i.e., low content of caffeine and pleasant aroma) produced by this cultivar (Raina *et al.* 1998). In the main, *C. canephora*'s beans are utilized to manufacture instant coffee (Steiger *et al.* 2002). Of the more than 80 species of *Coffea*, *C. arabica* is unique in being the only tetraploid (44 chromosomes) and self-fertile form (Raina *et al.* 1998; Steiger *et al.* 2002). In contrast, the remainder of the species, including *C. canephora*, are self-sterile and diploid (22 chromosomes; Raina *et al.* 1998; Steiger *et al.* 2002). Though distributed in several areas of Ethiopia, it is likely that *C. arabica* originated in this country's southwestern highlands (Anthony *et al.* 2001). From this source, humans have moved *C. arabica* throughout the world. This movement began with its initial transport to Yemen (possibly as early as 575 AD) followed subsequently by importation into Java, Reunion Island, the Amsterdam Botanical gardens and, finally, the New World (Anthony *et al.* 2002). In the process of its cultivation and human-mediated migration, two varieties originated; these were termed "Typica" and "Bourbon" (Anthony *et al.* 2002).

Many studies have indicated the importance of natural hybridization in the evolution of genus *Coffea* and, in particular, the cultivars. Several data sets have detected signatures of introgressive hybridization involving various species, including genetic exchange between the cultivar and its diploid progenitors (Cros *et al.* 1998; Mahé *et al.* 2007). However, the most significant class of hybridization—in terms of the origin of the cultivar—was the allopolyploid formation of *C. arabica*. The origin of this taxon is now known to be from a *natural* allopolyploid speciation event (see discussion by Ruas *et al.* 2003). Though the allopolyploid derivation of this species is well established, identifying the diploid progenitors for this species has been problematic. Raina *et al.* (1998) applied *in situ* hybridization to test for the genomic components of *C. arabica*. Results from this analysis led to the inference of *Coffea eugenioides* and *Coffea congensis* as the progenitors of this tetraploid. In contrast, though Lashermes *et al.* (1999) also identified *C. eugenioides* as one parent of the cultivar, they identified the other major cultivated species, *C. canephora*, as the second progenitor lineage.

Recently, Ruas *et al.* (2003) applied inter-simple sequence repeat markers to test the earlier, conflicting hypotheses. Consistent with the findings of Lashermes *et al.* (1999), Ruas *et al.*'s (2003) data indicated the closest genetic associations between *C. eugenioides*, *C. canephora*, and *C. arabica*. Ruas *et al.* (2003) indicated additional support for *C. eugenioides*' contribution by reporting its status as the only wild, diploid species shown to produce a beverage with "fine aroma and flavor." They argued that one effect of the genome from this species in the allotetraploid *C. arabica* might be reflected in the presence of these aroma and flavor characteristics. However, the reticulation that led to the derivation of *C. arabica* also resulted in ecological novelty relative to its parents. Specifically, *C. arabica* is adapted to environmental conditions different from either *C. eugenioides* or *C. canephora*. This conclusion is supported by *C. arabica*'s native range encompassing an area outside of the distribution of the diploid species (Lashermes *et al.* 1999). It thus appears likely that the allopolyploid speciation event resulting in *C. arabica* produced a lineage adapted to both a unique ecological setting and its use as a caffeine delivery system for humans.

5.6.7 Reticulate evolution and *Cannabis*

Ironically, some of the major areas for the production of illegal drugs in the United States are Federal land holdings. For example, between 1997 and 2003 over 3,000 metric tons—reflecting >3,000,000 plants—of *Cannabis sativa* (source of the drug, marijuana) were collected/destroyed from US public lands (Gaffrey 2003). The major production areas were located in the states of Arkansas, California, Hawaii, Kentucky, Missouri, and Tennessee, with most of the cultivation of *C. sativa* occurring in the least accessible sites within the Federal forests, parks, and refuges (Gaffrey 2003).

As with the production of any illegal substance, additional criminal activities (including assaults and homicides) are associated with the cultivation of *C. sativa*; violent incidences associated with this cultivation more than doubled on California public lands in 2003, relative to previous years (Gaffrey 2003). Though the cost in human injury and death is the most serious of the possible consequences from the production of *C. sativa*, significant environmental perturbations also occur. These include: (1) stripping of area of native vegetation; (2) diversion of water for irrigation from streams utilized by flora and fauna; (3) spread of toxic substances such as animal poisons and herbicides; and (4) poaching of local animal life to feed those tending the gardens (Gaffrey 2003).

Though probably best known as the source of intoxicating compounds (Small and Cronquist 1976), *C. sativa* has also provided fiber, seed, and seed-oil to humans for many millennia (Gilmore *et al.* 2007). Recently, there has been a renewed interest in the utilization of this plant for these purposes (e.g., see Struik *et al.* 2000). Notwithstanding the end products derived from this lineage, there has been considerable debate concerning the taxonomic status of various forms. Small and Cronquist (1976) provided a comprehensive, systematic treatment based on morphological characteristics. Their conclusions were to place all variants within the single species *C. sativa*, but with two subspecies, *sativa* and *indica*. These authors described the partially human-generated differences in the two subspecies in the following manner (Small and Cronquist 1976): (1) *C. sativa sativa* was "a group of generally northern plants of relatively limited intoxicant potential, influenced particularly by selection for fibre and oil agronomic qualities"; and (2) *C. sativa indica* included "a group of generally southern plants of considerable intoxicant potential, influenced particularly by selection for inebriant qualities." In contrast to Small and Cronquist's (1976) treatment, those interested in the plant as a producer of the intoxicant recognize two "species," *Cannabis indica* and *C. sativa*. *C. indica* and *C. sativa* normally refer to plants used to produce resin (for *hashish*) and leaves/inflorescences (for marijuana), respectively (de Meijer and van Soest 1992; Gilmore *et al.* 2007).

Much of the recent development of genetic markers used to differentiate the various categories of *C. sativa* has derived from the needs of forensic scientists (Gilmore *et al.* 2003, 2007). Significantly, in terms of the topic of this section, these studies have confirmed that the lineages within *C. sativa* have evolved in a reticulate manner. This conclusion is supported by variation at both nuclear and cytoplasmic loci. The pattern of genetic variation at nuclear loci, like the use of cannabinoid content (Small and Cronquist 1976), failed to differentiate clearly those plants used for fiber (*C. s. sativa*) from those grown for their higher percentage of intoxicant (*C. s. indica*; Gilmore *et al.* 2003). Likewise, though cpDNA variation generally placed "fiber" and "intoxicant" plant lineages into separate clades, there was significant intermixing of the lineages derived from the two subspecies of *C. sativa* (Gilmore *et al.* 2007). These findings, and those from previous morphological analyses (Small and Cronquist 1976) thus support the characterization of this complex as a "highly variable, highly hybridised and introgressed" species (Gilmore *et al.* 2007).

5.7 Summary and conclusions

My goal for this chapter was to emphasize the wide array of biological products from which humans have benefited that derive from lineages marked by reticulate evolution. That *H. sapiens* has not realized, until recently, that these plants, animals, and microorganisms reflect genetic exchange is immaterial to the significant role they have assumed in the cultural development and the survival of human populations. The degree to which the processes of hybrid speciation and introgressive hybridization have produced the biological characteristics exploited by humans is not known for many of the examples. However, those for which strong inferences of hybrid origin of the traits can be made (e.g., cotton, coffee, and chocolate) indicate the potential for reticulate origin of adaptations important for both the organism itself and *H. sapiens*.

In Chapters 6 and 7, I will continue the discussion of organisms that benefit our species. However, in these chapters I will review only those organisms from which we derive food products. As I stated previously, some of the examples presented in Chapter 5 have also been the source of food (e.g., dogs, cats, and water buffalo). Likewise, many of the examples presented in the following chapter have uses besides as a food source (e.g., sheep provide material for clothing). However, they have been chosen to highlight our use of reticulate lineages as sources for our dietary requirements.

CHAPTER 6

Reticulate evolution and beneficial organisms—part II

This hybrid zone is highly unusual in the biological world, because the mating events in catadromous eels presumably take place thousands of kilometers from where the hybrids are observed as maturing juveniles.

(Avise *et al.* 1990)

In Scandinavia, hybridization and gene introgression is so extensive that no individuals with pure *M. trossulus* genotypes have been found. However, *M. trossulus* alleles are maintained at high frequencies in the extremely low salinity Baltic Sea for some allozyme genes.

(Riginos and Cunningham 2005)

To explain the discordance between Y-chromosome and mtDNA phylogenies, several hypotheses are considered. We suggest that a plausible scenario involves mtDNA introgression between ancestral taxa before the relatively recent colonization of Western Europe, the Caucasus Mountains, and East Africa by *Capra* populations.

(Pidancier *et al.* 2006)

...these species diverged between 210,000 and 5,200,000 ybp, which did not overlap the predicted time for incomplete lineage sorting. These analyses also suggested that ancient introgression (~14,000 ybp) has resulted in the widespread distribution and high frequency of falcated-like mtDNA (5.5% of haplotypes) in North America.

(Peters *et al.* 2007)

6.1 Reticulate evolution and the formation of organisms utilized by humans

In Chapter 5, I reviewed data reflecting the breadth and depth of categories of human-utilized organisms. In keeping with the subject of this book, I focused on those lineages that both benefit *H. sapiens* and which demonstrate the affects from genetic exchange (e.g., allopolyploid speciation and introgression) in their evolutionary histories. In this chapter, and in Chapter 7, I continue this discussion of beneficial organisms, but I will restrict my attention to animals and plants utilized by humans for food. Though the categories of utilization are restricted to that of "food," the diversity of animals and plants falling into this descriptor is enormous.

Before proceeding with examples of animal and plant protein sources, I should state that though I have chosen to restrict my discussion of "microorganisms" to other categories (i.e., for beneficial organisms, placed in the class of drug producers—see Chapter 5; for harmful organisms, placed in the class of disease organisms—see Chapter 8); it should be recognized that many are utilized for food production. For example, the group known as "lactic acid bacteria" has been utilized in the fermentation of countless food products and drugs (Makarova *et al.* 2006; van de Guchte *et al.* 2006). Makarova *et al.* (2006) defined lactic acid bacteria, or "LAB," as "a group of microaerophilic, Gram-positive organisms that ferment hexose sugars to produce primarily lactic acid." They go on to point out that this group of organisms includes numerous taxa (e.g., *Lactococcus, Enterococcus, Oenococcus, Pediococcus, Streptococcus, Leuconostoc,* and *Lactobacillus*) that have been used since the dawn of agricultural practices in producing and preserving foods and beverages through their fermentative capabilities. In particular, the LAB are key components in the production of fermented dairy, meat, and vegetable products and are also important elements in the creation of wine, coffee, silage, and cocoa (Makarova *et al.* 2006; van de Guchte *et al.* 2006). As with bacterial groups in general (Figures 6.1 and 6.2—Ochman 2005; Ochman *et al.* 2000, 2005), the evolutionary history of the LAB has been characterized by both gene loss and gene acquisition (Makarova *et al.* 2006). For example, van de Guchte *et al.* (2006) examined the genome of the *Lactobacillus bulgaricus* strain ATCC11842 that was originally isolated from Bulgarian yogurt. Adaptations to the environmental aspects for the fermentation of milk products were evident both in the large number (270) of pseudogenes (i.e., genome reduction) and in the acquisition of key genic elements through horizontal transfer events (i.e., genome expansion). Specific signatures of genome expansion were detected in the analyses of the *L. bulgaricus* genome. First, numerous transposases, including two types not described previously, were detected. Second, genes that provide tolerance to oxidative stress, but that are rarely found in bacteria, were identified as likely acquired through lateral transfers from other species (van de Guchte *et al.* 2006).

As with the plant (and bacterial) derivatives, the lineages of animals to be discussed also come from an extremely wide array of biological types, including species of shellfish, fish, birds, and mammals. Similarly, some of these taxa provide major sources of animal protein (e.g., cattle and sheep), while others are more limited in terms of the numbers of humans for which they are a staple (e.g., African antelope and peacock bass).

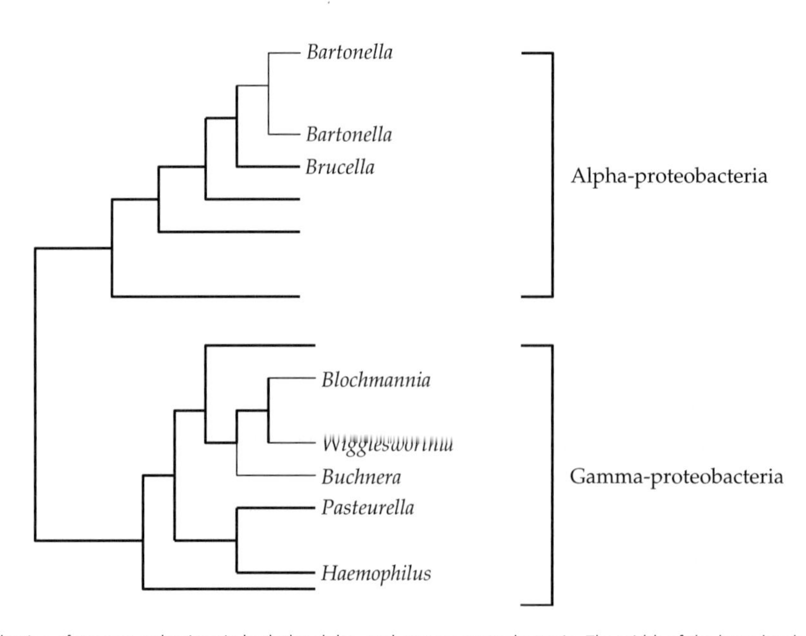

Figure 6.1 Reflection of genome reductions in both the alpha- and gamma-proteobacteria. The width of the branches leading to the ends of lineages indicates the relative sizes of the genomes. Those bacteria with reduced genome sizes are indicated by their generic designations (from Ochman 2005).

However, as with plant species that are utilized by limited numbers of humans (Chapter 7), these "less-utilized" animal assemblages and lineages still provide key resources to the populations who harvest and consume them. Furthermore, many of the species discussed (e.g., *Mytilus*, cattle, goats) provide examples of the effect that humans have had on the reticulate evolution of their food sources through the bringing together of related species into sympatric associations thereby facilitating introgressive hybridization.

The number of animal lineages to be discussed in this chapter is almost certainly an underestimate of those species utilized by *H. sapiens* as a protein source that have reticulate origins. However, the examples included illustrate clearly the degree to which the web-of-life metaphor is descriptive of the evolutionary patterns and processes associated with the origin of our animal food sources.

6.2 Reticulate evolution and sources of animal protein

Animal reticulate evolution has, until recently, been downplayed as a source of new lineages and novel adaptations (see Arnold 1997, 2006 for discussions). This trend continues somewhat (e.g., Coyne and Orr 2004), but the paradigm is shifting due to the wealth of animal assemblages that reflect the important, evolutionary effects from hybrid speciation and/or introgressive hybridization (see Arnold 1997, 2006; Seehausen 2004; Mallet 2005). The following examples reflect this same conclusion for animal lineages from which humans derive nourishment. However, unlike the wide assortment of examples of animal lineages not utilized by humans, the lineages discussed later also provide illustrations of the "practical" (from a *H. sapiens*' centric viewpoint) results that can accompany reticulate evolutionary processes. As with some of the examples discussed in Chapter 5, a portion of the reticulate animal lineages on which we feed may have gained adaptations due to genetic exchange events. For others, this hypothesis remains untested. Regardless of their adaptive potential, each of these lineages provides a benefit for humans and thus likely contributes to the fitness of individual *H. sapiens*.

6.2.1 Reticulate evolution and shellfish

The mussel genus, *Mytilus*, not only contains numerous, widely exploited, edible forms (e.g., *Mytilus edulis*; Bierne *et al.* 2003), but also provides numerous biological characteristics that illuminate important ecological, genetic, and evolutionary processes. For example, female *Mytilus*

	Utilize lactose	Utilize citrate	Produce H_2S	Produce indole	Produce urease	Lysine decarboxylase
Escherichia coli	+	–	–	+	–	+
Shigella flexneri	–	–	–	+	–	–
Salmonella enterica	–	+	+	–	–	+
Klebsiella pneumoniae	+	+	–	–	+	+
Serratia marcescens	–	+	–	–	+	+
Yersinia pestis	–	–	–	–	–	–

Figure 6.2 Phylogenetic relationships among various bacterial lineages. The sharing of physiological attributes (e.g., "utilize lactose") is often reflective of gene acquisition via horizontal transfer between species (from Ochman *et al.* 2000).

normally possess one mtDNA type inherited from the female parent while male *Mytilus* normally carry two mtDNA types that are inherited from each of their parents. This system of "doubly uniparental inheritance" facilitates the retention of two divergent mtDNA genomes in a single species (Hoeh *et al.* 1991). It also provides the opportunity for differential gene flow of the two mtDNA lineages given a decoupling of dispersal by male and female larvae and/or differential selection on the two mtDNA forms (Riginos *et al.* 2004). Furthermore, the form of inheritance itself can be disrupted (e.g., in cases of interspecific hybridization; Wood *et al.* 2003) thus leading to females with multiple mtDNA lineages and males with only a single mtDNA type.

In addition to their utility as models for understanding the biparental inheritance of cytoplasmic (i.e., mtDNA) genomes, these mussels have also allowed tests for selection-mediated reproductive isolation. Specifically, *Mytilus* has allowed analyses of (1) the proteins that facilitate recognition and fertilization of the egg by sperm that are broadcast into the open ocean and (2) the genetics of postzygotic selection deriving from incompatibilities between the genomes of the hybridizing species. In terms of point (1), adaptive divergence at gamete-recognition loci has been inferred, and its occurrence explained by selection against hybrids in areas of sympatry between various *Mytilus* species (Riginos and McDonald 2003; Springer and Crespi 2007). Adaptive changes were thus argued to have resulted from selection for the ability to differentiate between conspecific and heterospecific gametes (Springer and Crespi 2007). With regard to the second point, estimates of postzygotic selection against interspecific hybrids have provided evidence (as with countless other animal and plant complexes; Arnold 1997, 2006; Coyne and Orr 2004) for the contribution of many loci spread throughout the genomes of the parental species (Bierne *et al.* 2006).

Most significantly for the present discussion is the observation that *Mytilus* is a paradigm for the affect that introgressive hybridization may have on the genetic, ecological, and evolutionary characteristics of species assemblages. This is particularly important since it has been shown to involve commercially important species such as *M. edulis*. Supporting this conclusion are numerous data sets that illustrate the population genetic, phylogenetic, and phylogeographic structures of European and North American *Mytilus* populations. Though genetic admixtures have been detected in southern hemisphere populations as well (e.g., Borsa *et al.* 2007), the best studied cases of reticulate evolution in this assemblage involve the genetic interactions between *M. edulis* and *Mytilus galloprovincialis* in a 2,000 km zone of overlap along the western European coast (Daguin *et al.* 2001). Rawson and Hilbish (1998) tested the alternate hypotheses of retained ancestral polymorphisms (i.e., incomplete lineage sorting) versus introgressive hybridization as explanations for the presence of mtDNA (both the maternally and paternally inherited lineages, see previous discussion) sequences characteristic of *M. edulis* in *M. galloprovincialis* populations. Their data provided support for a recent introduction of these sequences into *M. galloprovincialis* (i.e., through introgression from *M. edulis*) rather than their being retained through the period of diversification of these species (Rawson and Hilbish 1998). Two decades earlier, Skibinski *et al.* (1978) had come to a similar conclusion in a study of nuclear loci variation in these two species, and mussels "intermediate" for morphological characteristics. In particular, Skibinski *et al.* (1978) concluded that introgressive hybridization had taken place between these two species in areas of sympatry in southwest England.

Daguin *et al.* (2001) and Bierne *et al.* (2003) extended the earlier studies of genetic variation in the *M. edulis/M. galloprovincialis* complex by examining one and three nuclear loci, respectively. In both cases, the authors concluded the geographically widespread occurrence of genetic exchange. However, the region of overlap was found to be a "mosaic" hybrid zone (Howard 1982, 1986; Harrison 1986) in which some populations possessed the genetic makeup of *M. edulis*, while other populations were admixed with alleles from both species (Daguin *et al.* 2001; Bierne *et al.* 2003). Reflecting the use of *M. edulis* as a commercial source of human food, some of the admixture in this zone of hybridization apparently reflected recent transfer of mussels from "cultivated stocks of mussels located in the Bay of Biscay...to the Mont Saint-Michel bay culture area" (Bierne *et al.* 2003).

In their 1998 paper, Rawson and Hilbish also reported data indicating introgression between *M. edulis* and a third species, *Mytilus trossulus*. The analysis of samples from the Baltic Sea revealed animals with nuclear genomes consisting mainly of *M. trossulus* alleles, but with mtDNA sequences characteristic of *M. edulis* (Rawson and Hilbish 1998). A recent analysis has substantiated these results and indicated that both the male- and female-transmitted mtDNA lineages in *M. trossulus* are similar to one another and to the female-transmitted mtDNA of *M. edulis* (Burzynski *et al.* 2006). Riginos and Cunningham (2005) reviewed data concerning the genetic structure of hybrid zones between these two species located on both sides of the Atlantic Ocean. From their extensive survey, they identified significant differences between the Canadian maritime and Scandinavian zones of overlap (Figure 6.3). In particular, they found that the Canadian coastal region was characterized by genetic discontinuities between *M. edulis* and *M. trossulus*; this pattern was consistent with the maintenance of the "independent genetic integrities" of these two taxa (Riginos and Cunningham 2005; Figure 6.3). In contrast, the Scandinavian hybrid zone was characterized by such a high degree of introgression that no *M. trossulus* animals were detected; only hybrid mussels with portions of the genome characteristic of this species were present (Riginos and Cunningham 2005; Figure 6.3).

Though the Canadian maritime zone was typified as reflecting reproductive isolation between *M. edulis* and *M. trossulus*, it would be inaccurate to conclude that there was no introgression. In fact, this zone contained introgressed forms, but also provided evidence for the maintenance of some degree of reproductive isolation (Figure 6.3; Toro *et al.* 2004; Riginos and Cunningham 2005). Riginos and Cunningham (2005) suggested two hypotheses to explain the different degree of introgression in the western and eastern Atlantic hybrid zones. The first hypothesis suggested a causal role for the different ecological settings resulting in selection for or against genetic exchange in the alternate hybrid zones. The second hypothesis invoked different times since the initiation of secondary contact between the two species as the determining

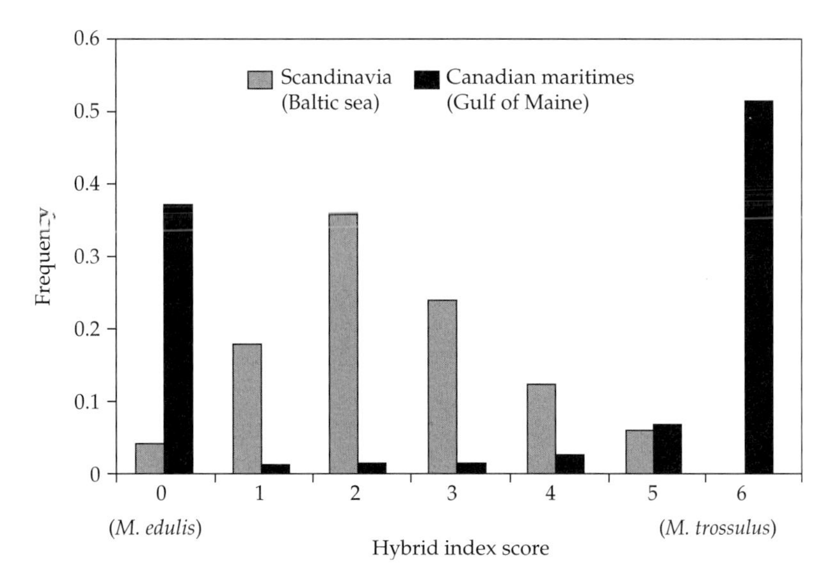

Figure 6.3 The frequency of *M. edulis*, *M. trossulus*, and natural hybrid genotypes in two hybrid zones (i.e., "Scandinavia" and "Canadian maritime"). A score of "0" or "6" indicates the presence of only *M. edulis* or *M. trossulus* genetic markers, respectively. Note that the Scandinavian zone contains almost all hybrid genotypes, while the Canadian maritime zone contains relatively few introgressed mussels (Riginos and Cunningham 2005).

factor. Neither hypothesis was strongly supported by the available evidence. In particular, the lack of accurate estimates of times since hybrid zone formation prevented a robust test of the alternate hypotheses (Riginos and Cunningham 2005).

6.2.2 Reticulate evolution and fish: trout

Trout species represent a significant, worldwide industry that produces large amounts of revenue for numerous countries and local municipalities. For example, 20 states within the United States produced US$74.9 million from sales of trout (eggs, fingerlings, etc.; National Agricultural Statistics Service 2007b). Various private and governmental outlets used these trout for restoration, conservation, and recreation (National Agricultural Statistics Service 2007b). Similarly, trout are utilized extensively in Europe as well. A study by Champigneulle and Cachera (2003) reflects, in a microcosm, their large-scale usage. These authors examined the effects from releasing ~500,000 trout annually on replenishment for angling along a 24 km stretch of the River Doubs. Champigneulle and Cachera (2003) concluded that natural recruitment, rather than artificial stocking, provided the majority of fish caught through angling.

Numerous trout clades present excellent examples of the significance of introgression in the evolutionary history of fishes (e.g., rainbow and cutthroat lineages; Williams *et al.* 2007; Gunnell *et al.* 2008). In addition, the affect that contemporary reticulation can have on genetic variation is also well illustrated by this assemblage (see Allendorf *et al.* 2001 for a discussion). However, I will limit my discussion to one exemplar of the role that reticulate evolution can play in the origin and development of trout assemblages. This is the brown trout, *Salmo trutta*. I will limit further this discussion to studies of introgression between divergent forms of this single species. This omits the wide array of introgression events known to occur between this species and other species of trout (e.g., Susnik *et al.* 2007).

S. trutta occurs naturally in Europe, specifically in the "Atlantic, Mediterranean, and southeastern drainages (Black, Caspian, and Aral seas)" (Presa *et al.* 2002). A common observation made in population-level and phylogeographic studies

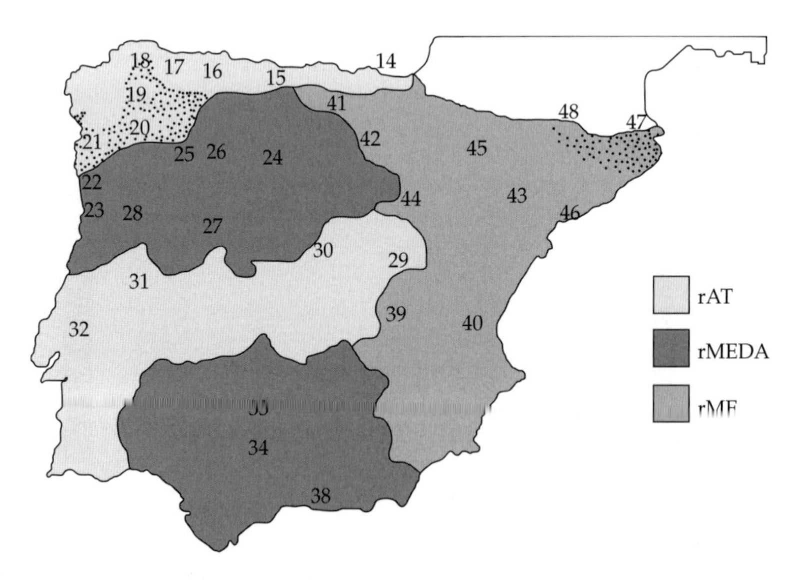

Figure 6.4 Genetic variation at the nuclear ribosomal RNA gene in brown trout, *S. trutta*, populations in the Iberian Peninsula. Three different classes of genotypes (i.e., rAT, rMEDA, and rME) are illustrated. The stippling near populations 18–21 and 47, 48 indicate areas of introgression between rAT/rMEDA and rME/rMEDA, respectively. However, additional hybrid individuals were detected in other regions as well (Presa *et al.* 2002).

is that this species is characterized by a complex pattern of genetic variation. Analyses of both mtDNA and nuclear loci have revealed that a portion of this complexity is due to introgressive hybridization. For example, Figure 6.4 illustrates admixtures of divergent ribosomal RNA gene lineages detected in the Iberian Peninsula. In total, Presa *et al.* (2002) found indications of introgression in ~18% of the 86 populations sampled across the range of the brown trout.

The conclusion of widespread genetic exchange between *S. trutta* lineages (specifically, in the Iberian Peninsula; Figure 6.4), thus leading to the concern that these forms will be lost through introgression, was emphasized by the subsequent analyses of Almodóvar *et al.* (2006) and Bouza *et al.* (2007). In the first of these analyses, stocking of waterways with hatchery grown progeny resulted in high levels of introgression with native trout. In the second study, contact between different lineages, characterized by divergent mtDNA genotypes, also led to introgression between lineages. In this latter case, some of the admixtures likely originated during the period of the last glaciation (Bouza *et al.* 2007). Regardless of proximate cause, introgression between divergent lineages of *S. trutta* has yielded numerous natural experiments of introgression-mediated evolutionary novelty (Figure 6.4; Presa *et al.* 2002; Almodóvar *et al.* 2006; Bouza *et al.* 2007).

6.2.3 Reticulate evolution and fish: peacock bass

The peacock bass (genus *Cichla*) belongs not to the bass clade, but rather is a member of the New World cichlid assemblage. Belonging as it does to the freshwater tropics, it forms a part of the "greatest diversity of freshwater fishes in the world" (Willis *et al.* 2007). Significantly, a portion of the South American tropical freshwater fishes has originated in the context of introgression between diverging lineages (Moritz *et al.* 2000). Peacock bass are so-named because of their brightly colored exterior. The attractive appearance of the species belonging to this genus, along with their size and behavior (i.e., aggressiveness when caught on a hook), has resulted in sportsmen and sportswomen paying several thousand dollars per week for the opportunity to angle for peacock bass. In fact, the popularity of sport fishing for this species complex has even led to the establishment of an "association" for communications concerning this sport fishing industry (http://www.peacockbassassociation.com).

Along with their beauty, and their importance in generating revenue from recreational sport fishing, the peacock bass clade also represents a resource for both subsistence and commercial fishing enterprises (Willis *et al.* 2007). Given this role as a food source for numerous indigenous peoples, inferences concerning the effects from introgressive hybridization on the evolution of this species web are significant. In this regard, Willis *et al.* (2007) examined the genetic diversity of *Cichla* lineages by examining over 450 individual fish representing all of the known and some putative, undescribed taxa.

In addition to corroborating previously defined phylogenetic relationships among certain species (e.g., see Farias *et al.* 2001; Renno *et al.* 2006), Willis *et al.* (2007) detected numerous cases of nonconcordance between the morphologically based assignment of specimens and their placement in the mtDNA phylogenies. Two groups of *Cichla* lineages were defined in this analysis, Clades "A" and "B." Among the samples assigned to these clades were four sets—identified as (1) *Cichla intermedia,* (2) *Cichla monoculus,* (3) *Cichla* sp. "Amazonas," and (4) *Cichla orinocensis*—that demonstrated closer evolutionary associations with lineages other than those belonging to their own species (Willis *et al.* 2007). Of these surprising associations, (1), (2), and (4) were inferred to have resulted from introgressive hybridization, while that of *Cichla* sp. "Amazonas" was argued to have resulted from the retention of ancestral polymorphisms (Willis *et al.* 2007). Furthermore, it was concluded that the introgression separately impacting *C. intermedia* and *C. monoculus* had been relatively recent, while that affecting the genetic structure of *C. orinocensis* had been ancient. In the latter case, the formation of the ancient, hybridization-derived lineage was termed an ancient "hybrid speciation" event (Willis *et al.* 2007). However, both the recent and ancient hybridization reflect genetic exchange that has affected lineages utilized for food (and recreation) by indigenous and foreign *H. sapiens* populations.

6.2.4 Reticulate evolution and fish: cod

The enormous importance of marine fish populations as a human food source has resulted in their endangerment from overharvesting. *Gadus morhua*, or cod, is one such species. For example, "Fisheries and Oceans Canada," the governmental agency providing oversight for Canada's fishing industry deemed it necessary to announce a moratorium on the cod harvest in 1992. This was followed (in 2003) by a total closure of not only many industrial cod harvest areas, but recreational fishing opportunities for this species as well (http://www.dfo-mpo.gc.ca/kids-enfants/map-carte/map_e.htm). Similarly, a report published by The Fisheries Secretariat of Sweden concluded that, "Scientific advice from the International Council of the Exploration of the Sea (ICES) has indicated that the two cod stocks in the Baltic Sea are suffering from unsustainable exploitation levels" (ORCA-EU 2007).

The state of cod fisheries in the Baltic Sea region is of particular importance to the present discussion because a hybrid zone has been identified between divergent lineages of *G. morhua* endemic to either the Baltic Sea or North Sea. In a series of analyses, Nielsen *et al.* (2001, 2003, 2005) defined the genetic variation of cod populations within these two seas, along with the transition zone between these two regions (Figure 6.5). Indeed, this zone of overlap and introgression appears to be a genetic/ecological transition zone for a variety of marine fish species (Nielsen *et al.* 2004). The ecological differences that may account for the genetic differentiation between the Baltic and North Sea, and thus the changeover in allele frequencies in this region, are likely correlated with the "steep salinity gradient from near oceanic salinities in the North Sea to the brackish Baltic Sea" (Nielsen *et al.* 2003).

The presence of alleles (at nuclear loci; Figure 6.5) characteristic of both the Baltic Sea and North Sea populations in the intermediate geographic region could have resulted either from the immigration of both types of *G. morhua* or introgressive hybridization in this area of overlap. Their detailed genetic analyses allowed Nielsen and his colleagues (Nielsen *et al.* 2003, 2005) to test these alternative hypotheses and thus formulate the following conclusions: (1) the region of admixture between the two lineages was characterized by recombinant individuals; (2) hybrid offspring were being produced within the transition zone, rather than from areas outside the zone of overlap;

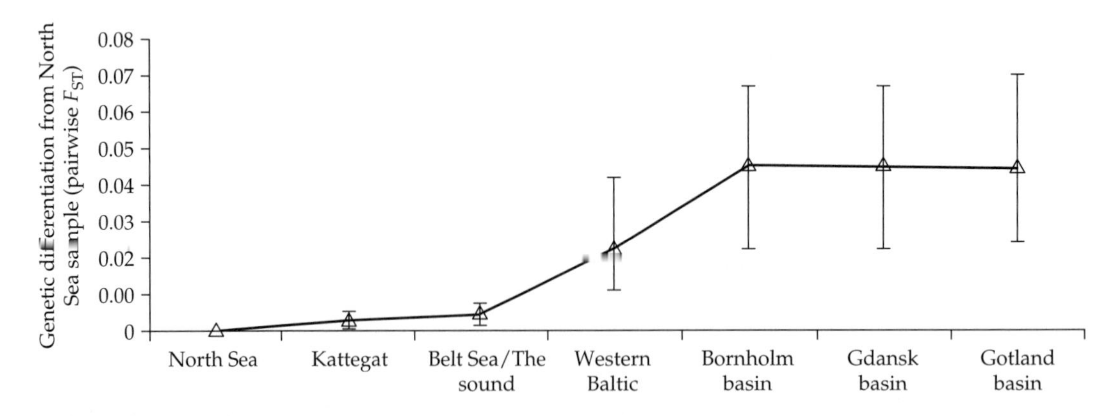

Figure 6.5 Transition of nuclear allelic variation in populations of cod, *G. morhua*. Nielsen *et al.* (2003) estimated the levels of genetic differentiation between the North Sea sample and all other populations following the procedures of Weir and Cockerham (1984; i.e., resulting in the "pairwise F_{st}" values on the *Y*-axis). The transition in genetic differentiation is coincidental with a changeover in salinity concentrations from the higher oceanic levels in the North Sea to the brackish water of the Baltic Sea (Nielsen *et al.* 2003).

and (3) the positioning of the hybrid zone in an area of steep ecological transition (Figure 6.5) could reflect either selection for hybrids that were more fit within this nonparental habitat, or selection against recombinants that are being continually replenished by hybridization between Baltic and North Sea fish migrating into the area of overlap.

6.2.5 Reticulate evolution and fish: eels

The European and American eels have been used to explore processes as diverse as panmixia, speciation, migration, and life history evolution. Avise *et al.* (1986) described a lack of mtDNA divergence for populations of the American eel (*Anguilla rostrata*) collected along a 4,000 km transect of the North American coastline. This led to the hypothesis that *A. rostrata* was a panmictic species. Likewise, Lintas *et al.* (1998), also through mtDNA analyses, arrived at the same conclusion for the European eel (*Anguilla anguilla*). In contrast, Avise *et al.* (1986) detected significant genetic divergence *between* the two *Anguilla* species, falsifying the hypothesis that they belonged to a single evolutionary lineage (Williams and Koehn 1984). Though the conclusion that there was a lack of genetic differentiation within the European and American eel lineages has now been shown to be incorrect (Wirth and Bernatchez 2001; Mank and Avise 2003) and thus the panmixia model falsified, the life history of these species does suggest unique conclusions concerning genetic exchange within and between *A. anguilla* and *A. rostrata*. Albert *et al.* (2006) have defined this life cycle as being characterized by the following: (1) both species begin their cycle in the North Atlantic (i.e., the Sargasso Sea) from which they disperse as larvae through either the Gulf Stream (to America) or the North Atlantic drift (to Europe; Albert *et al.* 2006); (2) once the larvae reach a certain size they go through metamorphosis, passing into the "glass eel" stage; (3) the "yellow eel" stage is reached after some 20≥160 days, at which time the eels can utilize aquatic habitats ranging from coastal to headwaters; and (4) following a period of 3–20 years, they reach maturation and metamorphose into silver eels, at which stage they migrate back to the Sargasso Sea to reproduce and die. During this life cycle, introgression is made possible by the temporal and spatial overlap of *A. anguilla* and *A. rostrata* in the Sargasso Sea spawning area (Albert *et al.* 2006).

Like cod and other human-utilized marine species, the European and American eels have been heavily depleted. For example, the yearly harvest of *A. anguilla* decreased by more than 40% during the decade 1988–1998 (Wirth and Bernatchez 2003). In addition, "a major decline in recruitment in the St. Lawrence River basin has been observed for *A. rostrata*" (Wirth and Bernatchez 2003). In regard to reticulate evolution, these food resources, also like other similarly utilized marine organisms, reflect signatures of introgressive hybridization. In particular, Avise *et al.* (1990) used genetic variation at nuclear and mtDNA loci along with morphological characters to test for interspecific matings between the American and European lineages. They sampled eels from Iceland, a region previously shown to contain animals with morphological characteristics thought to be diagnostic for *A. rostrata* (and not the resident *A. anguilla*; Williams and Koehn 1984). Their estimate of 2–4% hybrid individuals in this population was consistent with introgression from American into European eels having affected not only the genetic variation at the nuclear and mtDNA loci sampled by Avise *et al.* (1990), but also that underlying the morphological characteristics.

Albert *et al.* (2006) extended greatly the study of Avise *et al.* (1990) by sampling a total of 1,127 eels—748 from Iceland and 379 additional individuals from 1 North African, 3 European, and 4 North American populations. Because of the expanded population samples, and the surveying of many more loci (373 nuclear loci), Albert *et al.*'s (2006) study provided the opportunity to test for (1) the hybrid identity (e.g., F_1 hybrids versus second generation or later hybrids) of individual eels and (2) differences in the frequency of introgression across the eels' geographic distribution. Of the hybrids detected in this study, ~30% possessed genotypes indicating hybrid generations past the F_1 stage, with 5% of the Icelandic hybrids falling into the advanced generation category. Albert *et al.* (2006) detected a significantly higher mean frequency (15.5%) of recombinant eels in Iceland. Furthermore, different Icelandic sampling sites and different life cycle stages yielded estimated hybrid proportions ranging from ~7% to 100%. Interestingly, in terms of the geographic distribution of hybrids, no recombinant eels were detected in the European samples and few were identified from the North American populations. Almost the entire hybrid sample comes from Iceland. It was suggested that the occurrence of numerous hybrids in Iceland could be due to intermediate growth rates of hybrids, thus increasing their likelihood to colonize freshwater habitats in this "intermediate" geographic location (i.e., between North America and Europe; Albert *et al.* 2006). Regardless of the cause(s) for the formation and distribution of hybrids, the harvest of *Anguilla* from these populations reflects yet another example of the reticulate formation of a human food source.

6.2.6 Reticulate evolution and fish: Dolly Varden

As reflected by the following quote, the salmonid known as "Dolly Varden" (i.e., *Salvelinus malma*) fills a significant niche in the sport fishing industry in the state of Alaska:

> Dolly Varden is one of Alaska's most important and sought-after sport fish. The fish is unique, as it is the only member of the family Salmonidae, excluding salmon, that has readily adapted to the numerous small- to medium-size nonlake streams that enter our saltwater areas. Its importance and popularity can only increase as our population increases and further restrictions are placed on heavily used salmon streams. (Hubartt 1994)

Indeed, individual anglers spend thousands of dollars on 3–4 day excursions to collect this and other "arctic" fish species (e.g., http://www.cloverpassresort.com/alaska_fishing_packages.html).

The use of Dolly Varden both for food and recreation is not limited to the United States; *S. malma* is distributed from northwest North America to Siberia and Japan (Yamamoto *et al.* 2006). This species is only one of several belonging to the genus *Salvelinus*, with distributions encompassing the North Pacific rim. Members of this genus are all commonly referred to as "char." However, the shared nomenclature belies a great deal of taxonomic and evolutionary confusion concerning relationships within this assemblage (Brunner *et al.* 2001). One of the likely contributing factors to this uncertainty is widespread introgressive hybridization. Though genetic exchange has been documented between many species of *Salvelinus* (e.g., *Salvelinus alpinus* and *Salvelinus fontinalis*—Glémet *et al.* 1998; *S. fontinalis, Salvelinus confluentus,* and *Salvelinus namaycush*—Crespi and Fulton 2004), I will limit my discussion to studies that tested for introgression between Dolly Varden and three other char species, *Salvelinus leucomaenis, S. alpinus,* and *S. confluentus.*

Radchenko (2004) and Yamamoto *et al.* (2006) analyzed genetic variation in *S. malma* and *S. leucomaenis* in the Russian Far East and the Japanese Island of Hokkaido, respectively. Both studies identified patterns of variation indicative of recent and ancient introgressive hybridization between these two species. In the case of the Russian study, mtDNA genotypes suggested unidirectional introgression from *S. leucomaenis* into Dolly Varden populations located in rivers associated with the Okhotsk Sea basin (Radchenko 2004). Similarly, some of the Hokkaido Island populations of *S. malma* possessed both mtDNA and nuclear alleles that were otherwise diagnostic for *S. leucomaenis* (Yamamoto *et al.* 2006). In addition, like the Russian findings, the Hokkaido introgression was restricted to certain drainages, in this case to several rivers in the Shiretoko Peninsula (Yamamoto *et al.* 2006). Brunner *et al.* (2001) utilized mtDNA as well to examine within- and between-species genetic variation in *Salvelinus.* They were particularly interested in defining the phylogeographic distributions within the arctic char, *S. alpinus.* However, their study incorporated an analysis of mtDNA sequence variation in other char species as well, including the Dolly Varden. The phylogenetic distribution of samples reflected a signature consistent with mtDNA introgression from arctic char into *S. malma.* Specifically, Brunner *et al.* (2001) defined mtDNA genotypes within *S. malma* that were most closely related to those of *S. alpinus,* rather than to other Dolly Varden.

The final example of introgressive hybridization involving *S. malma* relates to genetic exchange with the "bull trout," *S. confluentus*. Two studies by Taylor and his colleagues (Taylor *et al.* 2001; Redenbach and Taylor 2002) illustrate well the affect from introgressive hybridization between these two species. Both studies identified a geographically extensive hybrid zone between *S. malma* and *S. confluentus* in northwestern North America. Taylor *et al.* (2001) examined genetic variation in 12 populations from six drainages. Eight of these populations contained recombinant individuals as assayed by nuclear loci (Taylor *et al.* 2001). In addition to the genetic divergence between bull trout and Dolly Varden at the nuclear loci, both studies detected highly divergent mtDNA lineages within *S. malma*. Though this could have reflected intraspecific lineage diversity, Redenbach and Taylor (2002) determined that one of these lineages had been introduced through hybridization with *S. confluentus*. Thus, like the hybridization between Dolly Varden and both arctic char and *S. leucomaenis*, the genetic exchange between *S. malma* and *S. confluentus* has left the genomes of the former pockmarked with genetic material of the latter.

6.2.7 Reticulate evolution and fish: tuna

> The high value of bluefin tuna (in terms of both ecology and economy) together with the serious concern regarding its conservation (risk of fisheries and stock collapse being recently stressed) led the SCRS to reiterate the critical necessity to develop an ambitious coordinated research program on this species. (International Commission for the Conservation of Atlantic Tunas 2007)

The annual income from this species complex (including Atlantic, Pacific northern, and southern bluefin tunas; *Thunnus thynnus, Thunnus orientalis,* and *Thunnus maccoyii*, respectively; Alvarado Bremer *et al.* 1997, 2005) has been estimated at greater than US$0.5 billion (International Commission for the Conservation of Atlantic Tunas 2007). Thus, the bluefin tuna has an enormous impact as both a food and economic resource for worldwide human populations. However, the reduction of stocks due to excessive human harvesting has caused the Director of the United States Marine Fisheries Service to call for a moratorium on bluefin fishing (Hogarth 2007; http://www.nmfs.noaa.gov). Similar to cod, bluefin species are in danger of being reduced to the point-of-no-return, at least in regard to their use as a human food source (International Commission for the Conservation of Atlantic Tunas 2007).

As with other key marine species utilized as food for humans, numerous data sets have been used to infer reticulate evolution in the histories of the assemblage containing bluefin tunas. Three studies, in particular, reflect the importance that introgressive hybridization has had in the genetic structuring of present-day populations. The earliest of these analyses (Chow and Kishino 1995) involved analyses of both nuclear (i.e., ribosomal RNA gene) and mtDNA sequences for each of the bluefin species (then considered subspecies) as well as albacore (*Thunnus alalunga*). Though the nuclear gene sequences defined relationships in agreement with morphology—placing the bluefin tuna together—the mtDNA sequences paired the Pacific northern bluefin with albacore samples (Chow and Kishino 1995). This finding led to the conclusion of extensive mtDNA introgression from albacore into Pacific northern bluefin populations, resulting in the widespread replacement of the mtDNA originally present in *T. orientalis* (Chow and Kishino 1995).

Two subsequent studies by Alvarado Bremer *et al.* (1997, 2005) supported the conclusions made by Chow and Kishino (1995). The first of these studies detected the phylogenetic pairing of *T. orientalis* with *T. alalunga* in a clade separate from the other bluefin tunas (Alvarado Bremer *et al.* 1997). This finding suggested the possibility of introgression of albacore genomic sequences into this lineage. However, it was also concluded that incomplete lineage sorting could account for the observation. Given that the latter study only involved the analysis of mtDNA sequences, the earlier study combining as it did both mtDNA and nuclear data reflected a more rigorous test of these alternate hypotheses.

A second analysis of mtDNA variation by Alvarado Bremer *et al.* (2005), one in which they examined over 600 Atlantic bluefin tuna, provided a great degree of resolution of the population genetic variation in this species and thus a stringent test for introgressive hybridization involving this lineage as well. Like the analyses involving *T. orientalis*, the data collected from the 600+ *T. thynnus* individuals defined patterns consistent with introgression of mtDNA from *T. alalunga* into bluefin tuna. Two highly divergent mtDNA sequence types were identified in the sample of *T. thynnus*, with ~3% of the individual fish possessing a mtDNA sequence that placed them with albacore (Alvarado Bremer *et al.* 2005). In contrast, nuclear sequences placed these same individuals with members of their own species. Thus, like *T. orientalis*, the Atlantic bluefin tuna has apparently been introgressed with mtDNA from *T. alalunga*. However, unlike Pacific bluefin tuna populations, this introgression has not led to a replacement of the original Atlantic bluefin tuna mtDNA form, but instead resulted in the presence of both the ancestral bluefin and the albacore mtDNA sequences (Alvarado Bremer *et al.* 2005). Notwithstanding the differences, both the Atlantic and Pacific branches of the bluefin tuna assemblage, and thus a major source of food for *H. sapiens*, possess signatures of reticulation. In the following sections, I will consider evidence for the same processes impacting food sources derived from avian species.

6.2.8 Reticulate evolution and birds: ducks

Grant and Grant (1992) estimated the overall frequency of hybridization between bird species to involve ~1 out of every 10 species. However, they found great variation in frequency between different bird orders. The highest frequency of hybridizing pairs was found for the order Anseriformes (ducks and geese) with almost half (i.e., 42%) of all species found to be hybridizing (Grant and Grant 1992). Given this propensity to mate outside the boundaries of their own species, and their widespread utilization by *H. sapiens*, ducks (and geese) illustrate well the effect of reticulation on a human food source. Furthermore, the economic value of these reticulating lineages can be considered from a variety of perspectives. These include the fact that one of the species that is widely cited for its role in introgressive hybridization, the mallard (*Anas platyrhynchus*), is one of the major sources for domesticated ducks (Hird *et al.* 2005). Another avenue for illustrating the economic value of the duck assemblage comes from estimates of expenditures by waterfowl hunters. For example, Henderson (2005) reviewed the economic benefit from these sportsmen and sportswomen and concluded that their expenditures during 2001, in the United States alone, generated "$2.3 billion in total output in the United States. Total output includes the direct, indirect, and induced effects of the expenditures associated with waterfowl hunting." Thus, both domesticated and wild duck populations reflect not only a widely used food resource for humans, but also an enormous economic benefit for numerous societies.

I have reviewed elsewhere (Arnold 2006; pp. 161–163) studies that illustrate the conservation issues arising from introgressive hybridization between *A. platyrhynchus* and a relatively rare species, the black duck (*Anas rubripes*). Specifically, Mank *et al.* (2004) demonstrated that the lack of monophyly for the black duck versus the mallard was due to the genetic assimilation, through introgression, of *A. rubripes* by *A. platyrhynchus*. In this instance then, the mallard is likely causing the extinction of another species through hybridization. However, the genome of the mallard is also a mosaic of elements derived from other species. For example, Avise *et al.* (1990) defined two mtDNA lineages in mallards; these lineages were named "Groups A and B." Recently, Kulikova *et al.* (2004, 2005) have invoked introgressive hybridization as at least a partial explanation for the presence of multiple mtDNA types in *A. platyrhynchus*. Figure 6.6 illustrates a portion of the mallard's geographic distribution and the mtDNA constitution of its populations indicating the role of introgression with other species. The cline in the frequency of the A and B mtDNA haplotypes is consistent with introgression of the Group A mtDNA into mallards from related species of ducks from Asia (that only possess Group A haplotypes; Kulikova *et al.* 2005). Furthermore, introgression into mallards has not only involved mtDNA but involved nuclear loci as well. Kulikova *et al.* (2004) detected introgression of Eastern Spot-billed duck

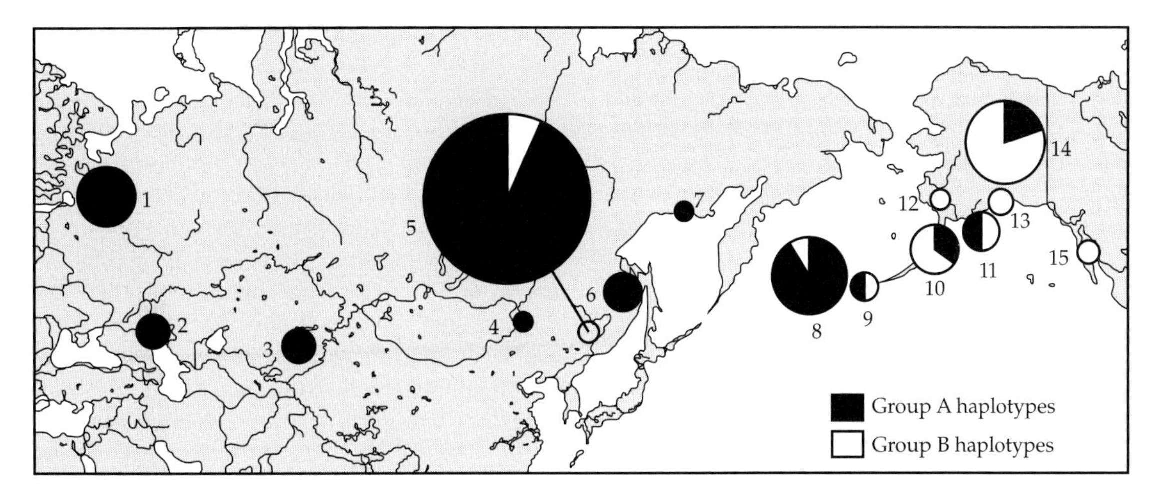

Figure 6.6 Frequency of two mtDNA haplotypes ("Group A" and "Group B") in the mallard, *A. platyrhynchus*. Each pie diagram indicates the proportion of individuals with either haplotype (as indicated by the filled and unfilled portions). The sizes of the pie diagrams reflect the relative sample sizes of each population. The Group A haplotypes, at high frequency in the easternmost samples, have been inferred to have originated through the introgression into mallards of mtDNA from other species such as the Eastern Spotbilled duck, *Anas zonorhyncha* (Kulikova *et al.* 2004, 2005).

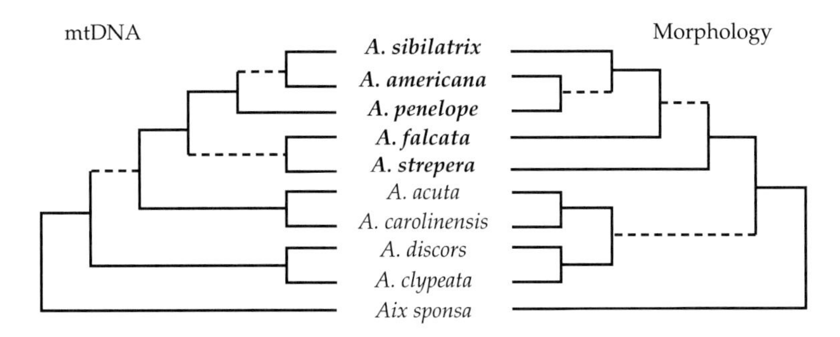

Figure 6.7 Phylogenetic relationships among (1) a subset of *Anas* species known as wigeons (indicated by bold type) based on mtDNA and morphology (Peters *et al.* 2005) and (2) their sister taxa. Dashed lines indicate the three regions of the phylogenies that show discordances in species placement.

alleles at the ornithine decarboxylase locus into mallard populations.

Genetic interactions between *Anas* species are not limited to cases involving mallards. The general interfertility between members of *Anas*, and thus the likelihood of introgression in nature, has been well established for decades (Johnsgard 1960). Figure 6.7 (Peters *et al.* 2005) reflects the discordant relationships among a series of *Anas* species known as the wigeons. The discordances illustrated in Figure 6.7 involve the placement of species based on morphology versus mtDNA sequences. However, similar levels of discordance were also detected between morphologically based phylogenies and phylogenies constructed from nuclear loci (Peters *et al.* 2005). From these analyses it is clear that, at the least, the sharing of the mtDNA sequence types likely results from introgressive hybridization between the various species. In addition, though introgression could not be conclusively inferred from the allelic variation at nuclear loci, it could not be rejected as an explanation (Peters *et al.* 2005). Indeed, nuclear (and mitochondrial) loci were utilized to infer introgression as the cause of the discordance indicated in Figure 6.7 between the gadwall and the falcated duck (*Anas strepera* and *Anas falcata*, respectively) (Peters and Omland 2007; Peters *et al.* 2007). Like the mallard lineage, other *Anas* species utilized extensively as a human food source are marked by past reticulate evolutionary processes.

6.2.9 Reticulate evolution and birds: partridge

As with the ducks and geese (and hares and rabbits—see later), members of the genus *Alectoris* (i.e., "partridges") are a primary species of interest for hunters in many countries. For example, *Alectoris chukar cypriotes* (the Chukar partridge found on the island of Cyprus) "...is the most important game species in Cyprus where several hundred thousand birds are shot down by hunters" (Tejedor *et al.* 2005). In addition, like species such as the mallard (Mank *et al.* 2004), human-mediated changes have apparently affected patterns of introgression.

In the case of partridge species, one of the largest human-based factors has been the release of large numbers of captive-reared animals to supplement the populations being hunted. Though reintroductions can be beneficial for genetic enrichment of critically small populations (O'Brien and Mayr 1991), for the *Alectoris* species the release of captive birds has often involved different species relative to the taxa present in the natural populations. This practice has facilitated widespread and genetically extensive introgression involving, for example, (1) native populations of *Alectoris rufa* (red-legged partridge) and either introduced rock partridges, *Alectoris graeca* (Negro *et al.* 2001) or chukar partridges (*Alectoris chukar*; Baratti *et al.* 2004; Barbanera *et al.* 2005; Tejedor *et al.* 2007) and (2) native populations of rock partridges and introduced chukar partridges (Barilani *et al.* 2007). Even when the restocking efforts have involved natural populations of what was thought to be a single species, the stocks used have sometimes been shown to be instead hybrid animals (e.g., Baratti *et al.* 2004; Barbanera *et al.* 2005).

Though introductions of foreign species and subsequent hybridization with native partridges have played a significant role in the genetic structuring of contemporary populations, non-human-mediated introgressive hybridization also occurs in this assemblage. Randi *et al.* (2003) and Barbanera *et al.* (2005) discussed evidence of such introgression between *A. graeca* and both *A. rufa* and *A. chukar*. In particular, the study by Randi and Bernard-Laurent (1999) elucidated evolutionary processes within one hybrid zone between the rock and red-legged partridge located in the southern range of the French Alps. Using nuclear loci, Randi and Bernard-Laurent (1999) defined a transition between diagnostic allele frequencies for the two species that spanned between 70 and 160 km and from this inferred secondary contact of the two partridge lineages following the glacial recession in the Alps (<6,000–8,000 ybp). These authors also concluded that the stability of the zone of overlap might be "sustained by the high reproductive success of hybrids" (Randi and Bernard-Laurent 1999). Thus, through secondary contacts caused by both human-mediated reintroductions and climatic variations, introgression has been pervasive in the species of *Alectoris*. The same can also be said for numerous mammalian species utilized by *Homo* as sources of animal protein.

6.2.10 Reticulate evolution and mammals: sheep

The importance of the origin and evolution of the domestic sheep is illustrated by the following quote from Chen *et al.* (2006): "Domestic sheep (*Ovis aries*) have played important roles in diverse human societies as a source of food, hide, and wool, and are one of the major components of agro-pastoral societies since the Neolithic." One example of this importance can be drawn from the present-day economic value of the domestic sheep industry. For example, Australia (one of the top sheep-producing countries) contained just over 100,000,000 million sheep and lambs in 2005. The gross value—just from the production of sheep and lamb meat—during 2004–2005 totaled ca. AU$2 billion (Australian Bureau of Statistics 2007). As an aside, and once again to emphasize the multiple benefits individual species of animals and plants can provide to humans, the value of the Australian wool market during this same time period was also ca. AU$2 billion (Australian Bureau of Statistics 2007).

In addition to their value as food (and clothing) producers for *H. sapiens*, *Ovis aries* also illustrates well a complex evolutionary history that includes introgressive hybridization. This is apparent from several different classes of data. First, multiple, divergent genetic lineages have been identified in this species. Meadows *et al.* (2007) reviewed the history of analyses of mtDNA from populations of sheep from various geographic regions—beginning with the work of Wood and Phua (1996) who detected two lineages in a limited sample (50 animals), and continuing through to several analyses that identified the presence of 3–5 mtDNA classes in various geographic regions (three in Asia and the Near East, Chen *et al.* 2006, Pedrosa *et al.* 2005; four in the Caucasus, Tapio *et al.* 2006; five in the Near East, Meadows *et al.* 2007). The identification of up to five divergent lineages reflects the contribution of independent domestication events to the genetic structure of contemporary *O. aries* populations. This level of diversity, caused by multiple domestications in various geographic regions, is now understood to be a common aspect for both animal and plant domesticates (e.g., Meadows *et al.* 2007; Saisho and Purugganan 2007).

In the case of sheep, the contemporary genetic diversity present in specific geographic regions (e.g., the Near East or Central Asia) is not only reflective of multiple points of origin for domestic sheep, but also likely indicates origins from multiple wild lineages. For example, Hiendleder *et al.* (2002) argued for the origin of two divergent domesticated lineages from two different subspecies of "mouflon" sheep. However, Meadows *et al.* (2007) pointed out that even with the detection of five domestic lineages "no extant wild *Ovis* progenitor has been identified." Yet, they also concluded that independent events may have "sampled diverse Moufloniformic populations, which have subsequently ceased to exist."

Regardless of the taxonomic sources of the divergent mtDNA lineages, that they arose in different geographic regions and yet are now found together indicates admixtures of these genetic isolates (Meadows *et al.* 2005). It is almost certain that some of the observed introgressions have been due—just as for partridges—to human-mediated introductions of sheep carrying one mtDNA haplotype into populations containing another haplotype (e.g., Chinese and Turkish populations, Chen *et al.* 2006; Iberian populations, Pereira *et al.* 2006; European sheep breeds, Lawson Handley *et al.* 2007). However, reticulate evolution has impacted this assemblage through additional avenues as well. For example, Valdez *et al.* (1978) documented introgressive hybridization involving *Ovis orientalis*, one of the wild species suggested as a progenitor for the Asian domesticated sheep clades (Hiendleder *et al.* 2002; but see Meadows *et al.* 2007). In addition, it is well accepted that the "wild" European mouflon, *Ovis musimon*, is actually a feral population from a Neolithic introduction (Hiendleder *et al.* 2002). The phylogenetic association of this taxon with *O. aries* is thus likely reflective of its origin from a domesticated population and intermittent introgression with this progenitor. Evolution preceding, during, and following the domestication of sheep was apparently marked by introgressive hybridization between divergent genetic lineages.

6.2.11 Reticulate evolution and mammals: goats

According to Pedrosa *et al.* (2005) "Sheep and goat domestication played an important role in the phenomenon of neolithization occurring in the late prehistory of the Near and Middle East... giving rise to a sedentary way of life." The positive contribution of sheep to this transition, and for present-day human cultures, was discussed earlier. Domesticated goats, *Capra hircus*, have at least an equivalent importance in many societies. MacHugh and Bradley (2001) discussed the twin facts that the goat is ridiculed as the "poor man's cow," yet the worldwide population at that time of 700 million goats provided small farmers (often in developing countries) with meat, milk, skins, and fiber. *C. hircus* may have been the first, domesticated, wild herbivore (MacHugh and Bradley 2001) and demonstrates the broadest ecological tolerance of any livestock species. Joshi *et al.*

(2004) supported this latter assertion by pointing to *C. hircus'* presence in Indian environments as diverse as the high elevations in the Himalayas, the desert regions of the state of Rajasthan, and the extremely wet coastal environments. Indeed, the importance of the domestic goat can also be illustrated through an analysis of Indian agriculture. In particular, Joshi *et al.* (2004) described India as containing "123 million goats comprising 20 recognized breeds...which together make up approximately 20% of the world's goat population."

In addition, to their highlighting the crucial cultural and economic status of the Indian populations of *C. hircus*, Joshi *et al.* (2004) collected mtDNA sequence data that indicated a multifaceted origin for this important domesticate (Figure 6.8). mtDNA sequence variation was analyzed for 363 individuals belonging to 10 separate breeds (nine named, and one termed "local") that were distributed throughout the Indian subcontinent (Figure 6.8). Once again like sheep, partridges, and so on the phylogeographic patterning suggested human-mediated genetic admixture. Though not perfectly correlated with current-day distributions of major peoples groups in India, the mtDNA variation in the goat populations (Figure 6.8) was inferred to have arisen partially through the separate migrations of Dravidian and Indo-Aryan language groups (Joshi *et al.* 2004).

Reticulation among the five, mtDNA lineages found in Indian *C. hircus* would have occurred as the goats accompanied the human migrants. Several lines of evidence also indicate the interaction, and introgression, between both wild *Capra* species and between wild and domesticated lineages. Figure 6.9 illustrates the results from an analysis designed to infer the ancestor of *Capra* (Ropiquet and Hassanin 2006). Instead of the mtDNA and nuclear sequences leading to the definition of a single ancestral lineage, discordant phylogenies were resolved from the two data sets (Figure 6.9). Of most interest for the question of *Capra*'s origin, it was concluded that mtDNA had introgressed from the Himalayan tahr (i.e., *Hemitragus*) lineage into that leading to the goat assemblage (Figure 6.9; Ropiquet and Hassanin 2006). This inference led Ropiquet and Hassanin (2006) to the conclusion that the origin of the goat clade occurred via hybrid speciation. They also hypothesized that the introgression of the mtDNA was adaptive. Specifically, they argued that the introgression of the mtDNA from the high altitude-adapted tahr into the proto-*Capra* would have "conferred a great selective advantage on goats."

Phylogenetic analyses of the entire *Capra* species complex have also detected signatures of introgression and hybrid lineage formation. Manceau *et al.* (1999) sequenced the cytochrome *b* (mtDNA) gene from the domestic goat and from the wild taxa, *Capra ibex ibex, Capra ibex nubiana, Capra ibex caucasica, Capra ibex sibirica, Capra pyrenaica, Capra aegagrus, Capra cylindricornis,* and *Capra falconeri.* The resulting phylogenetic analysis resolved discordances that were consistent with numerous reticulate events that affected the evolution of the genus, as a whole, and the domestic goat in particular. These discordant patterns included: (1) sequence variation demonstrating *C. ibex* to be polyphyletic; (2) the "Chiltan wild goat" (a type of *C. aegagrus*) was genetically related to *C. falconeri*; and (3) some *C. aegagrus* haplotypes also clustered with *C. cylindricornis.* In accounting for all of the cases of nonmonophyly, Manceau *et al.* (1999) reviewed the results of studies in which interspecific hybridization had been detected involving *C. aegagrus* and *C. ibex.* Significantly, *C. aegagrus* is thought to be the progenitor of the domestic goat. In addition to the evidence for hybridization between the two wild species, Manceau *et al.* (1999) also cited results indicating hybridization between *C. hircus* and "many other *Capra* spp."

A similar pattern to that detected by Manceau *et al.* (1999; i.e., of paraphyly/polyphyly among *Capra* species) was also described by Pidancier *et al.* (2006) from an analysis of both Y-chromosome and mtDNA sequences. The observation of numerous instances of introgressive hybridization involving *C. hircus* and its wild progenitor and sister taxa led Pidancier *et al.* (2006) to propose both recent and ancient genetic exchange to explain the discordances between the DNA-based phylogenies. Of particular significance was their hypothesis of ancient hybridization between the "bezoar-type" (i.e., the likely progenitor lineage of *C. hircus*) and the "ibex-type" of *Capra.* Ancient hybridization between maternal and paternal types that carried

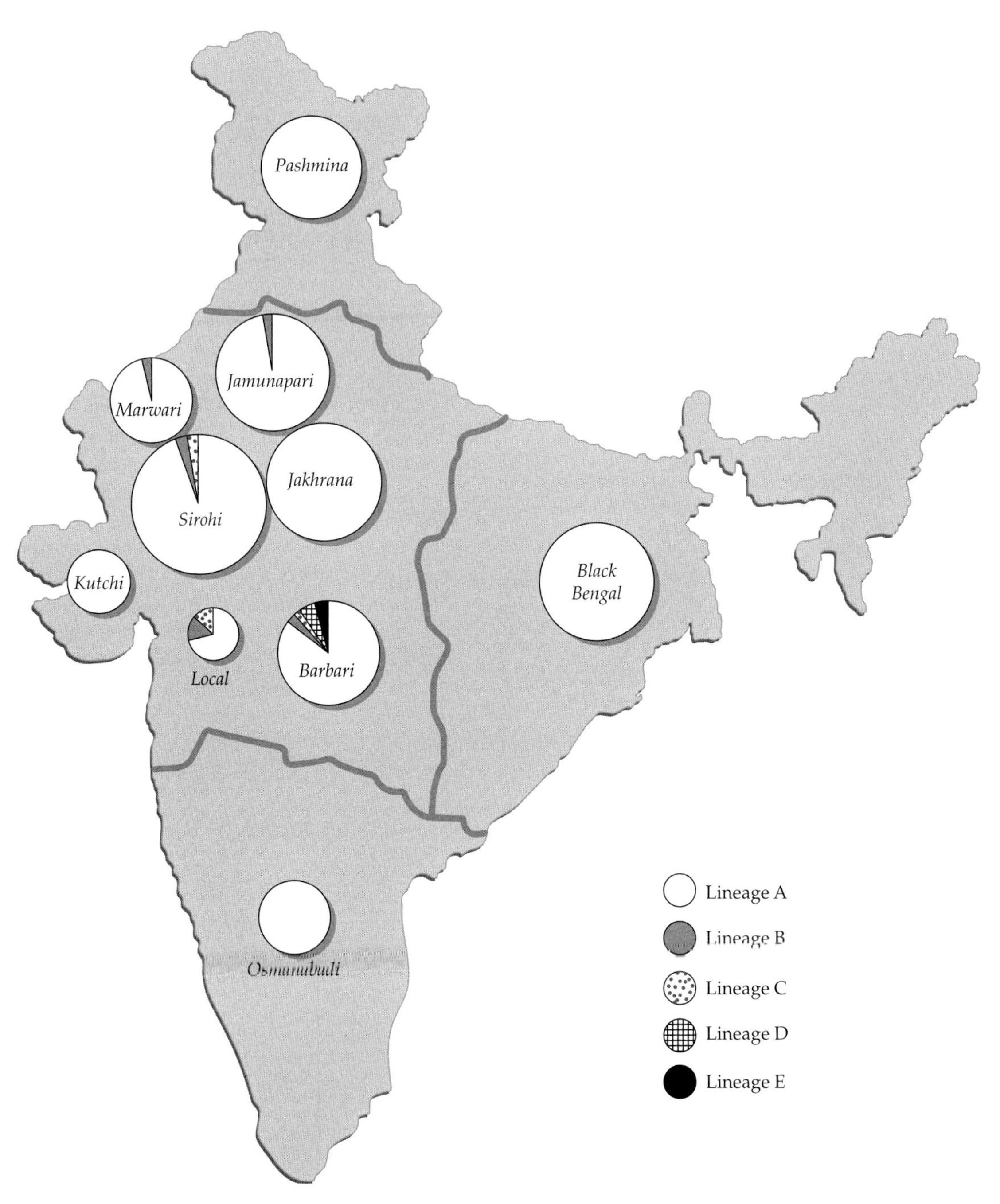

Figure 6.8 mtDNA variation in 10 breeds (9 named and 1 termed "local"—the latter consists of "nondescript goats distributed over the Barbari home tract"; Joshi *et al.* 2004) of domesticated goats from India. The diameter of each circle reflects the relative sample size, and the filled or unfilled areas reflect the proportion of the five mtDNA lineages in each breed (Joshi *et al.* 2004).

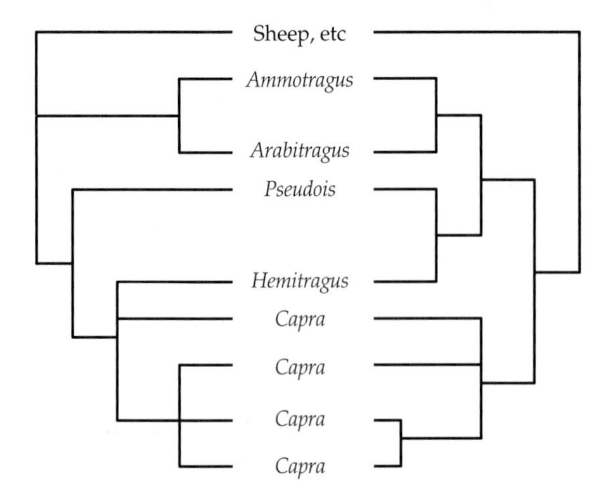

Figure 6.9 Phylogenies constructed from mtDNA (left-hand) and nuclear (right-hand) sequences to determine the most likely ancestor for the goat (*Capra*) assemblage. Possible progenitors and sister taxa for this clade were *Hemitragus* (Himalayan tahr), *Arabitragus* (Arabian tahr), *Pseudois* (bharals), and *Ammotragus* (aoudad). The discordant phylogenies are consistent with the hypothesis of mtDNA introgression from the ancestor of *Hemitragus* into that of *Capra*, making this genus the product of hybrid speciation (Ropiquet and Hassanin 2006).

the bezoar and ibex loci, respectively, would thus account for the admixture of mtDNA from bezoar and Y-chromosome variants from ibex detected in extant *Capra* species (Pidancier *et al.* 2006).

6.2.12 Reticulate evolution and mammals: chamois

Chamois, *Rupicapra pyrenaica* and *Rupicapra rupicapra*, are alpine-adapted ungulates that occur in mountainous regions of Europe and the Middle East (Schaschl *et al.* 2003). Archaeozoological findings, in particular, have indicated the preference of *Rupicapra* for the heavily forested and steep slopes of their alpine habitats (Baumann *et al.* 2005). Chamois have been a source of food for resident *Homo* populations and recently (i.e., during historical times) went through a severe bottleneck due to overexploitation (Baumann *et al.* 2005) and presumably also due to habitat loss. The utilization of chamois for food extends back hundreds of thousands of years. For example, Rivals and Deniaux (2005) examined tooth wear in fossils of *Rupicapra* ranging in age from ca. 440,000 to 35,000 ybp collected at paleolithic sites in southern France. The presence of these fossils (along with those of other species known to be utilized for food by humans) indicates the long-term dependence of archaic and anatomically modern *Homo* on these ungulates. Chamois meat continues to be utilized as a protein source by contemporary *H. sapiens* as well (e.g., Hofbauer *et al.* 2006).

Various loci—both nuclear and mtDNA—have been utilized to determine the population genetic and phylogeographic structure of *R. pyrenaica* and *R. rupicapra*. Each of these data sets has contributed to the dual inferences of significant genetic structuring between geographic regions, and yet introgression at some loci between both (1) lineages within species and (2) species (Pérez *et al.* 2002; Schaschl *et al.* 2003; Rodríguez *et al.* 2007). Pérez *et al.* (2002) assayed allelic variation at 23 nuclear (microsatellite) loci in each of the 3 subspecies of *R. pyrenaica* and 5 of 7 *R. rupicapra* subspecies. Phylogenetic analyses detected two assemblages consisting, respectively, of either *R. pyrenaica* or *R. rupicapra* samples; within each species, geographical and genetic distances were correlated. Both of these observations were consistent with an evolutionary scenario "...dominated by expansions and contractions within limited geographic regions, leading to alternate contact and isolation of contiguous populations" (Pérez *et al.* 2002). Similarly, Schaschl *et al.* (2003) detected patterns of genetic variability indicative of both isolation and introgression between divergent lineages within a sample of 443 *R. rupicapra* individuals collected in the Eastern Alps. In this latter study, genetic structuring consistent with limited gene flow was detected for mtDNA loci. Four different mtDNA lineages were detected (Schaschl *et al.* 2002). In contrast, divergence at nuclear loci did not increase with increasing geographical separation between populations sampled. Schaschl *et al.* (2002) hypothesized that variation in the dispersal behaviors of the two sexes—with the mtDNA-contributing females moving less than males—could account for this observation. This hypothesis was supported by telemetry studies demonstrating that (1) >90% of females were philopatric and (2) males dispersed more than females, especially before reproducing (Loison *et al.* 1999).

In addition to the data indicating extensive genetic exchange within the *Rupicapra* species, evidence of introgression involving different species also exists. Rodríguez *et al.* (2007) inferred such a past introgression event from sequence data gathered for both the mitochondrial-localized, cytochrome *b* gene and a nuclear pseudogene derived from this locus. These authors estimated an ancient origin of the lineage leading to the nuclear pseudogene. This estimate, coupled with estimates of the rate of evolution of the pseudogene and the functional (mitochondrial) locus, led to a derivative inference—the highly divergent pseudogene introgressed into *Rupicapra* from another species (Rodríguez *et al.* 2007). These authors further postulated that the divergent mtDNA lineages would have likely arisen in allopatry and were brought together through introgression as the ranges of the animals expanded until they overlapped (Rodríguez *et al.* 2007).

6.2.13 Reticulate evolution and mammals: banteng/kouprey

The banteng and kouprey are members of the Bovidae, and are further grouped in the genus *Bos*, an assemblage sometimes referred to as the "True Cattle" (Walker *et al.* 1975; p. 1429). Belonging to the category of "cattle" indicates that *Bos javanicus* and *Bos sauveli* (banteng and kouprey, respectively) have been utilized (possibly through domestication—Hassanin *et al.* 2006) as food sources for *H. sapiens* populations in their native, Southeast Asian, habitats. Indeed, this utilization, along with the trade in horns, likely led to the extermination of native populations of the kouprey (Hassanin *et al.* 2006) and the endangered status of the banteng (Hedges 2000). However, the value of banteng as a sport-hunting trophy (with excursions to collect a single bull costing over US$10,000; http://www.cabelas.com) has likely led to the numerical increase of *B. javanicus* in transplanted populations such as those in northern Australia.

One of the longest standing illustrations of the impact from introgression on the evolution of the banteng and kouprey involves the status of *B. sauveli*. This species has been alternately suggested to be (1) phylogenetically related to banteng and gaur, or to aurochs and domestic cattle, (2) a species belonging to a genus other than *Bos*, (3) a geographically separated banteng population, (4) a feral derivative of domesticated cattle, or (5) a hybrid derivative from crosses between banteng and either domesticated cattle, gaur, or water buffalo (Walker *et al.* 1975, p. 1431; Hassanin and Ropiquet 2004, 2007a, b; Galbreath *et al.* 2006). This final hypothesis concerning the origin of kouprey postulates its formation through hybrid speciation. Ironically, instead of the kouprey being a hybrid derivative of the banteng and another species, it now appears almost certain that some lineages of banteng are introgressive hybrids containing genes from the kouprey.

Studies by Hassanin and Ropiquet (2007a, b) have detected genetic variation suggestive of mtDNA introgression from *B. sauveli* into Cambodian populations of *B. javanicus*. The best documentation of this introgression comes from a combined analysis of mitochondrial and nuclear loci. Hassanin and Ropiquet (2007a) thus demonstrated that the mitochondrial genome of Cambodian banteng was more similar to kouprey sequences than to other banteng lineages. In contrast, all the banteng analyzed possessed identical nuclear sequences. This discordant pattern was consistent with the asymmetric introgression of mtDNA, without the transfer of nuclear alleles, resulting in the hybrid constitution of the Cambodian *B. javanicus* lineage. Estimates for the time period during which this introgression event took place fell within the Pleistocene (1.31 ± 0.45 mya) and thus did not implicate human-mediated genetic exchange (Hassanin and Ropiquet 2007a). In addition, supporting the likelihood of introgression between these species were the observations of mixed herds of kouprey and banteng in nature (e.g., Edmond-Blanc 1947).

Hassanin and Ropiquet's (2007a) findings also led to the rejection of the longstanding hypothesis mentioned earlier that *B. sauveli* is a hybrid lineage. Specifically, they did not detect elevated heterozygosity at the kouprey nuclear loci (i.e., expected if kouprey possessed a mixture of alleles from banteng and another species). Instead, Hassanin and Ropiquet (2007a) detected alleles diagnostic for *B. sauveli*. In total then, the findings from each of these studies supported one hypothesis of reticulation that affected banteng and kouprey evolution, but rejected another.

6.2.14 Reticulate evolution and mammals: cattle

That cattle possess great value for humans is a truism. Li *et al.* (2007) reflected the depth and breadth of this value by recognizing the past and contemporary roles of cattle as an agricultural, economic, cultural, and religious mainstay. Their fundamentally important impact as a *H. sapiens'* food source can be illustrated by considering that residents of Uruguay, United States, and Argentina—three societies that produce, and eat, large quantities of beef—consume an average of 87, 95, and 141 lb of beef per capita, respectively (Mathews and Vandeveer 2007).

I have previously (Arnold 2006, pp. 179–180) reviewed data concerning the evolution of European lineages of domesticated cattle (i.e., those most often utilized in the cattle industries of countries such as Argentina, Uruguay, and the United States). In particular, I discussed evidence of introgression into the cattle lineage domesticated in the Near East (i.e., *Bos taurus*; Bollongino *et al.* 2006) from male aurochs (*Bos primigenius*; the species from which cattle were domesticated) as the former was brought into contact with the latter in Europe (Götherström *et al.* 2005). Furthermore, though some studies have estimated a much lower frequency of female-mediated, relative to male-mediated, introgression (Edwards *et al.* 2007), Beja-Pereira *et al.* (2006) detected introgression between *B. taurus* introduced into Europe during the Neolithic period and native, wild animals.

A second domestication of cattle, also from wild aurochs (resulting in the *Bos indicus* lineage), took place on the Indian subcontinent, most likely in what is today Pakistan (see references in Beja-Pereira *et al.* 2006). Reticulate evolution, like that which affected *B. taurus*, has marked the evolutionary path of *B. indicus* as well. Some of the genetic exchange events have been due to human-mediated transplantations leading to *B. taurus* × *B. indicus* hybridization (Kikkawa *et al.* 2003). It has been demonstrated that expansions of *B. indicus* into Africa, Europe, and the Near East resulted in admixture ranging from nearly zero (i.e., little or no introgression) to ca. 50:50 ratios of alleles from the two lineages (Freeman *et al.* 2006).

Introgression impacting domestic cattle populations has not been restricted to genetic exchange between *B. taurus* and *B. indicus*; genes have been repeatedly transferred into domesticated *Bos* from a variety of wild species. Kikkawa *et al.* (2003) reported the presence of yak (i.e., *Bos grunniens*) mtDNA in native Nepalese cattle populations. These authors went on to review all of the data suggesting hybrid ancestries for Asian cattle and from this concluded that it was likely that as many as four wild species had introgressed with the domesticated lineage. Similarly, Nijman *et al.* (2003) detected the introgressive origin of some Southeast Asian (i.e., Malaysian and Indonesian) cattle from crosses between *B. indicus* and *B. javanicus* (banteng). Reticulate evolution involving domesticated cattle lineages has also impacted the genomes of wild relatives. Numerous studies have thus documented introgression from the domesticated lineages into wild taxa as diverse as the yak and the European and North American bison (*Bison bonasus* and *Bison bison*, respectively; Ward *et al.* 1999; Halbert and Derr 2007). For example, ~64% of North American bison populations sampled by Halbert and Derr (2007) demonstrated nuclear introgression from *B. taurus*.

The large degree to which domesticated and wild lineages in the cattle assemblage have been affected by reticulation has hopefully been well illustrated by the previous examples. Many of these instances have involved human intervention in the form of controlled crosses, or the transplantation of foreign populations into the habitats of native species. However, examples such as the hybrid derivation of the European bison (or wisent; Verkaar *et al.* 2004) belie the argument that *H. sapiens'* cultural practices alone can account for reticulate evolution within this complex.

6.2.15 Reticulate evolution and mammals: kob and waterbuck

"Three subspecies of the medium-sized kob antelope (*Kobus kob*)—found on the northern savannah plains across equatorial Africa—are distinguished by geographical distribution and variations in male coat colour" (Lorenzen *et al.* 2007). Similarly, "Two subspecies of waterbuck (*Kobus ellipsiprymnus*), common (*Kobus ellipsiprymnus ellipsiprymnus*) and defassa (*Kobus ellipsiprymnus defassa*), are recognized based on differences in rump pattern, coat colour and geographical distribution" (Figure 6.10; Lorenzen *et al.* 2006). Sadly, the extended wars in countries like Uganda, in conjunction with increasing human population sizes, have resulted in poaching that has driven clades of African bovids such as *Kobus* to near extinction (Lorenzen *et al.* 2007). Thus, Boddington (1997, p. 197) would write upon his return to the Central African Republic—after an absence of only 2 years—that Sudanese poachers had reduced hundreds of waterbuck and kob to "just scattered survivors." Tragic as such occurrences are, they reflect accurately the key role of kob antelope in subsistence hunting by humans living in equatorial Africa.

Analyses of genetic variation in the *K. kob* subspecies have detected patterns indicative of both evolutionary divergence and reticulation. In terms of divergence, discrete clades have been identified based on both nuclear and mtDNA loci. However, the clades do not necessarily reflect subspecific delineations (Birungi and Arctander 2000). One analysis that included all three subspecies of *K. kob* (i.e., *Kobus kob kob, Kobus kob thomasi*, and *Kobus kob leucotis*) utilized both mtDNA and nuclear microsatellite loci to assay genetic variability across the geographic range of this species (Lorenzen *et al.* 2007). Surprisingly, animals from the Murchison Falls region in Uganda were morphologically *K. k. thomasi*, yet possessed genetic variation characteristic of the morphologically distinct *K. k. leucotis* animals from Sudan and Ethiopia (Lorenzen

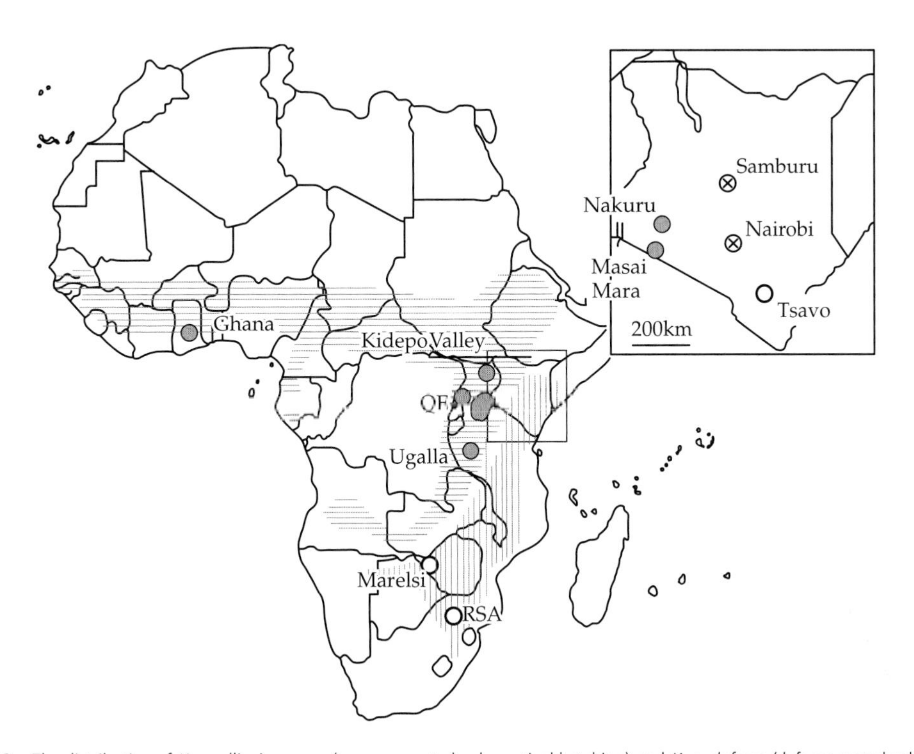

Figure 6.10 The distribution of *K. e. ellipsiprymnus* (common waterbuck; vertical hatching) and *K. e. defassa* (defassa waterbuck; horizontal hatching) with 11 sampling localities. Inset illustrates sampling localities in an area of overlap between the two subspecies in Kenya. Unfilled and filled circles indicate populations characterized by common and defassa waterbuck morphological traits, respectively. Circles containing "X" indicate populations found to consist of admixtures of nuclear and mtDNA variants from *K. e. ellipsiprymnus* and *K. e. defassa* (Lorenzen *et al.* 2006).

et al. 2007). This finding was recognized as the first example of such a disagreement between the morphological and genetic characterization for an African bovid (Lorenzen *et al.* 2007). A portion of the explanation for this discordance involved the occurrence of male-mediated gene flow from *K. k. leucotis* into the *K. k. thomasi* lineage (Lorenzen *et al.* 2007).

As with the *K. kob* subspecies, introgressive hybridization has also been inferred for the waterbuck taxa. These two taxa are easily diagnosable from both morphological and cytogenetic characters (Kingswood *et al.* 1998; Lorenzen *et al.* 2006). However, the possibility for genetic exchange exists due to an area of overlap between *K. e. ellipsiprymnus* and *K. e. defassa* in Kenya and northern Tanzania (Figure 6.10). The first indication that admixture may have occurred in this region came from observations of morphologically "intermediate" populations (see Lorenzen *et al.* 2006 for a discussion). More recently, the hypothesis of reticulation has been tested using nuclear and mitochondrial loci. Examinations of mtDNA and microsatellite variation in 11 populations of waterbuck detected genetic divergence between the common and defassa subspecies (Lorenzen *et al.* 2006). Yet, two Kenyan populations reflected apparent introgression between *K. e. ellipsiprymnus* and *K. e. defassa* (Figure 6.10). In the "Nairobi" population (Figure 6.10), the phylogenetic placement of the genotypes was intermediate relative to the common and defassa clades (Lorenzen *et al.* 2006). Furthermore, the "Samburu" population (Figure 6.10) possessed the common waterbuck morphological characteristics, but contained an admixture of genetic markers from both subspecies (Lorenzen *et al.* 2006).

6.2.16 Reticulate evolution and mammals: oryx and sable antelope

The African oryx (*Oryx* spp.) and the sable antelope (*Hippotragus niger*)—two additional members of the African bovid clade—are also widely used as a foodstuff by both indigenous and immigrant *H. sapiens'* populations. For example, Godwin (2000) records the following interaction within a Bushmen village between a shaman and a sick villager's mother "The problem," announces the shaman, on behalf of a dead ancestor, "all began with the gemsbok [*Oryx gazella*]. The one found dead near the village. You ate the meat, but you threw away the intestines and the stomach and the hooves. This was wasteful, and it angered the spirit, so now he will kill the girl."

Oryx and *Hippotragus*, also like kob and waterbuck lineages, reflect genetic admixture via introgressive hybridization. Sable antelope and oryx possess multiple, highly divergent mtDNA lineages (Matthee and Robinson 1999; Pitra *et al.* 2002; Masembe *et al.* 2006) that sometimes co-occur in the same population. For example, Masembe *et al.* (2006) detected three mtDNA lineages in their samples of East African oryx. One of the mtDNA lineages was coincident with the distribution of a single subspecies. In contrast, the two remaining mtDNA classes were found in the same populations, within a second subspecies' range. Masembe *et al.* (2006) inferred ancient introgressive hybridization between two overlapping lineages that had evolved in allopatry, but which had subsequently expanded into a common geographic region.

A similar pattern of mtDNA divergence and admixture has also been described for sable antelope populations. In this case, Pitra *et al.* (2002) defined a "tripartite pattern of genetic subdivision representing West Tanzania, Kenya/East Tanzania and Southern Africa." Of particular significance was the detection of multiple, highly divergent mtDNA lineages in the West Tanzanian region. This finding was once again seen as the result of reticulate evolution involving introgression between various *Hippotragus* genotypic classes. The inference of introgressive hybridization between what could be considered different subspecies of *H. niger* is greatly bolstered by the observation of natural hybrids between sable antelope and the related *Hippotragus equinus* (i.e., roan antelope; Robinson and Harley 1995).

6.2.17 Reticulate evolution and mammals: elephants and wooly mammoths

For *H. sapiens'* hunter-gatherer populations, members of the Elephantidae have represented an extremely attractive, but also difficult and dangerous, windfall of food. For example, this source of animal protein is of primary significance for present-day inhabitants of the forested regions of Africa. Von Meurers (1999, pp. 121–122), a physician who made numerous excursions into these areas, reflected this fact in his account of the following interaction with the Baka people of Cameroon:

A pitiful, half-starved horde of Pygmies gather around us. They have been hunting...elephant with three worn-out single-shot shotguns....Out of each shotgun barrel protrudes the wooden handle of a lance with a fine small blade....The hunter sneaks up under the elephant's belly and fires his lance into it....Then he follows—sometimes for days—until the huge animal becomes sick and weak enough for the final kill.

However, human utilization of members of the family Elephantidae (in the above case involving the African forest elephant, *Loxodonta cyclotis*) for food is not considered to be a recent phenomenon only. Thus, Stuart *et al.* (2002) reviewed evidence for the contribution of ancient human populations to the extinction of the wooly mammoth (*Mammuthus primigenius*) and observed that "Persuasive theoretical models for mammoth extinction continue to be produced, invoking either climatic change, human hunting, or a combination of the two."

Once again, I have reviewed elsewhere (Arnold 2006, pp. 159–160) the evidence indicating a significant impact from introgressive hybridization on the evolutionary history of the two African elephant species, *L. cyclotis* and *Loxodonta africana* (the savannah elephant). Briefly, though these species are easily differentiated on the basis of morphological and genetic characteristics (Roca *et al.* 2001, 2005; Comstock *et al.* 2002), admixtures of both mtDNA and nuclear alleles thought to be diagnostic for one or other of the species have been detected (e.g., Roca *et al.* 2001, 2005). In fact, the level of admixture, and the degree of introgression that it reflects (particularly for mtDNA; Roca *et al.* 2005) has led some to recognize only a single species of *Loxodonta* (Debruyne 2005). Reticulate evolution has thus been a major factor in determining the evolutionary genetics of the African elephant lineages.

As mentioned earlier, the utilization of *extant* Elephantidae lineages by humans does not encapsulate the entire history of associations between our own species and this potential food source. Instead, the hunter-gatherer culture of *H. sapiens* has been invoked as a likely contributor to the extinction of mammoths. Furthermore, in the context of the present discussion, it is significant that lineages of *Mammuthus* also demonstrate patterns of morphological and genetic variation likely caused by introgressive hybridization. One of the first reports of such data (at least to my knowledge) came from an analysis of fossil remains by Lister and Sher (2001). These workers mapped the transitions of fossil forms between the three recognized species, *Mammuthus meridionalis, Mammuthus trogontherii*, and *M. primigenius*. Their analysis led to the following conclusions (Lister and Sher 2001) concerning the transition from *M. trogontherii* to *M. primigenius*: (1) contrary to previous conclusions it was not gradual; (2) this transition involved sympatry between the two species; (3) the sympatry was then followed by replacement of the earlier form; (4) the transition began in Siberia and spread to Europe; and (5) there was a retention of some morphological characteristics of the former species in the transitional period and in *M. primigenius*, consistent with introgressive hybridization (this was also true for the fossils spanning the transition from *M. meridionalis* to *M. trogontherii*).

Higher-order phylogenetic analyses based on genetic information, which included the woolly mammoth and the extant Elephantidae, were also consistent with a role for introgression hybridization. Sequencing of both the entire mitochondrial genome and a set of nuclear genes have made these tests for introgression possible (Capelli *et al.* 2006; Krause *et al.* 2006; Rogaev *et al.* 2006). The analyses of the mtDNA were primarily designed to test the relationships of the African and Asian elephants relative to the wooly mammoth. In this regard, Krause *et al.*'s (2006) and Rogaev *et al.*'s (2006) finding that the Asian elephant (*Elephas maximus*) and the wooly mammoth shared the most recent common ancestor relative to *Loxodonta* seemed to falsify the hypothesis of a closer evolutionary relationship between the African elephant and the wooly mammoth.

Interestingly, Krause *et al.* (2006) emphasized the similarities in the short length of time between the divergence within the *Loxodonta/Elephas/Mammuthus* trichotomy and that within the chimpanzee/human/gorilla clade. This analogy was then extended to suggest that the previous difficulties in resolving the Elephantidae relationships might, like those of the primates, be partially due to a sharing of ancestral alleles across the brief periods between the different divergences (Krause *et al.* 2006). However, as we have already seen (Chapter 3), the *Homo/Pan/Gorilla* clade is also typified by a sharing of allelic variation inferred to be due to genetic exchange via introgression. This is likely the case for the Elephantidae taxa as well. Indeed, a subsequent analysis of nuclear allelic variation found some loci that supported (1) a closer relationship between the Asian elephant and the mammoth, (2) a closer relationship between the African elephant and the mammoth, or (3) neither of these relationships (Capelli *et al.* 2006). As already reiterated for other organisms, a finding of such discordance, in concert with evidence of contemporary hybrid zones (in this case for the African elephant species) strongly supports an inference of reticulate evolution. Not only extant, but also extinct members of the Elephantidae, utilized as sources of food by archaic and anatomically modern *Homo* lineages, appear to have evolved in a reticulate manner.

6.2.18 Reticulate evolution and mammals: pigs

Mona *et al.* (2007) provided a review of the current systematic state of the pig genus, *Sus*. Though the genus has been classified as "Southeast Asian" in origin and posited to contain 7 extant species and over 20 subspecies, there has been continual debate over the evolutionary and taxonomic identity of various lineages (Mona *et al.* 2007). However, one factor that is not debatable is the dietary and economic impact on humans of the domesticated member of this complex, *Sus scrofa*. Indeed, it is well established that the original category of the wild boar (progenitor of the domesticated pig) as a "prey item" for *H. sapiens* switched to its current class of utilization of "human domestic" *c.*9,000 ybp (discussed in Larson *et al.* 2005). One example of the tremendous nutritional and economic impact of domesticated pigs on *H. sapiens'* societies comes from the United States. In 2006, the US ranked second and third worldwide in terms of pork consumption and production, respectively, with the population of domesticated *S. scrofa* consisting of ~60 million animals (USDA 2006).

The confusion over the systematic and evolutionary relationships mentioned above result largely from the phylogenetic discordances resulting from analyses of various morphological and molecular data sets. For example, analyses of the origin of domesticated populations have determined multiple geographic sources for many of the present-day breeds of *S. scrofa* (e.g., Giuffra *et al.* 2000; Kijas and Andersson 2001). Introgression between lineages domesticated from different wild populations has apparently characterized pig husbandry leading to the expected phylogenetic signature of paraphyly (e.g., Robins *et al.* 2006). However, wild and feral species/populations of *Sus*, many of which have likely provided the basis for domesticated lineages, demonstrate the same patterns of paraphyly (Groves 1997; Giuffra *et al.* 2000; Larson *et al.* 2005; Mona *et al.* 2007). Indeed, the introgression among lineages of domesticated and wild *Sus* populations seems to be both extensive and worldwide in its distribution. Once again, it can be concluded that human populations depend on a reticulate assemblage as a major source of nourishment.

6.2.19 Reticulate evolution and mammals: hares and rabbits

The utilization of rabbits and hares as a source of meat has a long history. This is reflected in evidence from ancient and contemporary human activities. Many archaeological sites in both the New and Old Worlds (dated from *c.*12,000 to 1,400 ybp) are replete with the bones from rabbits and hares (Hockett and Bicho 2000; Lubinski 2003), indicative of the importance of these species for the survival of ancient *H. sapiens* populations. However, these organisms continue to occupy a significant niche in present-day human cultures as well. For example, rabbits and hares are main objects of the hunting traditions in many countries. Indeed, concerns that this food resource may be overharvested in some areas of Europe have led to proposals of how and when hunting should take place (Calvete *et al.* 2005). Another indication of modern *H. sapiens'* dependence on this clade of animals as a producer of meat is reflected in its inclusion on the menus of some of the world's finest restaurants (e.g., Bourdain 2004, pp. 94–95, 196–197).

The inclusion of hares and rabbits in this section reflects not only their capacity as an example of a mammalian food source for humans, but also as organisms that have and continue to undergo reticulate evolution. Alves and his colleagues, along with other groups of researchers, have developed the European hare (*Lepus* spp.) into a now classic model system for defining the evolutionary consequences of introgressive hybridization

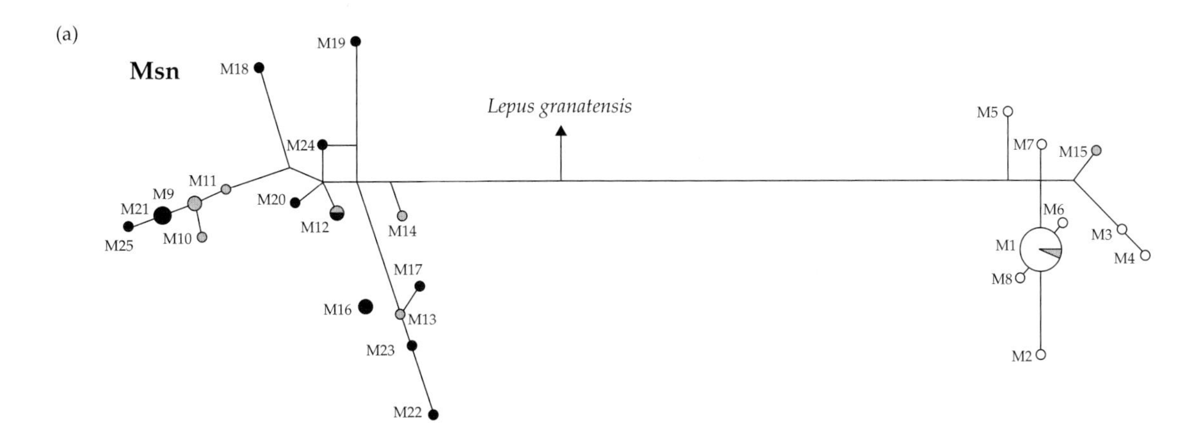

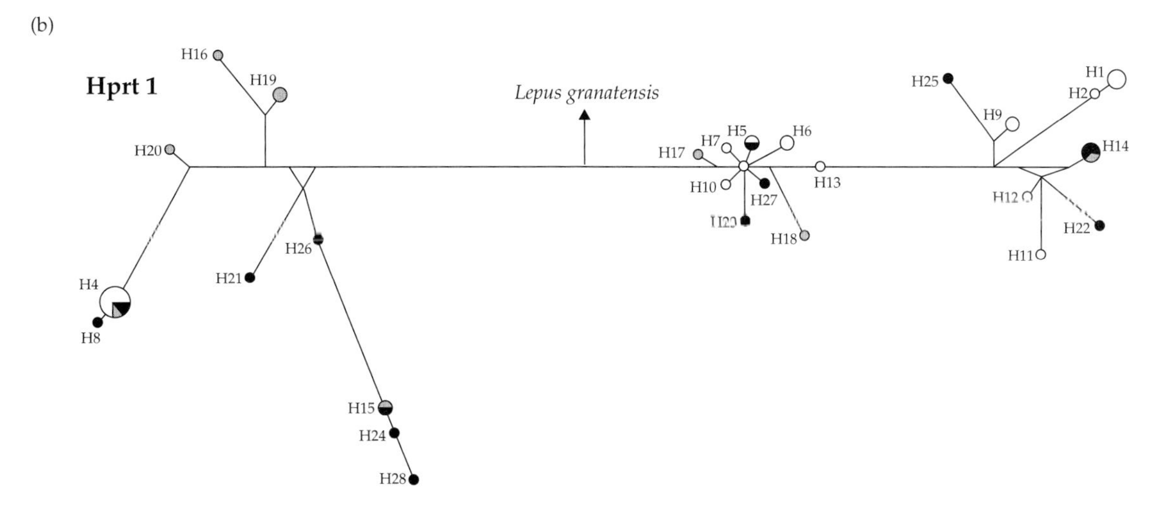

Figure 6.11 Allelic networks constructed from variation in the European rabbit subspecies, *O. c. cuniculus* and *O. c. algirus*. The (a) *Msn* and (b) *Hprt1* loci are located near the centromere and telomere of the X-chromosome, respectively. The size of the circles indicate the relative frequency of each allele. The solid, shaded, and open portions of each circle reflect animals collected from southwestern, central, and northeastern Iberian Peninsula populations, respectively. The shading of the circles also reflects the subspecific and genetic constitution of the populations given that solid = *O. c. algirus*, open = *O. c. cuniculus*, and shaded = the region of overlap (i.e., hybrid zone) between the two subspecies. Introgression of *Hprt1*, but not *Msn*, alleles is indicated by the admixture of alleles of the former, but not the latter, on both sides of the networks (Geraldes *et al.* 2006).

(e.g., Thulin *et al.* 1997; Thulin and Tegelström 2002; Alves *et al.* 2003, 2006, 2008; Melo-Ferreira *et al.* 2005, 2007). As I have reviewed this literature elsewhere (Arnold 2006, pp. 170–172), I will only summarize their findings here. These workers observed: (1) mtDNA introgressive hybridization between the native Swedish species, *Lepus timidus* (the mountain hare) and the introduced brown hare, *Lepus europaeus* (Thulin *et al.* 1997; Thulin and Tegelström 2002); (2) asymmetry in the mtDNA introgression between *L. timidus* and *L. europaeus* (i.e., *L. timidus* > *L. europaeus*); and (3) introgression of mtDNA (possibly reflecting adaptive trait transfer due to climate change) from *L. timidus* into *L. europaeus*, *Lepus granatensis*, and the Cantabrian endemic, *Lepus castroviejoi* in the Iberian Peninsula (Alves *et al.* 2003, 2006, 2008; Melo-Ferreira *et al.* 2005, 2007).

Similar to the European *Lepus* assemblage, the European rabbit (*Oryctolagus cuniculus*) demonstrates both evolutionary diversification and admixture of lineages. First, the two subspecies of the European rabbit—*Oryctolagus cuniculus cuniculus* and *Oryctolagus cuniculus algirus*—are divergent from one another at mtDNA, Y-chromosome and X-chromosome loci (Branco *et al.* 2000, 2002; Geraldes *et al.* 2005, 2006). Second, in areas of spatial overlap between these two divergent lineages genetic admixture has been detected (Branco *et al.* 2000, 2002; Geraldes *et al.* 2006). In regard to the second point, Geraldes *et al.*'s (2006) analysis of X-chromosome variability in the Iberian Peninsula hybrid zone between these two subspecies detected differential introgression of the sampled loci. They determined that the two genes located near the centromere and the two genes located near the telomeres of the X-chromosome introgressed rarely and frequently, respectively. Figure 6.11 illustrates the differential patterns seen at two of these loci—the centromeric locus, *Msn* and the telomeric locus, *Hprt1*. No admixture of empty and solid circles in the left- and right-hand (i.e., *O. c. algirus* and *O. c. cuniculus*) portions of the network, respectively, reflects the lack of introgression of *Msn* alleles between the subspecies (Figure 6.11). In contrast, *Hprt1* alleles from the two subspecies were extensively admixed (Figure 6.11), indicating genetic exchange between *O. c. algirus* and *O. c. cuniculus* (Geraldes *et al.* 2006). These contrasting results indicated that the rabbit genomes were semipermeable to the introgression of alleles from a divergent lineage (Key 1968). Furthermore, the loci affected by introgression indicated the reticulate, genetic constitution of yet another mammal widely utilized by humans for food.

6.3 Summary and conclusions

The major conclusion that can hopefully be drawn from this chapter is that many of the human foodstuffs generated from animals have evolved at least in part as a result of reticulation. Specifically, introgressive hybridization and hybrid speciation have affected our most widely used animal protein sources such as cattle, pigs, tuna, sheep, goats, and trout as well as some of the lesser-utilized species such as peacock bass, chamois, mammoths, and partridges. Indeed, there are few examples of animal species that *H. sapiens* has utilized, or is currently utilizing, that have *not* been impacted by reticulate evolution.

A second, and equally important, contribution to our understanding of the evolution of these food sources is the impact that humans and, in particular their movement of domesticated animals, have had on the catalysis of reticulate evolution. Thus, human migrations involving populations dependent on domesticated animals, included those animal food sources. This sometimes brought the accompanying animals into the range of other domesticated lineages with which the migrant was intentionally or accidentally introgressed. Likewise, the migrating animals were often brought into the ranges of wild taxa related to those lineages from which the domesticated forms were isolated. This too often resulted in genetic exchange. Regardless of the mechanism of these crosses (i.e., human-mediated or not), they enriched the genomes of the animals upon which our species relies. In Chapter 7, I will continue my discussion of the reticulate nature of human food sources, but with the focus turned to examples of plant products. The patterns seen for domesticated animal species are evident for domesticated plant lineages as well.

CHAPTER 7

Reticulate evolution and beneficial organisms—part III

> The original grapefruit biotype originated in the Caribbean, most probably by a natural hybridization between pummelo and sweet orange, perhaps followed by introgression back to pummelo.
>
> (Moore 2001)

> Sect. Petota contains many examples of reticulate evolution: both hybridization at the diploid level...and the formation of allopolyploid species (from triploids to hexaploids), have been reported.
>
> (Volkov *et al.* 2003)

> Genetic diversity of crop species in sub-Sahelian Africa is still poorly documented. Among such crops, pearl millet is one of the most important staple species...Accessions of cultivated pearl millet showed introgressions of wild alleles in the western, central, and eastern parts of Niger.
>
> (Mariac *et al.* 2006)

> Our analyses of spatial transects and temporal series in the New World revealed differential replacement of alleles derived from eastern versus western Europe, with admixture evident in all individuals.
>
> (Whitfield *et al.* 2006)

> Our current view of rice domestication supports multiple domestications coupled with limited introgression that transferred key domestication alleles between divergent rice gene pools.
>
> (Kovach *et al.* 2007)

7.1 Reticulate evolution and the formation of organisms utilized by humans

The previous two chapters have reflected the wealth of examples of organisms that benefit humankind by providing clothing, tools, containers, building materials, food, and much more. In the present chapter, I will conclude the discussion of beneficial organisms by surveying a set of plant species and species complexes from which *H. sapiens* derives nourishment. Thus, I will highlight the reticulate nature of our food sources by discussing data for plants from which we harvest roots, seeds, fruits, stems, or leaves. Of the 20 lineages to be discussed, some reflect major, worldwide crop species such as rice and wheat, while others reflect much more geographically limited species, such as cassava and grapefruit. Similar to the animal examples in Chapter 6, a limited geographic distribution of a plant cultivar does not reflect an unimportant role. Instead, some of the more geographically restricted plant lineages are of primary importance for local populations of humans, in terms of income and/or nourishment.

Humans have not only affected the evolution of plant-derived food sources through the intentional use of native taxa, causing subsequent genetic exchange between wild and cultivated lineages, but they have also inadvertently provided avenues for the evolution of hybrid forms useful for domestication. For example, Hughes *et al.* (2007) tested the hypothesis that "Backyard gardens, dump heaps, and kitchen middens are thought to have provided important venues for early crop domestication via generation of hybrids between otherwise isolated plant species." Edgar Anderson and Ledyard Stebbins constructed the conceptual framework for this hypothesis in a series of publications (Anderson 1948, 1949; Anderson and Stebbins 1954). Specifically, Anderson and Stebbins hypothesized that disturbance, with or without human involvement, could lead to the sympatry of previously separated taxa and the production of [disturbed] habitats that the hybrid progeny could occupy. This "hybridization of the habitat," as Anderson (1948) termed it, could thus lead to both the origin and survival of introgressive genotypes.

Hughes *et al.* (2007) tested for the effects from human-mediated habitat disturbance on the evolution of a south-central Mexico food source, the leguminous tree genus, *Leucaena*. They accomplished this through a combination of genetical, geographical, archaeological, and ethnobotanical analyses. In regard to the role of introgression between wild species in the production of domesticated lineages, their analyses determined that: (1) predomestication cultivation and multiple transitions from wild to cultivated forms led to the sympatry of 13 *Leucaena* species; (2) introgressive hybridization among the various species—brought into sympatric associations through human-mediated disturbances—resulted in numerous recombinant lineages; (3) the degree of introgressive hybridization estimated for this cultivar/wild complex exceeded any other domesticated complex in Mexico; however, (4) other plant species complexes utilized in Mexico, such as *Agave* and *Opuntia*, reflected exactly the same avenues (i.e., "backyard hybridization") through which domestication had proceeded. Therefore, anthropogenic hybridization of the habitat (Anderson 1948) may have played a significant role, not only in Mexican agriculture, but also throughout the history of Mesoamerican crop development (Hughes *et al.* 2007). Indeed, the fundamentally important role that human-mediated genetic exchange has played in the origin of plant lineages utilized as food sources will be illustrated repeatedly in this chapter.

7.2 Reticulate evolution and plant-based food sources

I stated in the introduction to Chapter 6 that reticulate evolution involving animal lineages had been historically treated as a process of minor (if any) relevance (e.g., Mayr 1942, 1963). I also mentioned that this viewpoint is losing ground owing to data. In contrast, with regard to plant lineages, reticulate evolution has nearly always been presented as having a major role in diversification and adaptation (but see Wagner 1970; Mayr 1992). In particular, natural hybridization between divergent plant taxa is well recognized as having produced countless examples of hybrid lineages—both those recognized as "species" and those termed "introgressed" (see Arnold 1997, 2006; Rieseberg 1997; Soltis *et al.* 2003 for reviews). Furthermore, adaptive evolution has similarly been inferred, both in the context of hybrid speciation and introgressive hybridization (e.g., Anderson 1949; Stebbins and Anderson 1954; Arnold 1997, 2006; Rieseberg *et al.* 2003). The outcomes from plant reticulate evolution, like those for microorganisms and animals, are thus seen to include both of the major evolutionary processes—lineage formation and adaptation. Most importantly, for the present discussion, these outcomes have also formed the basis for foodstuffs on which all *H. sapiens* societies depend.

7.2.1 Reticulate evolution and honey

The American Honey Producers Association—in testimony given before the United States Senate Committee on Agriculture, Nutrition and Forestry (AHPA 2007)—provided the following details concerning the crucial effect of *Apis mellifera* (i.e., the honeybee) on the production of human plant-based foods: (1) income from sales of honey during 2006, in the top six producing states alone, exceeded US$80 million; (2) pollination services provided by captive hives are essential for the multibillion dollar agricultural industry; (3) honeybees pollinate >90 crop species as diverse as melons and broccoli (and including nonfood crops such as cotton); (4) crops such as almonds, which are the major agricultural export of California (this state provides 100% of almonds used in the United States and 80% of those used worldwide), depend entirely on *A. mellifera* for pollination; and (5) as of 2006, >140 billion honeybees (i.e., 2 million colonies) were needed to service US crop plants. Applying this formula to global crop production illustrates the vast dependency—for both food and income—of human populations on *A. mellifera*.

The evolutionary origin of honeybees has been pinpointed to the continent of Africa, with subsequent spreading of populations (both with and without human-mediated transport) to Eurasia and the New World (Whitfield *et al.* 2006). Evidence from numerous analyses suggests that the evolutionary history of *A. mellifera* has involved extensive reticulation. With regard to European populations, a phylogenetic analysis by Arias and Sheppard (1996) detected numerous instances of discordance that they attributed to intersubspecific introgression. For example, Italy was a geographic region identified by Arias and Sheppard (1996) as providing "substantial evidence of subspecific introgression." Consistent with this hypothesis, a subsequent population genetic analysis of *Apis mellifera ligustica* (continental Italian samples) and *Apis mellifera sicula* (Sicilian samples) confirmed the hybrid origin of both subspecies (Franck *et al.* 2000).

The best-documented instance of reticulate evolution of *A. mellifera* involves introgressive hybridization in the New World between the earlier introduced European honeybees with populations of the African lineage, *Apis mellifera scutellata*. At least seven European subspecies were introduced into North America alone (beginning in 1622 and continuing for several hundred years; Whitfield *et al.* 2006). Similarly, introductions of subspecies from both western and eastern Europe were made into South America. The introductions of multiple lineages of European origin resulted in admixed populations. However, the greatest genetic exchange was catalyzed by the introduction of African *A. m. scutellata* into Brazil in 1956 (Whitfield *et al.* 2006).

The spread of the African *A. mellifera* lineage into the New World has involved extensive introgression with European taxa (Figure 7.1). So extensive has been the genetic assimilation of the European subspecies by *A. m. scutellata* that the process has been termed "Africanization" and the resulting populations "Africanized" (e.g., Clarke *et al.* 2002; Pinto *et al.* 2004, 2005). For example, of 775 samples from Brazil and Uruguay, >90% possessed mtDNA variants from *A. m. scutellata* (Figure 7.1; Collet *et al.* 2006). Similarly, Mexican honeybee populations sampled by Kraus *et al.* (2007) possessed African mtDNA at frequencies ranging from 67% to 95% and African nuclear alleles at a frequency of ~58%. Finally, a feral honeybee population in southern Texas demonstrated a changeover in genetic composition indicating that a "panmictic European population was replaced by panmictic mixtures of *A. m. scutellata* and European genes within 5 years." (Pinto *et al.* 2005).

7.2.2 Reticulate evolution and root crops: potato

The dependence of human populations on the cultivated potato (*Solanum tuberosum*) can be seen from both a historical and contemporary frame of reference. Historically, one of the best-known migration events had its impetus from a famine caused by the loss of potatoes as a food source. Specifically, the Irish potato famine beginning in the mid-1800s was caused by the fungus, *Phytophthora infestans*, and resulted in >1 million deaths and the immigration of a similar number of people to the United States (Gómez-Alpizar *et al.* 2007). The present-day utilization of potatoes by *H. sapiens* reflects, if anything, an even greater dependence. Thus, in 2006, US farmers produced 43.5 billion pounds of potatoes with a value of US$3.2 billion (Lucier and Jerardo 2007). These values represent record amounts (Lucier and Jerardo

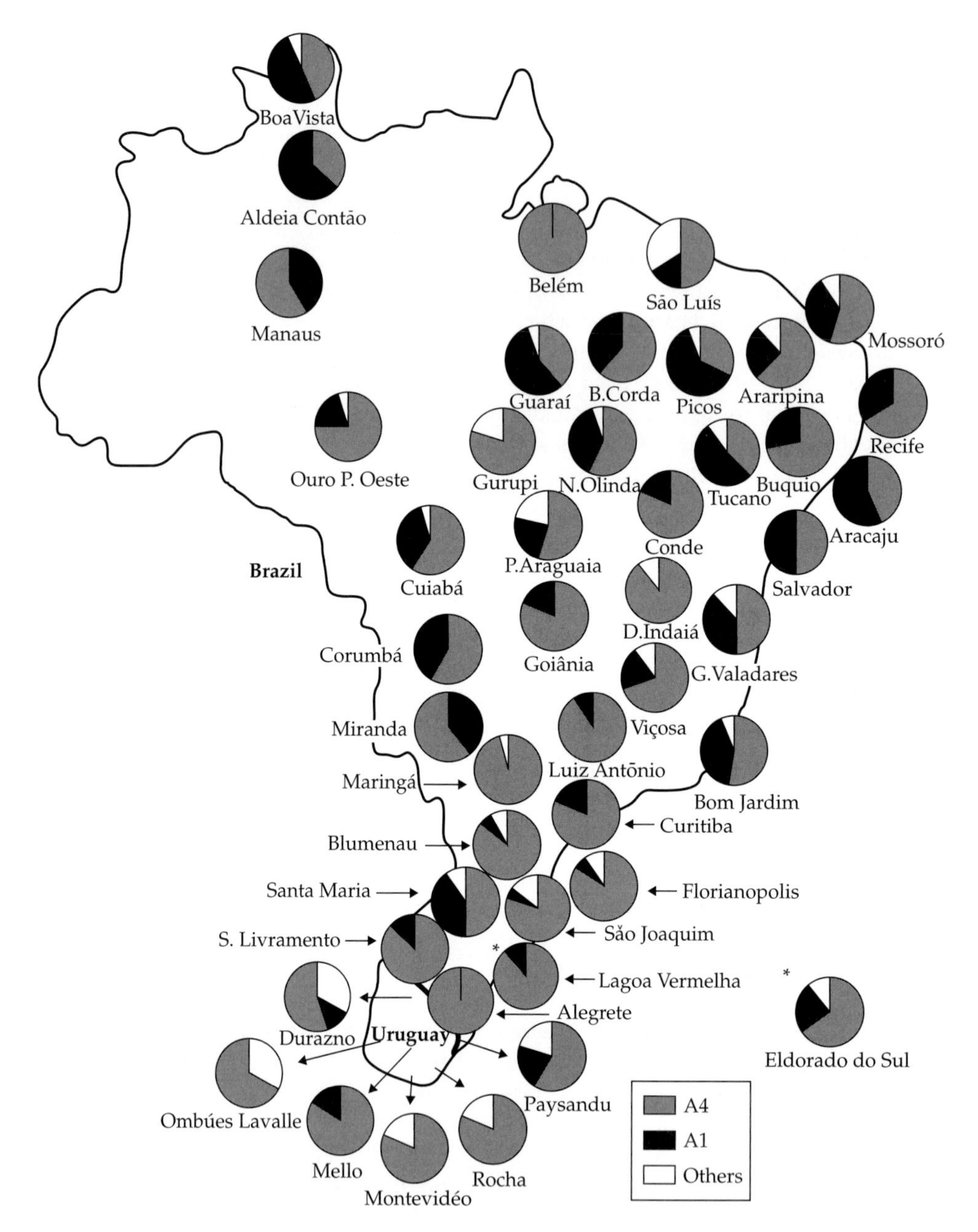

Figure 7.1 Frequency of the African honeybee mtDNA variants, "A1" and "A4", and variants from European lineages (i.e., included in the category "others") in Brazilian and Uruguayan populations. Note that the African subspecies, *A. m. scutellata*, was not present until it was introduced into Brazil in 1956. Thus, the nearly complete replacement of European mtDNA by that of African lineages has occurred in c. 50 years (Collet *et al.* 2006).

2007) and are likely predictive of future increases in worldwide consumption.

The reticulate nature of the clade containing the cultivated potato is apparent from numerous studies. For example, Castillo and Spooner (1997), in their discussion of the difficulties in the taxonomic placement of species belonging to *Solanum* series *Conicibaccata* section *Petota* recognized the potential contribution of genetic admixture due to introgression. Similarly, Rodriguez and Spooner (1997) detected patterns of genetic variability and phylogenetic placement of subspecies of *Solanum bulbocastanum* and *Solanum cardiophyllum* consistent with introgression of cpDNA variants between these species. Volkov *et al.* (2003), as reflected by the quote at the begining of this chapter, also concluded that this assemblage illustrated various outcomes of reticulate evolution. In particular, they reported evidence for both introgressive hybridization (e.g., in the "*Solanum brevicaule* complex") and allopolyploid speciation (including *S. tuberosum*). Huamán and Spooner (2002) provided a review of the evidence supporting the extensive impact of introgressive hybridization and allopolyploid speciation on the origin and evolution of wild and cultivated potato lineages. They envisioned this impact as being facilitated by the (1) growing of divergent cultivars in the same field, (2) co-occurrence of wild and cultivated species, and (3) production of gametes with different ploidy levels in the same species (providing the opportunity for introgression between diploid and polyploid taxa; Huamán and Spooner 2002). The presence of such extensive genetic exchange, and the weak morphological support for the recognition of different lineages resulted in Huamán and Spooner's (2002) classification of all cultivated potatoes into the single species, *S. tuberosum*.

Spooner and his colleagues have provided further tests of the hypothesis that the evolutionary trajectory of the cultivated potato reflects the web-of-life, rather than tree-of-life metaphor. In an analysis designed to ascertain whether the European potato derived from an "Andean" or "Chilean" lineage, Ríos *et al.* (2007) determined the genetic variation in landraces at nuclear and chloroplast loci. Specifically, they examined 19 Canary Island (i.e., the first "European" record of potato introductions), 14 Andean, and 11 Chilean landraces. Though Ríos *et al.* (2007) concluded that the *early* European potato likely had a Chilean origin (but see Ames and Spooner 2008), they also pointed out that the Canary Island source populations were an admixture of both Andean and Chilean lineages. This complex genetic source on the Canary Islands is illustrative of the observation that "modern advanced cultivars have germplasm from over 16 wild species" (Ríos *et al.* 2007). Indeed, the most extensive analysis of landraces to date by Spooner *et al.* (2007) detected limited numbers of definable genetic and taxonomic groupings. This too was reflective of the extent to which introgression and hybrid (allopolyploid) speciation have helped to structure wild and cultivated forms of the potato.

7.2.3 Reticulate evolution and root crops: cassava

In their introduction to a targeted issue of the journal *Plant Molecular Biology*, focusing on cassava (*Manihot esculenta* ssp. *esculenta*; also known as tapioca, manioc, and yuca; Olsen 2004), Fauquet and Tohme (2004) stated that this crop: (1) is a daily staple of >700 million people living in the tropics; (2) is one of the most important calorie sources for these peoples (listed as fourth by Cock 1982 behind rice, sugarcane, and maize); (3) is produced mainly by poor farmers; and thus (4) represents a vital source of both food and income for these producers. Globally, *M. e.* ssp. *esculenta* ranks fourth as a supplier of calories for tropical *H. sapiens* populations, and in Africa it ranks second in this category. Yet, despite the dependence of hundreds of millions of people on this crop, "investment in research to improve cassava has lagged behind that of other staple crops such as rice, wheat, maize, and potatoes" (Fauquet and Tohme 2004). It is tempting to infer that this relative lack of investment reflects not the relative importance of this crop *per se*, but rather a dearth of interest by investors in either the countries or the peoples of low socioeconomic status who depend on this species for sustenance.

Two of the questions of greatest significance for crop species, both from an evolutionary as well as an "improvement" standpoint, is where did they originate, and from what taxon(a)? These two questions are related in that answers to them will reveal possible causes of evolutionary diversification and domestication and possible sources for additional genetic variation for crop improvement. In the case of cassava, the hypothesized sites of origin and the possible progenitor species are

Neotropical (Olsen and Schaal 1999; Olsen 2004). Though postulated by many to be a hybrid derivative from multiple wild taxa, it now appears that *M. e.* ssp. *esculenta* instead may have been domesticated from the wild species, *Manihot esculenta* ssp. *flabellifolia* in the southern Amazon basin. Olsen and Schaal (1999) and Olsen (2004) arrived at this inference from surveys of genetic variation at nuclear loci.

Notwithstanding the single point of origin and nonhybrid derivation of *M. e.* ssp. *esculenta*, introgressive hybridization postdomestication between this crop and other *Manihot* lineages has been detected in numerous geographical locales. For example, Lefèvre and Charrier (1993) examined nuclear gene variation (using isozyme analysis) within 365 *M. esculenta* cultivars and 109 samples from "wild" (i.e., consisting of *Manihot glaziovii* and natural hybrids between this species and the crop species) populations from Africa, South America, and India. The results from this analysis identified "spontaneous" hybrids between the two *Manihot* species in Ivory Coast (Lefèvre and Charrier 1993). This introgression was human mediated in that both species were introduced into Africa—*M. esculenta* as a food source and *M. glaziovii* as a source for rubber production (Lefèvre and Charrier 1993).

In addition to the introgression detected in the Old World, genetic exchange between cassava and wild *Manihot* taxa has also taken place very frequently within the Neotropics. For example, a study of microsatellite and morphological variation of populations in French Guiana determined that (1) hybrid zone formation between wild and cultivated forms was relatively common and (2) exchange had often been asymmetric and largely in the direction of crop > wild (Duputié *et al.* 2007). Similarly, Nassar (2003) also inferred crop to wild gene flow between cassava and a number of naturally occurring *Manihot* species. Though there may be a bias in directionality, the formation of hybrid zones, as with any taxa, provides a bridge for genes to move from wild taxa into the cultivar as well. This is particularly to be expected when farming practices include the incorporation of volunteer seedlings into the cultivated fields; some of the seedlings will almost certainly have originated from crosses between divergent cultivar lineages, or between cultivars and nearby wild species (Elias *et al.* 2001). This was indeed the inference drawn from a genetic analysis of 117 accessions of cassava from Brazil and Guyana (Elias *et al.* 2004). The detection of alleles in the cultivars, found previously only in wild species, resulted in the conclusion that there had been wild > crop introgression. Elias *et al.* (2004), as in their earlier study, inferred the causal role of farming practices that included the use of volunteer seedlings in this pattern of introgression. Thus, for cassava, it is likely that enrichment from its wild relatives will continue to occur at least in geographic regions inhabited by human populations with relatively "unmanaged" agricultural practices (Elias *et al.* 2001; Duputié *et al.* 2007).

7.2.4 Reticulate evolution and root crops: yams

Yams, genus *Dioscorea*, evolved in three subhumid/humid tropical regions—Southeast Asia, West Africa, and the New World Tropics (Tamiru *et al.* 2007). Sub-Saharan Africa accounts for > 95% of yam production, and *Dioscorea* is the second most important African tuber crop after cassava (Scarcelli *et al.* 2006a; Tamiru *et al.* 2007). Indeed, in the densely populated areas of countries such as Ethiopia, this crop, like cassava in other areas, is a crucial factor in the livelihood of the resident human populations (Tamiru *et al.* 2007). Though produced throughout Sub-Saharan Africa, more than 90% of world yam production occurs in West Africa (Scarcelli *et al.* 2006a). The West African cultivars are recognized as belonging to the "*Dioscorea cayenensis–Dioscorea rotundata* complex," and are found to be in geographic proximity with wild species such as *Dioscorea abyssinica, Dioscorea praehensilis*, and *Dioscorea burkilliana* (Mignouna and Dansi 2003; Scarcelli *et al.* 2006a).

Two factors have facilitated a significant role for introgressive hybridization between cultivated and wild *Dioscorea* species in the evolution of the domesticated yam. The first, mentioned above, is the sympatry between the cultivars and numerous native taxa. Such sympatry has provided the opportunities for wild × domesticate hybrid formation

(Scarcelli *et al.* 2006a, b). However, the importance of these hybrids in affecting reticulate evolution of yams has been enhanced by particular agricultural practices by local farming communities. This practice, termed "ennoblement" by Mignouna and Dansi (2003) involves the incorporation of tubers from wild yams into the cultivated fields. The tubers selected from the wild and planted into the cultivated fields are referred to as "pre-ennobled" (Figure 7.2; Scarcelli *et al.* 2006b). One example of this process involves the Fon and Nago agriculturalists in Benin. These farmers collected tubers, either close to their villages or on hunting trips to more distant areas, which they transplanted into their fields. While these pre-ennobled plants were sometimes genetically and morphologically indistinguishable from known landraces (Mignouna and Dansi 2003), in some cases they were novel with regard to both genotype and phenotype (Mignouna and Dansi 2003).

Scarcelli *et al.* (2006a,b) tested the hypothesis that a proportion of the *Dioscorea* pre-ennobled plants utilized by indigenous farmers in Benin were actually crop x wild hybrids. Both of their studies did indeed provide evidence for the utilization by farmers of natural hybrids between the cultivated and wild species (Figure 7.2). In the first study, this was reflected by the nuclear genotypes of the wild-collected plants being equally closely related to the cultivar lineage and to wild *D. praehensilis* (Scarcelli *et al.* 2006a). The second, more detailed analysis (Scarcelli *et al.* 2006b), involved genotyping of wild and cultivated plants within villages and across different geographic regions. At the village level, >50% of the pre-ennobled yam samples were found to be of hybrid origin (Scarcelli *et al.* 2006b). In addition, a paternity analysis estimated that 77.4% of seeds collected from cultivated plants had interspecific (i.e., wild species) fathers, indicating the high frequency occurrence of spontaneous interspecific hybridization onto cultivated plants (Scarcelli *et al.* 2006b).

Scarcelli *et al.* (2006b) detected an equally large effect from introgression when all samples of the various wild and cultivated (both pre-ennobled and previously domesticated plants) lineages were analyzed. Figure 7.2 illustrates the genetic assignment of the *Dioscorea* samples identified before genotyping as belonging to *D. abyssinica*, *D. praehensilis*, the cultivar (*D. rotundata*), or the class of pre-ennobled yams. The genotypic constitution of these plants indicated that 7.5% of all individuals assigned morphologically to one of the three species had conflicting genotypes (i.e., hybrid

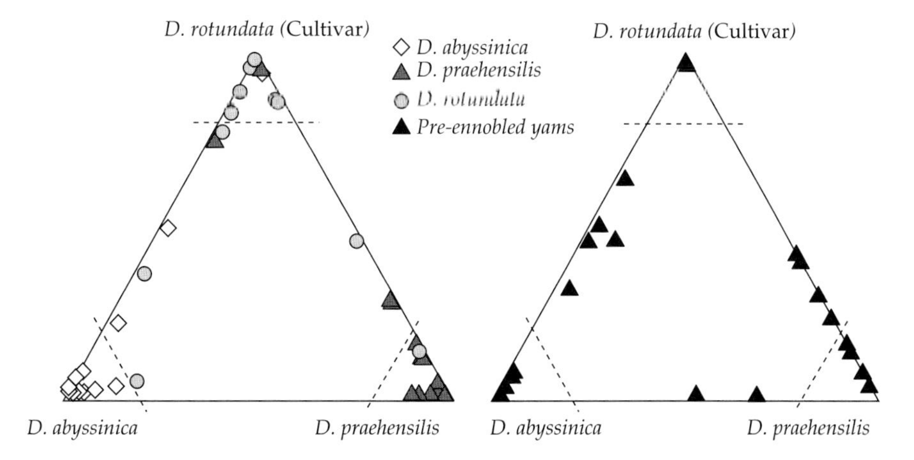

Figure 7.2 Genetic assignments of wild/previously domesticated yams (left-hand diagram) and pre-ennobled yams (right-hand diagram) using nuclear microsatellite loci. *D. abyssinica* and *D. praehensilis* are wild species and *D. rotundata* is the cultivated lineage. The dashed lines indicate the cutoff for assignment to one of the three categories. The samples that fall outside of the corners demarcated by these lines (in both the left- and right-hand diagrams) are inferred to be various hybrid genotypes (Scarcelli *et al.* 2006b).

or assigned to an alternate species) and that 37% of the pre-ennobled samples were hybrids (Figure 7.2; Scarcelli *et al.* 2006b). Both the sympatry of the cultivated and wild *Dioscorea* species and the West African indigenous farming practices have allowed an enrichment of a key foodstuff. However, this type of process is not isolated to yam cultivation. Jarvis and Hodgkin (1999), in a review of numerous examples of domestication, indicated not only the prevalence, but also the fundamental importance of wild > crop introgression when they stated:

> The information reviewed here has shown that introgression between crop cultivars and their wild relatives is an ongoing process affecting the genetic diversity of crops today…The availability of large gene pools becomes even more essential as farmers need to adapt over time to changing conditions that result from new population pressures, land degradation, and environmental change.

7.2.5 Reticulate evolution and seed crops: pearl millet

Pearl millet, *Pennisetum glaucum*, is one of the most important crop species of the Sahel region of Africa. The Sahel is primarily savanna-like habitat extending for ~2,400 miles from the Atlantic Ocean to the Horn of Africa and crossing the countries of Senegal, Mauritania, Mali, Burkina Faso, Niger, Nigeria, Chad, Sudan, and Eritrea (Mariac *et al.* 2006; http://en.wikipedia.org/wiki/Sahel). The significance of this crop can be illustrated by the fact that in Niger alone, over 65% of the cultivated land is occupied by *P. glaucum* (Mariac *et al.* 2006). Its importance is also apparent from the antiquity of the domestication process leading to the present-day cultivars. Specifically, (1) by "the early second millennium BC widely dispersed populations in West Africa were cultivating morphologically domesticated pearl millet" and (2) "pearl millet had dispersed to India by approx. 1700 BC" suggesting "that this process began by the third millennium BC and saw rapid dispersal of cropping across the northern savannas" of Africa (Fuller 2007). The final portion of this quote indicates that human populations located in the semiarid Sahel region of Africa are not the only *H. sapiens* to depend on pearl millet for sustenance. Indeed, *P. glaucum* is "the staple food of the semiarid state of Rajasthan in northwest India" (vom Brocke *et al.* 2003). In addition to its crucial role as a plant-derived food source for humans, *P. glaucum* also provides fodder for domesticated animals in dryer climates (e.g., Namibia; http://www.fao.org/ag/agp/agpc/doc/counprof/namibia/namibia.htm) thus indirectly providing sources of animal protein as well.

Though some degree of postzygotic, reproductive isolation has been demonstrated in crosses between wild (*P. glaucum* subsp. *monodii*) and cultivated (*P. glaucum* subsp. *glaucum*) pearl millet (Amoukou and Marchais 1993), bidirectional introgression has also been detected. For example, Busso *et al.* (2000) tested the effect of farmer behavior—in terms of crop management—on the levels of cultivar genetic diversity. Using nuclear loci, these workers examined the diversity within and among the plants grown by individual farmers in two villages in northeastern Nigeria. Interestingly, Busso *et al.* (2000) concluded from their data that the individual cultivated plots were relatively isolated from one another due to the local agricultural practices. In contrast, genetic exchange was detected between wild *P. glaucum* subsp. *monodii* and the cultivar, *P. glaucum* subsp. *glaucum*. This was caused by the farmers' practice of harvesting seeds from volunteer plant types known as "Lafsir." This category included plants known to be wild × cultivar hybrid derivatives (Busso *et al.* 2000).

A second analysis, by Mariac *et al.* (2006), also provided estimates of the diversity in wild and cultivated populations of pearl millet. In this case, the country from which samples were collected was Niger. The patterns of genetic variation at 25 nuclear (microsatellite) loci for 46 *P. glaucum* subsp. *monodii* and 421 *P. glaucum* subsp. *glaucum* accessions led to the following conclusions: (1) introgression was bidirectional between cultivated and wild populations; (2) between 4.3% and 1.4% of wild and cultivar samples, respectively, possessed introgressed alleles; and (3) cultivated populations from which introgressed genotypes were collected occurred in western, central, and eastern Niger (i.e., throughout the entire area sampled; Mariac *et al.* 2006).

7.2.6 Reticulate evolution and seed crops: wheat

"With 620 million tons produced annually worldwide, wheat provides about one-fifth of the calories consumed by humans...Roughly 95% of the wheat crop is common wheat, used for making bread, cookies, and pastries, whereas the remaining 5% is durum wheat, used for making pasta and other semolina products." This was the description given by Dubcovsky and Dvorak (2007) of the essential role of *Triticum aestivum* as a plant-based food generator for *H. sapiens*. The crucial economic importance of this crop species can be illustrated as well by the following figures for the 2006/2007 wheat harvest of the United States (one of the major producer countries—exceeded only by China, the European Union and, in some years, India; http://www.ers.usda.gov/Briefing/Wheat). Just over 57 million acres of wheat were planted, from which ~1,800 million bushels of wheat were harvested, resulting in an income of ~US$8 billion (http://www.ers.usda.gov/Data/Wheat/YBtable01.asp).

The origin and continued evolutionary trajectory that resulted in *T. aestivum* reflects well the processes of reticulation and the web-of-life metaphor. Furthermore, reticulate processes continue to affect the evolution of the species complex that includes common wheat. Returning to the review of Dubcovsky and Dvorak (2007), the following were listed as steps in the evolution of allopolyploid (i.e., allohexaploid) *T. aestivum*: (1) ~10,000 ybp, in southeastern Turkey, "einkorn" (*Triticum monococcum*) and "emmer" (*Triticum turgidum* ssp. *dicoccon*) wheat were domesticated; (2) both of these species still exist and their genomic constitutions are known to differ, with the former being a diploid (genome—"AA") and the latter an allotetraploid (genome—"BBAA"); (3) these species of "hulled" wheat were spread by humans through Asia, Europe, and Africa; (4) the spread of emmer wheat brought it into contact with a second, wild, subspecies (*Triticum turgidum* ssp. *dicoccoides*) resulting in introgression; (5) sympatry of emmer wheat with *Aegilops tauschii* (genome—"DD") occurred through the Northeastern spread of emmer from its center of domestication; and (6) "5" led to the allopolyploid formation of *T. aestivum* (genome—BBAADD) from hybridization between emmer wheat and *A. tauschii*.

As outlined above, the steps leading to the evolution of common wheat were marked by numerous instances of allopolyploid hybrid speciation (see Kilian *et al.* 2006 for an additional description of this process). Two recent studies, also by Dvorak's group, illustrate the extensive effects from introgressive hybridization as well. First, as mentioned above, tetraploid (i.e., emmer wheat) lineages of domesticated *Triticum* contributed to the formation of hexaploid common wheat through hybridization with *A. tauschii*. Yet, these emmer wheat lineages had undergone previous introgression. In particular, Luo *et al.* (2007) found that domesticated emmer wheat possessed a genetic structure like that found in wild emmer wheat populations—that is, they detected two diverged lineages of domesticated emmer just as is found in the wild populations. However, in addition to the patterns of genetic variation indicating two divergent lineages, Luo *et al.* (2007) also detected genomic mosaicism indicative of introgression between the wild and cultivated tetraploid wheat. Similarly, Dvorak *et al.* (2006), in an analysis of nuclear gene variation in tetraploid wheat lineages and hexaploid, common wheat, detected not only the contribution of an entire genome to the formation of *T. aestivum*, but also postorigin introgression from wild tetraploids into the hexaploid domesticate. This latter finding indicated that both domesticated and wild emmer wheat had contributed to the evolution of *T. aestivum* (Dvorak *et al.* 2006).

7.2.7 Reticulate evolution and seed crops: Eragrostis

Eragrostis tef (commonly known as "tef") is a cereal crop species utilized for making bread termed "taita" or (in Amharic) "injera." This food product is a staple throughout Ethiopia and Eritrea (Ingram and Doyle 2003). In Ethiopia, *E. tef* cultivation occupies the greatest area (~2.17 million hectares) and ranks third in total production of any cereal. Though tef is primarily used for flour, it also provides starting material for porridge and the alcoholic beverages, "tella" and "katikala" (Teklu and Tefera 2005). Tef's utility also includes other, food-related, cultural practices. For example, "Injera is not only a kind of bread—it's also an eating utensil. In Ethiopia and Eritrea, this spongy, sour flatbread is used to scoop up meat and vegetable stews. Injera also lines the tray on which the stews are served, soaking up their juices as the meal progresses. When this edible tablecloth is eaten, the meal is officially over" (http://www.exploratorium.edu/cooking/bread/recipe-injera.html). Finally, the production of injera from *E. tef* flour reflects not only the utilization of indigenous plant species, but also cultural constraints and the use of technologies specially adapted for the available plant products. Lyons and D'Andrea (2003) thus argued that the adoption of cooking techniques utilizing griddles, rather than ovens, was partially explained by the domestication of indigenous grains lacking in gluten (like tef); women applied the food preparation technology best suited to their species-of-choice (Lyons and D'Andrea 2003).

Domesticated *E. tef* belongs to a genus of ~350 species (Ingram and Doyle 2004). Though the monophyly of this clade has been questioned, Ingram and Doyle (2004) concluded that all species of *Eragrostis* did indeed share a most recent common ancestor. In addition, members from several other genera (i.e., *Acamptoclados, Diandrochloa, Neeragrostis,* and *Pogonarthria*) also fell within the *Eragrostis* assemblage. Of particular significance for the present discussion is the evidence for frequent reticulate evolution during the diversification of *Eragrostis*. Specifically, numerous taxa have arisen via allopolyploid speciation. *Eragrostis macilenta, Eragrostis minor, Eragrostis mexicana,* and *Eragrostis cilianensis* are allopolyploids that appear to be phylogenetically related (Ingram and Doyle 2003). However, the finding that *E. cilianensis* shares only one of its genomes with the other three species—with the other genome in this allotetraploid derived from an unrelated diploid progenitor—indicates the complexity of the evolutionary history of this set of *Eragrostis* allopolyploids (Ingram and Doyle 2003).

The most significant results for this section are those indicating the involvement of reticulate events in the evolution of *E. tef*. These include the following observations (Ingram and Doyle 2003): (1) this species has also been shown to be an allopolyploid derivative (i.e., allotetraploid; $2n$ = 40 chromosomes); (2) tef is evolutionarily closely allied with the wild allotetraploid, *Eragrostis pilosa*; and (3) one of the tef genomes is closely related to the tetraploid (likely allotetraploid) species, *Eragrostis heteromera*. Thus, the evolutionary history of a staple Ethiopian/Eritrean food and cultural element involved reticulate speciation.

7.2.8 Reticulate evolution and seed crops: peanut

Raina and Mukai (1999) list the peanut, *Arachis hypogaea*, as the third most important grain legume crop in the world. One indication of its utility as a food source for *H. sapiens* populations is reflected by the fact that Americans consume nearly 1.1 billion kg of peanuts per year (Barkley *et al.* 2007). Raina and Mukai (1999) supported their contention by citing *A. hypogaea*'s cultivation across divergent agricultural practices in Asia, Africa, and North, Central and South America (the latter being its place of origin; reviewed by Kochert *et al.* 1996 and Jung *et al.* 2003), reflective of its adaptation to a diverse array of tropical and subtropical environments.

A. hypogaea's fundamental importance for humans resides not only in its use as a direct source of food, but also because it functions as (1) a feedstock for domesticated animals, (2) a source of high-quality oil, and (3) a ground cover species (Raina and Mukai 1999). This species is also of economic significance for producer countries. For example, the United States is the third largest exporter of peanuts after China and Argentina (Dohlman and Livezey 2005). Peanuts represent a relatively small share of total agricultural production in the United States, yet in 2004, income from this crop was nearly US$800 million (Dohlman and Livezey 2005). Its importance can be seen particularly at the state level. (*A. hypogaea* cultivation takes place largely in the states of Georgia, Alabama, Florida, South Carolina,

Texas, Oklahoma, New Mexico, Virginia, and North Carolina.) Thus, although less than 1% of US crop income derives from *A. hypogaea* it produced the second highest amount of crop revenue in Georgia and was also a leading income producer for Alabama and Florida (Dohlman and Livezey 2005).

Section *Arachis*, of the genus *Arachis*—in which cultivated peanut belongs—contains 31 species (Fávero *et al.* 2006). Of these 31 species, 29 are diploid taxa and 2 are allopolyploid derivatives; the 2 allotetraploids are *Arachis monticola* and *A. hypogaea*. A variety of analyses have been performed to infer the diploid parents for *A. monticola* and *A. hypogaea*, but with most research directed at understanding the derivation of the cultivar (Kochert *et al.* 1996; Raina and Mukai 1999; Jung *et al.* 2003; Fávero *et al.* 2006). For example, Kochert *et al.* (1996) utilized chromosome markers along with patterns of genetic variation at both nuclear and chloroplast loci to decipher evolutionary relationships among the diploid and tetraploid species. These data led to the conclusion that the cultivated lineage is derived from hybridization between the diploid species *Arachis duranensis* and *Arachis ipaensis* (Kochert *et al.* 1996). Results from this analysis also suggested that the most likely geographic point of origin was northern Argentina or southern Bolivia (Kochert *et al.* 1996). Finally, the maternally inherited cpDNA was donated by plants from *A. duranensis*, indicating this lineage's role as the female parent in the original cross with *A. ipaensis* (Kochert *et al.* 1996). Subsequent studies that examined additional genetic loci and progeny from artificial crosses had, in the main, validated Kochert *et al.*'s (1996) conclusions concerning the progenitor genomes found in cultivated peanut. In contrast, it has recently been inferred that the allotetraploid *A. monticola* lineage was the direct progenitor of allotetraploid *A. hypogaea* (Seijo *et al.* 2007). Regardless of whether or not the cultivar was formed independently from crosses between diploid taxa, or arose directly from another allopolyploid species, the New World derivation of the plant that provides the highly nutritious peanut oil, butter, and seeds (Barkley *et al.* 2007) provides yet another example of processes that underlie the web of life.

7.2.9 Reticulate evolution and seed crops: rice

The rice genus, *Oryza*, contains 21 wild species and 2 cultivated species that possess 9 recognized categories of genomes (Ge *et al.* 1999). The rice cultivars are of immense importance as a plant food source for humans. The domestication of the most widespread cultivars, *Oryza sativa indica* and *Oryza sativa japonica*, occurred between 9,000 and 10,000 ybp somewhere in the region "extending from eastern China, through the foothills of the Himalayas in Vietnam, Thailand, and Myanmar, to eastern India" (Kovach *et al.* 2007). The progenitor for both of the *O. sativa* subspecies is now recognized as being "Asian common rice," *Oryza rufipogon* (Kovach *et al.* 2007). The domestication of the most commonly utilized rice lineage (i.e., *O. sativa*) thus involved at least two independent events (Londo *et al.* 2006).

Khush (1997) listed the following figures to substantiate his claim that cultivated rice (mainly *Oryza sativa*, but also including the African cultivar, *Oryza glaberrima*) is "the world's single most important food crop": (1) it was the primary food source for greater than one-third of humans; (2) greater than 90% of the rice harvested was in Asia where about 60% of humans resided; (3) rice provided 35–60% of the caloric intake of >3 billion Asians; (4) rice cultivation occurred on about 148 million hectares (i.e., ~11% of all cultivated land); (5) rice was the only significant crop of its kind that was consumed almost entirely by *H. sapiens*; and (6) in 1996 rice production worldwide exceeded 550 million tons, with the major producers being China, India, Indonesia, Bangladesh, Vietnam, Thailand, and Myanmar. Given an annual population increase of ~2% in rice consuming countries, Khush (1997) estimated a need for rice production to reach 850 million tons by the year 2025 (mainly through genetic improvement rather than increases in the area of cultivation).

Both allopolyploidization and introgressive hybridization characterize the 23 species that make up the *Oryza* assemblage (Ge *et al.* 1999). At least seven of the species are known to be allopolyploid derivatives from crosses between various diploid lineages (Figure 7.3). However, the

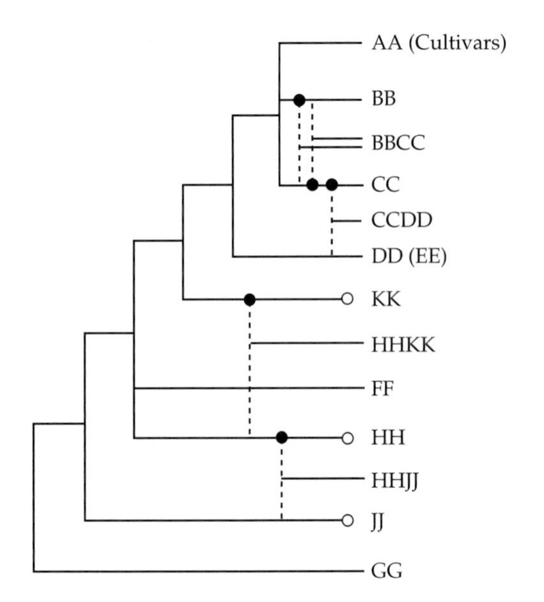

Figure 7.3 Phylogeny reflecting relationships among the various diploid species of *Oryza* and the derivation of the allotetraploid lineages. Dashed lines indicate the combinations of the diploid lineages leading to the origins of the allotetraploid species; ● indicates maternal parents; and ○ indicates diploid genomes not yet identified in extant species of *Oryza* (Ge *et al.* 1999).

clade that contains both Asian and African forms of cultivated rice, as well as their progenitors (Khush 1997), contains only diploid taxa possessing the "AA" genome constitution (Figure 7.3; Ge *et al.* 1999). Yet, the evolutionary trajectory of this clade has involved repeated bouts of introgressive hybridization. An analysis of genomic regions showing either elevated or reduced levels of sequence divergence led Tang *et al.* (2006) to infer extensive admixture among the *Oryza* species belonging to the progenitor/cultivar assemblage. In particular, these authors concluded that the AA-genome diploid complex demonstrated extensive signatures of introgression between divergent lineages (Tang *et al.* 2006).

Cheng *et al.* (2003) concluded that the interspersed phylogenetic placements of *O. sativa* and *O. rufipogon* lineages were likely due to "gene flow that may have occurred spontaneously among the strains of [these] closely related species." However, in addition to gene flow between progenitor (*O. rufipogon*) and derivative (*O. sativa*), there has also apparently been introgression between the different cultivars. In particular, Kovach *et al.* (2007) reviewed evidence indicating the transfer of allelic variation between *O. s. indica* and *O. s. japonica* at several key "domestication" loci. These included loci affecting grain shattering (*Sh4*), seed coat color (*Rc*), and amylose content (*Wx*). The introgression of these—and most likely other loci as well (Kovach *et al.* 2007)—between *O. s. indica* and *O. s. japonica* thus reflects the human-mediated incorporation of traits that increase the quality of this key source of food for *H. sapiens*. Intriguingly, Tang *et al.* (2006) suggested that this human-mediated process might have involved utilization of naturally occurring hybrids as the starting material for the domesticated lineages.

7.2.10 Reticulate evolution and seed crops: maize

Maize (*Zea mays mays*) domestication is thought to have occurred *c.*9,000 ybp in the southern Mexican highlands (Matsuoka *et al.* 2002). Cock (1982) listed maize as the third and fourth most important calorie source for tropical and worldwide human populations, respectively. Indeed, he listed 600 billion kilocalories/day as the amount of energy generated for *H. sapiens* from this crop alone. Global production of maize in 2006/2007 was greater than 700 million metric tons (World Agricultural Outlook Board 2007). Furthermore, it is predicted that the United States will export a record 2.45 billion bushels during 2007/2008 (World Agricultural Outlook Board 2007). For the United States, the extremely large economic benefit from *Z. m. mays*, is reflected in its 2006 value as a grain crop of ~US$34 billion (National Agricultural Statistics Service 2007a). Each of these facts reflects the integral role played by maize in human nutrition and economic development (including its capacity as a source of biofuel).

Reticulate evolutionary processes, as in many of our other examples of human-food-source evolution, affected both the origin and ongoing development of the maize lineage. The formation of *Z. m. mays* was affected *c.*11 mya by hybridization between two divergent taxa, leading to an allopolyploid derivative (Gaut and Doebley 1997; Wei *et al.* 2007). Interestingly, this allopolyploidization event was inferred to be "segmental" in that the progenitor genomes would have been similar enough to allow some of the homologous chromosomes to pair correctly during meiosis (Gaut and Doebley 1997). Furthermore, the number of chromosomes possessed by both progenitors has been estimated to be $n = 10$ (Wei *et al.* 2007). Given allo[tetra]polyploidization, the derivative lineage leading to *Z. m. mays* should have possessed $n = 20$ chromosomes (i.e., twice the number in each of its progenitors due to doubling during the allotetraploidy event). Instead, maize has $n = 10$ chromosomes like its putative parents. Wei *et al.* (2007) suggested that "the progenitor chromosomes underwent breakage and fusion and eventually reassembled into ten new mosaic chromosomes that are genetically diploid as opposed to being allotetraploid."

In addition to the ancient (segmental-allopolyploid) formation of the progenitor of *Z. m. mays*, further reticulate evolution is ongoing, as evidenced by natural hybridization between the cultivar and its Teosinte relatives (studies reviewed by Baltazar *et al.* 2005 and Ellstrand *et al.* 2007). Though barriers to introgression are present between Teosinte taxa and maize, these do not prevent some level of genetic exchange. Thus, Baltazar *et al.* (2005) and Ellstrand *et al.* (2007) detected hybrid formation in experiments in which *Z. m. mays* was paired with Teosinte plants. In particular, these studies resulted in the inference of greater potential for introgressive hybridization between maize and the Teosinte that was most likely its direct progenitor, *Zea mays parviglumis* (Clark *et al.* 2004). Furthermore, probably owing to divergent floral characteristics, Baltazar *et al.* (2005) concluded that it was much more likely for hybridization to be in the direction of Teosinte > maize, rather than the reverse. Consistent with this conclusion was the detection of *Z. m. parviglumis* alleles in *Z. m. mays* plants at the "domestication" locus, *tb1* (Clark *et al.* 2004). Given that this latter finding reflects introgression, it parallels the data for rice in which the spread of alleles important for cultivation was human mediated.

7.2.11 Reticulate evolution and seed crops: pea

Ofuya and Akhidue (2005) reviewed information concerning the production and nutritional value of the legume species known as "pulses"; this terminology refers to all of the species of legumes grown for the harvest of their dried seeds. In particular, the data reviewed by Ofuya and Akhidue (2005) showed the former Soviet Union to be the major producer (7,800 metric tonnes annually) of the common pea, *Pisum sativum*. This finding, though outdated in terms of present-day countries, reflects the ongoing, significant value of this crop as a food source from Europe to Eastern Asia. Indeed, the nutritional value of *P. sativum* is reflected in its ranking as one of the top pulses in terms of total carbohydrates and dietary fiber content (Ofuya and Akhidue 2005). Furthermore, once again using the US agricultural system as an indicator of crop value, the 2006 *P. sativum* harvest netted more than US$80 million (National Agricultural Statistics Service 2007a). Though small compared to receipts from such crops as wheat or maize, the income from pea cultivation still reflects a significant economic benefit to many local US communities. Given the ten-fold or higher production outlay of *P. sativum* in other parts of the world (Ofuya and Akhidue 2005), the cultivation of this species plays an even greater role in maintaining the health (both physical and economic) of European and Asian societies.

Weeden (2007) discussed data concerning the domestication history of pea. *P. sativum* is native to the region of the Middle East between Turkey and Iraq, with *Pisum sativum elatius* considered to represent the most likely wild progenitor for pea (Weeden 2007). From its geographic point of origin, more primitive lineages were transported throughout the world, with the last iteration in this domestication process reflected "by the modern cultivars of the fresh market pea" (Weeden 2007). Findings from both Vershinin *et al.* (2003) and Jing *et al.* (2005) are consistent with the hypothesis that the events associated with pea domestication included introgressive hybridization. Both of these studies utilized transposable elements to test for underlying processes affecting the origin and evolution of *Pisum* cultivars. From these analyses came inferences of reticulate evolution in the derivation of both *P. sativum* and *Pisum abyssinicum* (often called *Pisum sativum abyssinicum*; Weeden 2007). Specifically, the genomic locations of the transposable markers (Vershinin *et al.* 2003; Jing *et al.* 2005) suggested the following: (1) *P. abyssinicum* was a separate species relative to the other cultivar, *P. sativum*; (2) the two cultivars reflected independent domestication events; (3) *P. abyssinicum* was likely formed through hybridization, with one of its progenitors being *Pisum fulvum*; and (4) the high level of polymorphic loci in *P. sativum*, relative to *P. abyssinicum*, was consistent with a greater impact from introgression in its formation, particularly from the species, *P. elatius*.

7.2.12 Reticulate evolution and seed crops: cowpea

> Cowpea is the most economically important indigenous African legume crop . . . Cowpeas are of vital importance to the livelihood of several millions of people in West and Central Africa. Rural families that make up the larger part of the population of these regions derive from its production, food, animal feed, alongside cash income. (Gómez 2003; http://www.fao.org/inpho/content/compend/toc_main.htm)

Gómez (2003) listed the worldwide production figures for *Vigna unguiculata* ssp. *unguiculata* var. *unguiculata* (i.e., cowpea) as 68% from Africa, 17% from Brazil, 3% from Asia, 2% from the United States (so-called blackeyed peas), and 10% from the "rest of the world." Cowpea's significance for human populations is reflected in its utility as a rich source of protein for humans and their domesticated animals, and as a nitrogen fixing species for other crops (Gómez 2003; Ba 2004). Cultivation of cowpea occupies ~12.5 million hectares of land, with the annual worldwide production of this cultivar exceeding 3 million tons (Feleke *et al.* 2006). *Vigna unguiculata* is of primary importance for farmers who deal with regular droughts. For example, this crop is a critical legume for the Sahelian zone of Sub-Saharan Africa (Fang *et al.* 20007).

An African origin for the cowpea is well accepted (Feleke *et al.* 2006), with the centers of highest diversity for wild and cultivated species being in southeastern and West Africa, respectively (Fang *et al.* 2007). Given the level of diversity in cultivars from West Africa, this region has been hypothesized as the area in which domestication occurred. However, a northeastern African domestication event is also supported by some data sets (see Ba *et al.* 2004 and Fang *et al.* 2007 for a list of references and discussions of these alternate hypotheses). The likely wild lineage, from which *V. u. unguiculata* var. *unguiculata* was domesticated, has been identified as *Vigna unguiculata* ssp. *unguiculata* var. *spontanea* (Pasquet 1999).

The list of studies that have inferred the impact from introgressive hybridization (whether or not mediated by *H. sapiens*) in the evolution of wild and cultivated *V. unguiculata* is extensive (e.g., Rawal 1975; Pasquet 1999; Coulibaly *et al.* 2002; Ba *et al.* 2004; Feleke *et al.* 2006; Fang *et al.* 2007; Simon *et al.* 2007). Each of these data sets reflect the close tie between ongoing introgression between wild and cultivated forms, and the genetic and phenotypic structure of the plants utilized by humans for both their own food, and for fodder for their domesticated animals. The following points reflect a subset of the genetic analyses that led to inferences of reticulate evolution in the *V. unguiculata* wild-cultivated assemblage. First, a study of nuclear loci (i.e., as assayed from isozymes) indicated a large hybrid zone between wild and cultivated lineages in northern Nigeria and Niger (Rawal 1975). Second, nuclear genetic variation (also assayed through isozymes) detected introgression between several wild lineages distributed throughout southern African countries, including South Africa, Namibia, Zimbabwe, Zambia, and Botswana—the instances of introgression included ssp. *stenophylla* × ssp. *tenuis*, var. *spontanea* × ssp. *pubescens*, and var. *spontanea* × var. *dekindtiana* or ssp. *stenophylla* (Pasquet 1999). Third, variation at cpDNA loci reflected introgression between wild and cultivated *V. unguiculata* throughout a zone extending from Senegal to Tanzania through to South Africa (Feleke *et al.* 2006). Fourth, nuclear [AFLP] data indicated that genetic exchange between domesticate and progenitor, as well as among wild taxa, had occurred throughout the range of cowpea in Africa (Coulibaly *et al.* 2002). In summary, then, it is clear that extensive introgression has blurred "the distinction between the ancestral state and post-domestication evolution" (Coulibaly *et al.* 2002) leading to a web of relationships between wild and cultivated forms of *V. unguiculata*.

7.2.13 Reticulate evolution and seed crops: common bean

The total production value of the common bean, *Phaseolus vulgaris*, in 2006 for US farmers exceeded US$500 million (National Agricultural Statistics Service 2007a). However, of even greater importance is that this species is the most significant grain legume used directly as a food source by *H. sapiens* populations (Broughton *et al.* 2003). The critical role filled by the common bean is reflected in its cultivation on ~13 million hectares of land in Latin America and Africa, yielding some 7 million metric tonnes (Broughton *et al.* 2003). The significance of the agricultural statistics for Latin America resides in the fact that *P. vulgaris* is often the primary source of protein for the resident human populations (e.g., in Mexico and Brazil) and that it is a crucial component of small-scale or subsistence farms (i.e., 1–10 hectares in size; Broughton *et al.* 2003). Similarly, the >4 million hectares of common bean cultivation in eastern and southern Africa is primarily due to the work of "resource-poor farmers with very few inputs…primarily on small-scale, marginal farms" (Broughton *et al.* 2003). A high percentage of the world's population thus depends on *P. vulgaris* as a primary source of nutrition.

The *V. unguiculata* assemblage was domesticated from New World lineages. Common bean originated from wild populations of *Phaseolus* in both Meso- and South America. Using cpDNA loci, Chacón *et al.* (2005) inferred a single domestication event for the South American cultivars, located somewhere in the southern Andean region. In contrast, their data led to inferences of up to four separate domestication events in the establishment of the different Mesoamerican races (Chacón *et al.* 2005). The degree to which reticulate evolution, in the form of introgressive hybridization, has affected the *P. vulgaris* complex is similar to that described earlier for cowpea. For example, an equally parsimonious explanation for the close genetic similarity among the various Mesoamerican *P. vulgaris* cultivars with local wild species is that a single domestication event occurred with subsequent spread of the cultivar and introgression with the resident wild lineages (Chacón *et al.* 2005). Consistent with this hypothesis is the observation that cultivated and wild common bean lineages have formed extensive hybrid zones. These zones of introgression occur across the entire New World range of *Phaseolus* (Beebe *et al.* 1997, 2001; Papa and Gepts 2003; Islam *et al.* 2004; Blair *et al.* 2007). Though introgression seems to be asymmetrical, with gene flow occurring mainly from cultivated to wild populations, introgression has also impacted the common bean cultivars (Papa and Gepts 2003).

In addition, to introgression between wild and domesticated populations in the New World, genetic exchange between the Mesoamerican and South American cultivars has also been described in the Iberian Peninsula. Separate introductions into Iberia during the sixteenth century brought the Meso- and South American *P. vulgaris* lineages into contact (Santalla *et al.* 2002; Rodiño *et al.* 2006). This human-mediated association led to introgressive hybridization, which yielded numerous recombinant genotypes. Indeed, the extremely high level of variation in the Iberian accessions resulted in the description of this region as a "secondary center of genetic diversity for the common bean" (Santalla *et al.* 2002).

7.2.14 Reticulate evolution and fruit crops: breadfruit

The *Bounty* was now heading for Tahiti with no further landfalls. On that beautiful island the main objective of the expedition, the breadfruit, was growing in abundance on countless trees…On Saturday, April 4, 1789, the *Bounty* weighed anchor and sailed out from Toaroa harbor. On board were 1,015 breadfruit plants in 774 pots, 39 tubs, and 24 boxes. (Wahlroos 2001; as cited by Pacific Union College Library, puc.edu)

Thus was the genus *Artocarpus* and, most especially, the widely distributed tropical cultivar, *Artocarpus altilis* (i.e., breadfruit), tied to the "Mutiny on the *Bounty*." However, this was not the last expedition of the infamous "Captain Bligh" to collect this plant for transport to British colonies located in the Caribbean (Wahlroos 2001; as cited by Pacific Union College Library, puc.edu).

Although numerous cultivar lineages of *A. altilis*—developed over thousands of years—are currently used as a source of carbohydrates, proteins, and fiber across Melanesia, Micronesia, and Polynesia (Zerega *et al.* 2004), no significant commercial cultivation has developed (Roberts-Nkrumah and Badrie 2005). Indeed, only in the Caribbean has commercialization of this crop occurred, and that in response to the market for expatriate West Indians in North America and the United Kingdom (Roberts-Nkrumah and Badrie 2005). Some *A. altilis* cultivators utilize seeded varieties, while other farmers depend on seedless lineages (the latter necessitating vegetative propagation). Of the nearly 60 wild species of *Artocarpus*, the sister-taxa of *A. altilis* are most likely *Artocarpus camansi* and *Artocarpus mariannensis* (Zerega *et al.* 2004, 2005). Yet, the geographic point of domestication and the evolutionary trajectory of the lineage leading to the cultivar have been unresolved until recently (see Zerega *et al.* 2004 for a review). In terms of the possibility that reticulate evolution has contributed to the derivation of breadfruit, Fosberg (1960) hypothesized an introgression-based origin (involving *Artocarpus blancoi*) for the *A. altilis* variants found in the Philippines. He also inferred past introgressive hybridization between the cultivar and *A. mariannensis* in Micronesia (Fosberg 1960). Consistent with this latter hypothesis was Ragone's (2001) detection of putative interspecific hybrids between these two species in Micronesian samples.

Zerega *et al.* (2004, 2005) analyzed genetic variation at nuclear loci to infer the avenues of dispersal and modes of evolution in the wild and cultivated *Artocarpus* species. The patterns of genetic variation resulted in a two-step model to explain the evolution of breadfruit: (1) the majority (but not all; Figure 7.4) of Melanesian and Polynesian cultivars were derived from human-mediated asexual propagation and artificial selection of *A. camansi* variants (Zerega *et al.* 2004); and (2) the Micronesian

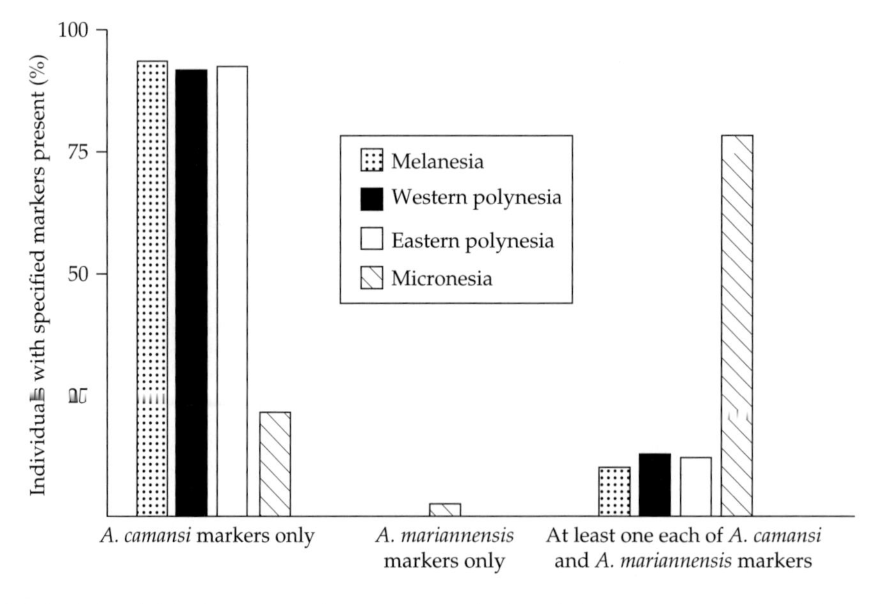

Figure 7.4 The distribution of alleles from *A. camansi* and *A. mariannensis* in breadfruit cultivars grown in Melanesia, Polynesia, and Micronesia. The presence of alleles from both species in all sampled regions indicates the presence of introgression, but with the highest levels of genetic exchange in Micronesian cultivars (Zerega *et al.* 2004).

A. altilis originated from introgressive hybridization between the introduced, *A. camansi*-derived lineage and the naturally occurring, *A. mariannensis* (Figure 7.4; Zerega *et al.* 2004). The authors summarized their findings that supported the inference of a reticulate evolutionary origin for the Micronesian cultivars by stating, "Polynesian cultivars have only *A. camansi*-specific markers present, while most Micronesian cultivars have both *A. camansi*- and *A. mariannensis*-specific markers present within individual cultivars" (Figure 7.4; Zerega *et al.* 2004). Thus, *H. sapiens* again inadvertently caused the reticulate evolution of a foodstuff by the introduction of a previously domesticated cultivar into the region of a related, wild taxon.

7.2.15 Reticulate evolution and fruit crops: peach palm

Mora-Urpí *et al.* (1997) reviewed data concerning the biology and nutritional/economic value of the peach palm, *Bactris gasipaes*. This species has been a staple food crop for its Amerindian cultivators since pre-Columbian times, with types of products ranging from fresh and dried fruit to beverages (Mora-Urpí *et al.* 1997). Indeed, the Native Amerindians supposedly valued only their wives and children more highly than they did *B. gasipaes* (see quotes in Mora-Urpí *et al.* 1997). Two main parts of the plant are either utilized by local populations or exported; these are the fruits and the "heart-of-palm." Mora-Urpí *et al.* (1997) described the "true" heart-of-palm as consisting of "the tender tubular part composed of immature leaves wrapped within the tender petiole sheaths."

Peach palm is still considered an underutilized tropical crop species, but it is gaining in worldwide popularity, especially due to the designation of heart-of-palm as a gourmet vegetable (Clement and Manshardt 2000). However, I have included this species under the heading of "Fruit Crops" to indicate its importance as such for its native farmers. Traditional and modern cultivation practices differ greatly for *B. gasipaes*, and reflect mainly its use as a source of either fruit or heart-of-palm. Fruits, partly due to their consistency and flavor, are utilized mainly as a subsistence farming product, rather than as an item for export (Mora-Urpí *et al.* 1997). As a subsistence plant species peach palm is usually sown at low densities (3–20 plants/hectare) as part of the farmer's home garden or agroforestry system (Mora-Urpí *et al.* 1997). In contrast, within a managed agricultural framework *B. gasipaes* is planted at much higher densities (400–500 plants/hectare for fruit and 3,000–20,000 plants/hectare for heart-of-palm production; Mora-Urpí *et al.* 1997). Annual profits from small-scale peach palm cultivation reflect a substantial revenue source for the farming communities involved. For example, sales of fruit alone make this the third and seventh most valuable crop for Peruvian farmers near Iquitos/Yurimaguas and Pucallpa, respectively (Mora-Urpí *et al.* 1997).

Reticulate evolution of *B. gasipaes*, particularly in the form of crop-wild introgression, has been inferred from a number of studies. Rodrigues *et al.* (2004) concluded—from an analysis of nuclear loci—that the frequency of wild × crop hybridization was limited, but not zero. In addition, they discussed data from previous studies substantiating such genetic exchange. Rodrigues *et al.* (2004) also detected gene exchange in a contact zone between the two landraces, Putumayo and Pará, resulting in the highly heterozygous Solimões cultivar. Similarly, Couvreur *et al.* (2007) suggested that introgression between wild and domesticated *B. gasipaes* might explain their inability to separate these lineages using either cpDNA or nuclear microsatellite variation. However, the most definitive evidence of crop-wild introgression came from an earlier analysis by Couvreur *et al.* (2006). Figures 7.5 and 7.6 illustrate the effects of introgression on both the genetic variation and fruit morphology within cultivated plants from northwest Ecuador. Specifically, these plants demonstrated a mixture of microsatellite alleles from cultivated and wild populations and fruit volumes intermediate between wild and cultivated lineages (Couvreur *et al.* 2006).

7.2.16 Reticulate evolution and fruit crops: apples

The apple is the most popular fruit from the temperate regions of the world, where they thrive as they generally require a distinct winter season.

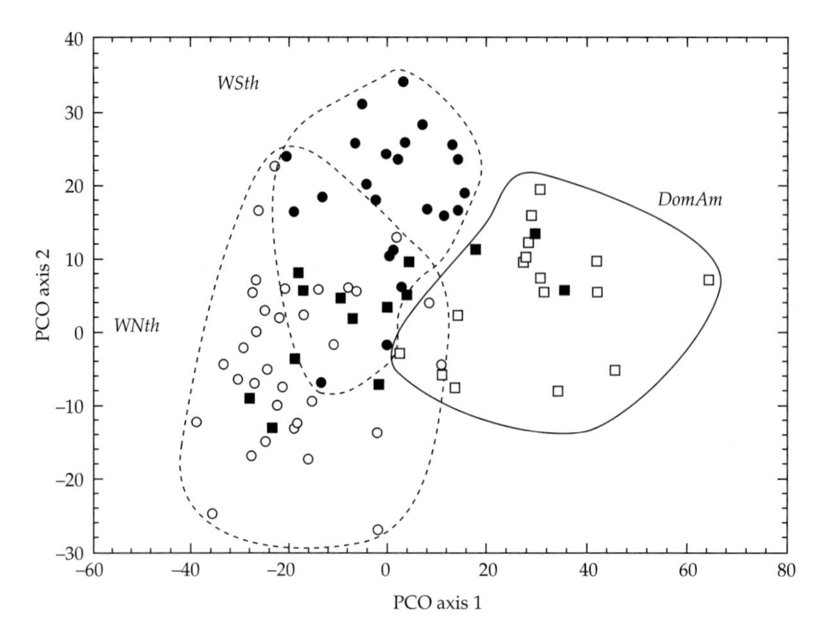

Figure 7.5 Distribution of principal coordinate scores for peach palm (*B. gasipaes*) samples: "*WSth*" = wild populations in southern part of distribution (filled circles); "*WNth*" = wild populations in northern part of distribution (open circles); "*DomAm*" = cultivated plants not in geographic/genetic contact with wild plants (open squares). The various lines near these designations enclose the genetic variation found in these three classes. Filled squares indicate individuals of peach palm cultivated in northwestern Ecuador. These plants possessed intermediate genotypes, indicative of introgression between wild and cultivated lineages (Couvreur *et al.* 2006).

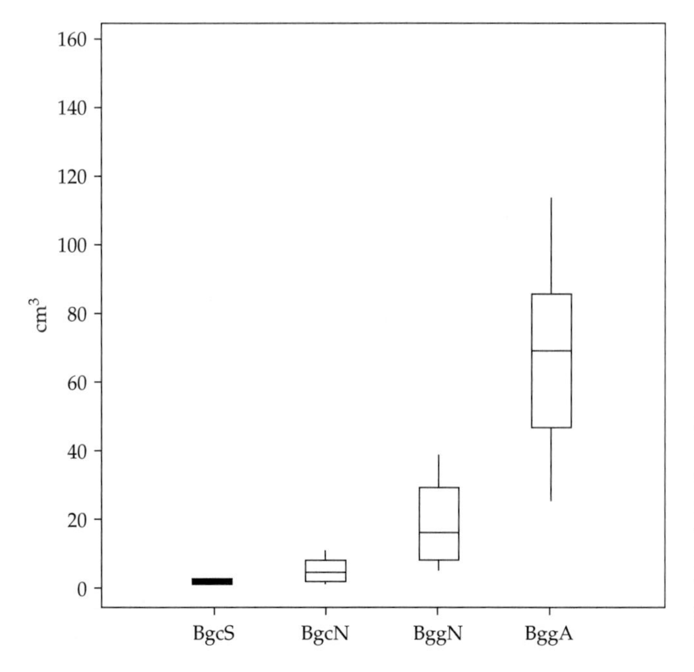

Figure 7.6 Volume of fruits produced by *B. gasipaes* (i.e., peach palm) plants from wild (BgcS and BgcN) and cultivated (BggN and BggA) populations. As expected, the wild populations have the smallest mean fruit sizes due to a lack of human-mediated selection. The intermediate fruit sizes of the BggN cultivars from northwestern Ecuador likely reflect introgression from wild genotypes into the cultivars (Couvreur *et al.* 2006).

According to FAO, apples are produced in substantial amounts in 93 countries around the world, with China accounting for 40% of world production. (http://www.croptrust.org/)

The US apple production during 2006 was ~5 million tons yielding > US$2 billion of income (Pollack and Perez 2007). Given that the United States contributes ~7% of the worldwide apple harvest (United States Department of Agriculture/Foreign Agricultural Service 2006), the total market is 60 million tons of apples, resulting in an annual worldwide income of close to US$30 billion.

Domestication of the apple (*Malus domestica*) likely occurred sometime before 8,000 ybp, possibly from Central Asian wild apple species (i.e., *Malus sieversii*; Harris *et al.* 2002). However, though the ancient association of this fruit with humans is not controversial, the geographic point and mode of its derivation are. Thus, Coart *et al.* (2006) found a close association between European *M. domestica* and the local wild species, *Malus sylvestris*, and a more distant relationship with *M. sieversii*. This led Coart *et al.* (2006) to conclude "This study hereby reopens the exciting discussion on the origin of *M. domestica*."

Regardless of the place of origin of domesticated apple lineages, numerous authors have pointed to reticulate events in the ancient and more recent evolution of *M. domestica*. For example, Xu *et al.* (2007) inferred the likely allopolyploid formation of the clade to which the apple lineages belong (i.e., Maloideae). In addition, both Hokanson *et al.* (2001) and Robinson *et al.* (2001) reviewed studies resulting in the hypothesis of a complex introgressive hybrid origin—involving several wild apple species—for *M. domestica*. Apple has also apparently undergone postdomestication reticulate evolution as well. Two studies by Coart *et al.* (2003, 2006) illustrate this inference well. In the first of these analyses, nuclear loci were sampled for over 120 wild and domesticated lineages. This study detected an ~2.5% frequency of admixed individuals, all of which were found in wild *M. sylvestris* populations (Coart *et al.* 2003). In contrast to the "low level" of introgression detected from the earlier screening of nuclear loci, assays incorporating a larger number of plants, and the addition of cpDNA loci, detected much higher admixture estimates (Coart *et al.* 2006). Specifically, the following observations were generated: (1) of the sampled individuals, 11% reflected nuclear introgression; (2) an even higher proportion of *M. sylvestris* samples possessed *M. domestica* cpDNA variants; (3) the bidirectional introgressive hybridization between wild and domesticated populations was likely due to the use of local wild populations for sources of domesticated lineages, with subsequent introgression of cpDNA from the cultivars into the resident *M. sylvestris* populations (Coart *et al.* 2006).

7.2.17 Reticulate evolution and fruit crops: coconut

The coconut palm, *Cocos nucifera*, exemplifies again observations made repeatedly in this and the two preceding chapters. First, this species has benefited numerous tropical, *H. sapiens* populations as a source of food and economic stimulus in countries such as India, Sri Lanka, Malaysia, Indonesia, Nigeria, Jamaica, and Brazil (Teulat *et al.* 2000; Manimekalai and Nagarajan 2006b). In 2003, 10.6 million hectares were utilized for *C. nucifera* plantations, yielding ~53 million nuts (Manimekalai and Nagarajan 2006a). The second observation that can be drawn from an analysis of coconut is that, though I have placed this species into the category of food source, coconut palm is a species with a multitude of uses. For example, parts of the coconut palm are utilized to produce drink containers, alcohol, furniture, activated charcoal, ropes, doormats, woven articles, thatch, and biodiesel (Bruman 1945; Corpuz 2004; Baudouin *et al.* 2006).

The domesticated and wild coconut palms are both placed within *C. nucifera*, but nonetheless reflect well-differentiated genetic lineages (Zizumbo-Villarreal and Piñero 1998; Zizumbo-Villarreal *et al.* 2002, 2005, 2006; Perera *et al.* 2003; Manimekalai and Nagarajan 2006a). Furthermore, their genetic differentiation is reflected in divergent morphologies, with the cultivars possessing the *Niu vai*-type and the wild populations the *Niu kafa*-type morphology (Zizumbo-Villarreal *et al.* 2006). A portion of the morphological differences between wild plants and cultivars likely reflects the enhanced ability of the *Niu kafa*-types to disperse by floating (Zizumbo-Villarreal *et al.* 2006) thus allowing oceanic colonization of islands and continents (e.g., the Cocos Islands in the Indian Ocean; Leach *et al.* 2003).

The ability of the wild coconut to disperse by floating could bring these lineages into contact with domesticated forms that have been introduced by humans thus facilitating introgressive hybridization. Furthermore, introgression between different cultivars brought together by the establishment of multiple plantations could also contribute to *C. nucifera* (wild and domesticate) genetic diversity. Indeed, it is well recognized that both wild–domesticate and domesticate–domesticate introgression have contributed to the genetic variation in contemporary wild and cultivar populations (Lebrun *et al.* 1998; Teulat *et al.* 2000; Perera *et al.* 2003; Manimekalai and Nagarajan 2006b). However, the avenue by which the cultivated and wild lineages have been brought together has more frequently involved the human-mediated transport of domesticated coconuts into areas previously occupied by wild *C. nucifera* (Zizumbo-Villarreal *et al.* 2006). Humans have catalyzed additional genetic exchange between "wild" and domesticated genomes by the importation and cultivation of both *Niu vai*-type and *Niu kafa*-type plants. This has been the case for Mexican populations of coconuts, which were introduced to both the Gulf of Mexico and Pacific coasts (*Niu kafa* and *Niu vai*, respectively; Zizumbo-Villarreal and Piñero 1998; Zizumbo-Villarreal *et al.* 2005). Subsequent to their importation, introgression occurred not only between different cultivars, but also between plants belonging to the *Niu vai*-type and *Niu kafa*-type lineages (Zizumbo-Villarreal *et al.* 2006). Numerous modes for genetic exchange have thus been identified among the worldwide *C. nucifera* populations.

7.2.18 Reticulate evolution and fruit crops: pepino

The pepino dulce (*Solanum muricatum*) is a common fruit in the markets of Colombia, Ecuador, Peru, Bolivia, and Chile. It comes in a variety of shapes, sizes, colors, and qualities. Many are exotically colored in bright yellow set off with jagged purple streaks. Most are about as big as goose eggs; some are bigger. Inside, they are somewhat like honeydew melons: watery and pleasantly flavored, but normally not overly sweet. (National Research Council 1989)

Given its common name of pepino dulce (Spanish for "sweet cucumber"), it seems surprising that *S. muricatum* is not well known outside of its Andean homeland (National Research Council 1989). For example, the pepino is much less utilized outside South America than are its close relatives, the tomato and the potato (Blanca *et al.* 2007). In contrast, in the Andes, this domesticate has been utilized since pre-Colombian times, as reflected by its representation on Peruvian pottery from this period (National Research Council 1989). However, this crop species has recently become more visible and attractive to other parts of the world and is now being cultivated on European, South Pacific, and North American farms (Blanca *et al.* 2007). It has been argued that the pepino would increase even more in attractiveness to non-South American consumers through genetic modification to increase the sugar content of its fruits (Rodríguez-Burruezo *et al.* 2003).

Solanum muricatum reflects not only ancient domestication, but also a reticulate history involving introgressive hybridization with various wild species. Thus, although pepino may have originated from a single species (possibly *Solanum caripense*; Stiefkens *et al.* 1999), genetic exchange within the clade to which it belongs has likely been extensive (Prohens *et al.* 2006). Most importantly, postdomestication introgression between a number of wild lineages and *S. muricatum* also appears to have been a significant determinant of the genetic structure of present-day pepino cultivars. The genetic exchange between pepino cultivars and wild taxa has apparently been facilitated by the proximity of cultivated fields to wild populations (Anderson 1975). Figure 7.7 illustrates the association of different *S. muricatum* DNA sequences with a divergent array of wild species. Derived from an analysis of nuclear loci by Blanca *et al.* (2007), this pattern is a strong indication that introgressive hybridization has brought together previously divergent lineages.

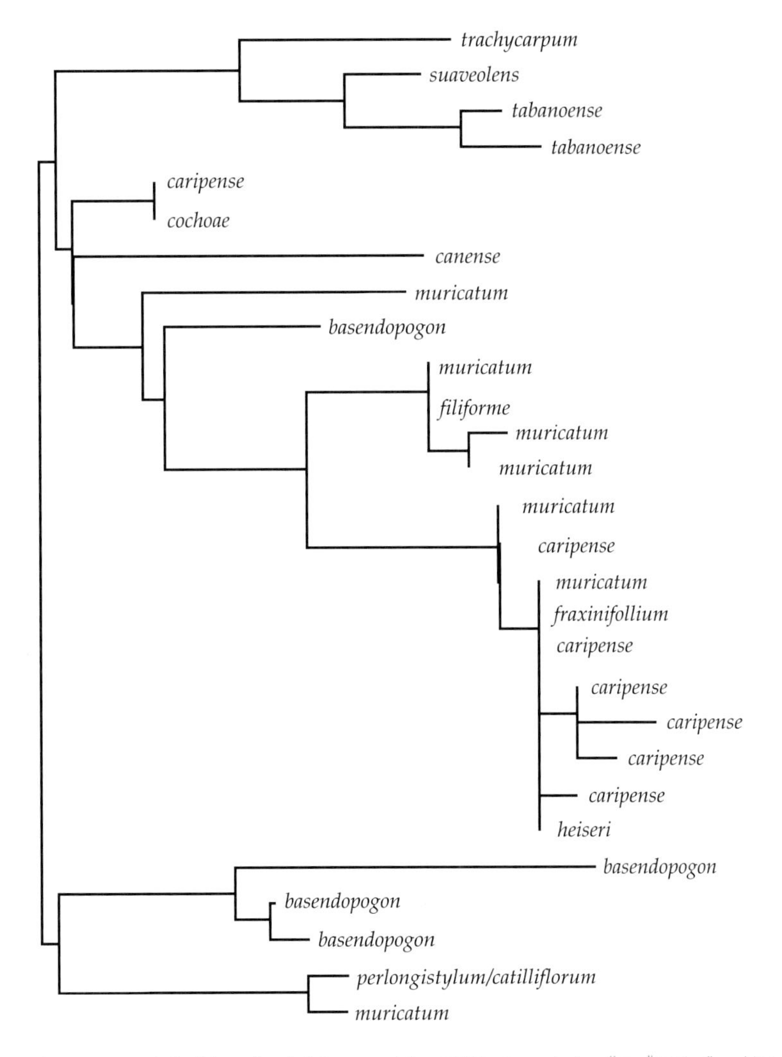

Figure 7.7 Phylogenetic arrangement of wild species of *Solanum* and the cultivar, *S. muricatum* (i.e., "pepino" and listed as "*muricatum*" in this phylogeny). Note that the distribution of *S. muricatum* throughout the phylogeny is likely mainly due to introgression between the cultivar and various native, wild taxa with which it has come into contact (Blanca *et al.* 2007).

These workers detected evidence for introgression affecting *S. muricatum*, involving *S. caripense*, *Solanum catilliflorum*, and *Solanum perlongistylum*. Blanca *et al.*'s (2007) survey also detected genetic variation inferred to have originated from additional, but as yet unidentified, wild species. Finally, postdomestication hybridization between the pepino and wild relatives was hypothesized to have given rise to at least two stabilized hybrid lineages recognized as *Solanum filiforme* and *Solanum cochoae* (Blanca *et al.* 2007).

7.2.19 Reticulate evolution and fruit crops: citrus

The genus *Citrus* contains numerous cultivars that are grown and utilized extensively throughout the world. Indeed, as the United Nations Conference on Trade and Development reported, "Citrus fruits are the first fruit crop in international trade in terms of value" (UNCTAD 2005). Some of the *Citrus* species of most value, both as a source of food and economic strength, are the grapefruit, lime, lemon, pummelo, citron, mandarin, and sour and sweet oranges (*Citrus paradisi, Citrus aurantifolia, Citrus limon, Citrus grandis, Citrus medica, Citrus reticulata, Citrus aurantium,* and *Citrus sinensis,* respectively; Moore 2001). Although 140 countries contribute to the production of these cultivars, over two-thirds of all citrus fruits originate in Brazil, various Mediterranean countries, the United States, and China (UNCTAD 2005). Of the *Citrus* producing countries, those in the Mediterranean provide over half of the fresh fruit exports, while Brazil and the United States are centers for processed fruits (mostly represented by orange juice; UNCTAD 2005). Notwithstanding the use made of citrus fruits by the various cultivators, the nutritional and economic gains from this farming industry are undeniable. Once again we use the United States as an example of economic benefit: 2006 receipts for grapefruits, lemons, and oranges alone totaled > US$2,500 million (Pollack and Perez 2007).

The origin of the genus, *Citrus,* is hypothesized to have occurred in the subtropical and tropical zones of Southeast Asia, followed by dispersal to other continents (see Nicolosi *et al.* 2000 for references). However, the evolutionary relationships among the cultivated species have been difficult to discern, likely due to extensive hybrid lineage formation and introgression (Moore 2001). In this regard, Barrett and Rhodes (1976) argued for the recognition of only three "true biological species," those being *C. grandis* (pummelo), *C. medica* (citron), and *C. reticulata* (mandarin). The remaining *Citrus* domesticated species were recognized as having a hybrid derivation (Barrett and Rhodes 1976; Nicolosi *et al.* 2000; Moore 2001; de Moraes *et al.* 2007).

Figure 7.8 illustrates the complicated, reticulate evolutionary history for grapefruit, lime, lemon, sour orange, and sweet orange. The complexity of these derivations is obvious, but one example is indicative of the vagaries of the cultivars' origins. This involves the formation of the (very) different hybrid species recognized as *C. aurantium* and *C. sinensis*—that is, the sour and sweet orange, respectively. Although both of these species formed through hybridization between the pummelo and mandarin lineages (with most of their genomes derived from *C. reticulata*; Barrett and Rhodes 1976; Nicolosi *et al.* 2000), the resulting cultivars produce fruit with significantly different flavor characteristics (Moore 2001). The hypothesis constructed to explain such divergent evolutionary endpoints involved their derivation from introgression of pummelo into different mandarin subspecies (Barrett and Rhodes 1976). Finally, and somewhat ironically, genetic analyses have shown that mandarin as well (i.e., one of the three "true" species) contains numerous recombinant genotypes (Filho *et al.* 1998). Similarly, Barrett and Rhodes (1976), while naming the mandarin as one of the three biologically relevant lineages, also discussed the formation of the heterogeneous *C. reticulata* variants from introgression both among mandarin subspecies as well as from *C. grandis*.

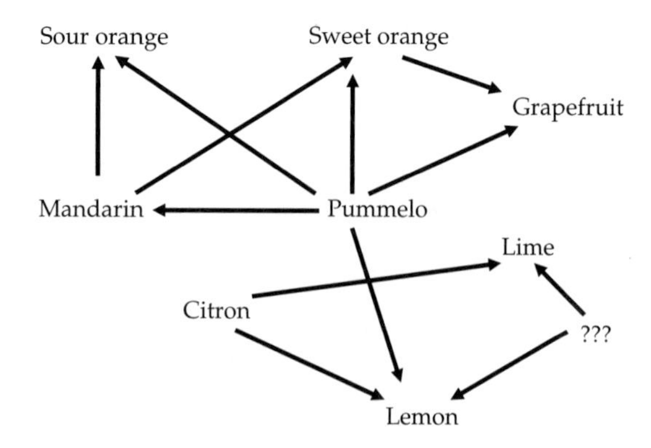

Figure 7.8 The reticulate evolutionary history of grapefruit, lime, lemon, pummelo, citron, mandarin, and sour and sweet oranges (*C. paradisi, C. aurantifolia, C. limon, C. grandis, C. medica, C. reticulata, C. aurantium,* and *C. sinensis,* respectively). ??? indicates the genetic contribution by unidentified species to the formation of both limes and lemons. Arrows indicate the genetic contributions in the origin/evolution of the various cultivars (adapted from information in Barrett and Rhodes 1976, Nicolosi *et al.* 2000 and Moore 2001).

7.2.20 Reticulate evolution and sugarcane

Sugarcane, *Saccharum officinarum,* exceeds all other sugar-producing crops, contributing ~75% of the sucrose utilized by humans (Dillon *et al.* 2007). This cultivar is capable of accumulating (in its stem) between 16% and 50% of its fresh and dry weight, respectively, as sucrose (Dillon *et al.* 2007). Brazil and India are the two largest producers of raw sugar from *S. officinarum*, with both countries devoting ~4 million hectares of land to its cultivation (Bolling and Suarez 2001; Singh *et al.* 2007). Each of the years during the period 1997–2001, Brazil produced 250–300 million metric tons of sugarcane, yielding ~20 million metric tons of raw sugar (Bolling and Suarez 2001). Brazil's population (fifth largest in the world) has also demonstrated a historically high utilization of sugar as a caloric source, annually averaging 9.5 million tons consumed countrywide and with each citizen consuming ~50 kg of sugar per year (Bolling and Suarez 2001). Furthermore, during the year 2000, Brazilian sugar exports totaled >7.5 million metric tons (Bolling and Suarez 2001). In contrast to the major contribution to *S. officinarum* cultivation and sugar production by India and Brazil, during the same time period the United States devoted <1 million acres to its cultivation, and produced only ~4 million tons of sugar from sugarcane (http://www.ers.usda.gov/Briefing/Sugar/Background.htm).

The domestication of sugarcane, and its subsequent transportation by humans, began thousands of years ago (Brandes and Sartoris 1936). Genetic analyses suggest that the geographic point of origin occurred somewhere in the Sahul region (i.e., the continental plate occupied by New Guinea, Australia, and Tasmania), most likely in Papua New Guinea (Lebot 1999). The complex nature of this derivation, and the role of introgressive hybridization (both with and without human intervention), can be illustrated well with the following observations drawn from numerous studies. First, the genus *Saccharum* consists of six species; the wild taxa, *Saccharum robustum* and *Saccharum spontaneum* and the cultivars, *Saccharum sinense, Saccharum barberi, Saccharum edule,* and of course, the most commonly used cultivar, *S. officinarum* (Takahashi *et al.* 2005). Second, though the evolution of this species complex includes polyploid formation from within the same lineage (i.e., autopolyploidy; Grivet *et al.* 1996), the two wild species, *S. robustum* and *S. spontaneum*, have also apparently been the progenitors of numerous hybrid derivatives (Lebot 1999). This latter process is the likely cause of the observed discordance between different phylogenies derived from various DNA sequence data sets (Takahashi *et al.* 2005). Third, within the past century and a half, the formation of present-day cultivars of *S. officinarum* has resulted from human-mediated crosses between *S. officinarum* and *S. spontaneum* and *S. sinense* and *S. barberi* (Dillon *et al.* 2007). This process, termed "nobilization" was in response to disease outbreaks and involved introgression between *S. officinarum* and the other species to introduce resistance to such epidemics into the cultivar (Brandes and Sartoris 1936; Jannoo *et al.* 2007). In total, the processes of polyploidy and introgressive hybridization in the formation of *S. officinarum* have resulted in "the most complex of all crop genomes studied to date" (Jannoo *et al.* 2007).

7.3 Summary and conclusions

The examples discussed in this chapter reveal the intimate relationship between reticulation and the origin and evolution of plant-based foods. As with the subjects covered in the previous two chapters, not all possible examples were included in the present chapter. For example, I did not detail the many hybrid species from the family Brassicaceae (e.g., from the genus *Brassica*; Yang *et al.* 2002) that provide critical foodstuffs for human populations. Yet, the taxa in the preceding sections reflect not only the diversity of plant foods isolated by *H. sapiens* from admixed lineages, but also the array of reticulate processes yielding these lineages. Postdomestication introgressive hybridization between cultivar lineages, or between cultivars and their wild progenitors, or between cultivars and unrelated wild taxa have contributed to the genetic and phenotypic structure of present-day plant domesticates. However, even the biological starting material—that is, the progenitors of our cultivars—exhibit admixed genomes. Some of these reflect allopolyploidization, while others are the products of introgression between divergent lineages. In sum, our plant-based foods, as with the animals that provide most of our meat protein, are of reticulate origin and thus demonstrate our continued dependence on products from the web of life for nourishment.

CHAPTER 8

Reticulate evolution of disease vectors and diseases

...there is a minimum of 9 mutational steps between these putatively introgressed haplotypes and the *macromelasoma* main grouping...that is strongly supportive of introgression from *brasiliensis* into *macromelasoma*.

(Monteiro *et al.* 2004)

Phylogenomic analysis identifies evidence for lateral gene transfer of bacterial genes into the *E. histolytica* genome, the effects of which centre on expanding aspects of *E. histolytica's* metabolic repertoire.

(Loftus *et al.* 2005)

The discrepancies between the mitochondrial and the nuclear trees, with the latter being more congruent with cytological data, could be due to mitochondrial gene flow within certain geographical areas.

(Krueger and Hennings 2006)

It is now generally agreed that there were at least two critical gene transfer events in the evolution of pathogenic *V. cholerae* from a nonpathogenic precursor strain...

(Faruque *et al.* 2007)

Based on data from two ZEBOV genes, we also demonstrate, within the family *Filoviridae*, recombination between the two lineages.

(Wittmann *et al.* 2007)

...admixture analysis detected evidence of contemporary interform gene flow...lack of differentiation between M and S forms likely reflects substantial introgression....

(Yawson *et al.* 2007)

8.1 Reticulate evolution and the development of human disease vectors and diseases

The preceding chapters (1–7) have focused attention on what could be termed "positive" (or possibly in some instances, better typified as "neutral") effects on the human lineage from reticulate evolution. On the one hand, our own species and those of our sister lineages have been directly affected by introgressive hybridization. The inference of the origin or transfer of adaptive traits, rather than simply the introgression of neutral variants, is more strongly supported for some loci (Hawks *et al.* 2008). Yet, the occurrence of either adaptive or neutral introgression between, for example, different *Homo* lineages has transitioned from an unrealistic idea, to the null hypothesis, for many workers in the field (Chapter 4). Similarly, the organisms that allow the founding, growth, and maintenance of *H. sapiens'* populations—including those that provide building materials, clothing, and food—possess genomes speckled with material from other evolutionary lineages (Chapters 5–7). Once again, in some cases the very act of bringing together genomic elements from lineages on different evolutionary trajectories has apparently provided the basis for human-mediated innovations (e.g., cotton; Jiang *et al.* 1998; Rong *et al.* 2007).

Unlike each of the examples from Chapters 1–7, the biological properties of the organisms included in the present chapter lead to destructive consequences when they come into contact with humans. In particular, human pathogens and the vectors of these pathogens contribute to untold annual suffering and death among members of our species. However, the interactions—though negative in the extreme for *H. sapiens*—are underlain by adaptations beneficial to the pathogens/vectors. Furthermore, these adaptations are sometimes the result of the same classes of reticulate events (i.e., hybrid lineage formation and introgressive hybridization) that have produced *H. sapiens* and those organisms that we count as benefactors for our survival. Also like the reticulate events discussed in the previous seven chapters, there is a complexity of mode and mechanisms that reflects the richness of the web-of-life metaphor. Indeed, some vectors that belong to lineages that have been affected by reticulate evolution similarly carry some human pathogenic organisms that are of admixed ancestry. For example, the vector and the causal agent of Chagas disease—*Triatoma* and *Trypanosoma cruzi*, respectively—reflect one instance of this. Furthermore, many human pathogens would not have disease-causing capabilities in the absence of genetic transfers from their own parasites; the most obvious examples of this process come from the interactions between bacteria and their parasites known as bacteriophage (see Wagner and Waldor 2002 for a review). Thus, the level of complexity resulting in the evolution of the disease vectors and their pathogenic cargo can result from reticulate evolution. The examples discussed below indicate the widespread occurrence of reticulate evolutionary processes in those organisms against which we must battle for individual- and species-level survival.

8.2 Reticulate evolution and human disease vectors

Broadly defined, the vectors of human diseases vary from mammals that transmit rabies to insects that transfer the causal agents of some of the most devastating of diseases such as malaria, Chagas, and river blindness. *H. sapiens* acts as its own vector when sexually transmitted pathogens are involved, including of course the most recent plague, human immunodeficiency virus (HIV). In the discussion of examples of vectors, I will concentrate on those organisms that provide the most widespread and destructive (in terms of human lives affected) pathogens the opportunity to invade our populations. Specifically, I will focus on insects (reduviid bugs—Chagas disease; blackflies—river blindness; mosquitoes—malaria, West Nile, etc.) that lead to the infection of over half a billion people annually (http://www.cdc.gov). Of course, the vectors chosen also exemplify the processes of reticulate evolution. However, additional examples could have been discussed (e.g., placing our own introgressed genome as an example of a "vector" species that evolved at least partially by introgressive hybridization). Yet, the species chosen reflect well the negative impact on humans by organisms that owe their evolutionary trajectory to the web of life.

8.2.1 Reticulate evolution and *Triatoma*, the vector for Chagas disease

Chagas disease (i.e., American trypanosomiasis), caused by the protozoan parasite *T. cruzi*, and named for the Brazilian physician Carlos Chagas who first described it in 1909, occurs throughout Central and South America (including Mexico; Moncayo 1999). It is estimated that 16–18 million people are infected with this parasite, with an additional 25% of Latin Americans (~120 million) at risk of infection (Scientific Working Group on Chagas Disease 2005). Of those infected, 25–30% will be affected with fatal cardiac, esophageal, digestive, and/or neurological pathologies (Scientific Working Group on Chagas Disease 2005; Gürtler *et al.* 2007). Blood-feeding insects, broadly termed "Assassin Bugs" (subfamily Triatominae; World Health Organization [WHO]: http://www.who.int/tdr/diseases/chagas/diseaseinfo.htm), are the vectors for the introduction of the disease into mammalian populations. Members of the genus *Triatoma*, the vectors of this disease, have been referred the more descriptive name of "Kissing Bugs" indicating their preference for feeding on the soft, fleshy eye lids, and lips of their [sleeping] victims (Ross personal communication). Though the initial introduction of the causal agent of Chagas requires *Triatoma* vectors, further transmission of the disease can also occur through blood transfusions and congenitally from infected mothers to their fetuses (WHO: http://www.who.int/tdr/diseases/chagas/diseaseinfo.htm).

Given that the development of a vaccine is unlikely, the most effective strategies to limit the disease involve providing quality housing that is not amenable to *Triatoma* infestations and killing the vectors in human dwellings in which the bugs already occur (WHO: http://www.who.int/tdr/diseases/chagas/diseaseinfo.htm). As an indication of the effectiveness of such control measures, Figure 8.1 illustrates the pattern of infections of *Triatoma*, dog and human populations in several villages in Argentina before and after the application of insecticides (Gürtler *et al.* 2007). Though the decrease in human (i.e., child) infection rate was slower than that of dogs or vectors, by the end of the sequence of three applications—accompanied by the modification of some dwellings to limit *Triatoma* infestations—human infection was negligible (Figure 8.1; Gürtler *et al.* 2007). In spite of this demonstration of the importance of adopting procedures for the control of Chagas disease, and the obvious advances made in certain regions (Moncayo 1999), there is great concern that *T. cruzi* infection will continue to impact millions of Latin Americans. At-risk populations occur in poor (rural or urban) areas and thus are often isolated from either control measures or medical attention (Scientific Working Group on Chagas Disease 2005).

The interfertility of *Triatoma* species, as demonstrated through laboratory crosses, and the occurrence of natural hybrids between various species is well established. In particular, Usinger *et al.* (1966) discussed data indicating introgressive hybridization among South American species belonging to the "*Triatoma infestans* complex." This is significant for the present discussion given that *T. infestans* is the main conduit for *T. cruzi* infections among Brazilian populations (Moncayo 1999). Similarly, populations of the most important vector in

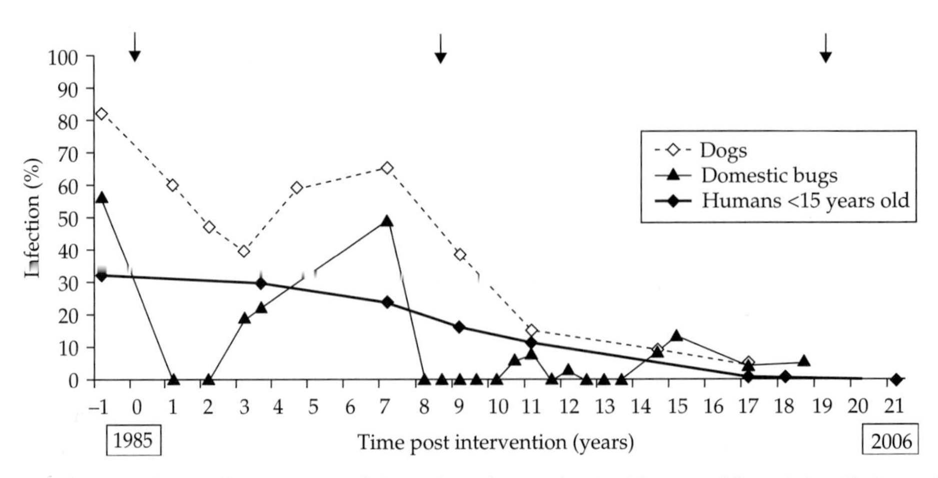

Figure 8.1 The frequency of *T. cruzi* (causative agent of Chagas disease) in populations of the vector ("domestic bugs"), dogs and humans in a series of Argentinean villages. The three vertical arrows indicate the timing of control measures that included the application of insecticides to human dwellings and the modification of some dwellings to make them less amenable to infestations by the vectors.

semiarid regions of Brazil, *Triatoma brasiliensis*, also demonstrate morphological and genetic variation consistent with introgression. Recent studies of *T. brasiliensis* have resolved the presence of at least four genetically, and ecologically, diverse lineages (Costa *et al.* 2002, 2003; Monteiro *et al.* 2004; Borges *et al.* 2005). Partial reproductive isolation has also been demonstrated via experimental crosses between individuals from each of these four lineages (Costa *et al.* 2003). These, and more recently collected, data have led to the suggestion that *T. brasiliensis* should actually be recognized as a complex of three species (Monteiro *et al.* 2004).

Notwithstanding the significant genetic differentiation and the partial reproductive discontinuity among the "*T. brasiliensis*" lineages, genetic exchange among naturally occurring populations has been predicted (Costa *et al.* 2003). This prediction has been borne out by analyses of both morphological and mtDNA variation across the geographic range of the *T. brasiliensis* complex. Specifically, Monteiro *et al.* (2004) reported evidence that introgressive hybridization had occurred between three of the four lineages—*brasiliensis* and *macromelasoma*, *brasiliensis* and *juazeiro*, *juazeiro* and *macromelasoma*. The definition of these divergent lineages was not only of significance for understanding the process of evolutionary diversification in this complex, but also for strategic planning of control measures that would target the predominant vector (i.e., the *brasiliensis* lineage; Monteiro *et al.* 2004).

8.2.2 Reticulate evolution and *Simulium*, the vector for river blindness

Onchocerciasis, known as river blindness "because the transmission is most intense in remote African rural agricultural villages, located near rapidly flowing streams" (Centers for Disease Control and Prevention 2004), is caused by the filarial (i.e., nematode) parasite, *Onchocerca volvulus* (Centers for Disease Control and Prevention 2004; Mansiangi *et al.* 2007). About 99% of the ~18 million infected humans live in Africa. The remaining populations occur in Yemen, and six New World countries (Mexico, Guatemala, Ecuador, Colombia, Venezuela, and Brazil; Centers for Disease Control and Prevention 2004). Onchocerciasis is responsible for blindness or visual impairment in ~800,000 of the infected persons (Centers for Disease Control and Prevention 2004). Tragically, though reflecting hope for future generations, these cases of blindness are preventable with annual treatment with the drug ivermectine (http://www.who.int/blindness/causes/priority/en/index3.html).

Approximately half of the 50 or so sibling species belonging to the "*Simulium damnosum* species complex" of blackflies act as vectors for *O. volvulus* (Mansiangi *et al.* 2007). This clade's remarkable pattern of evolutionary diversification represents a wonderful opportunity to test for evolutionary processes leading to such radiations (Krueger and Hennings 2006; Krueger *et al.* 2006b). In addition, knowledge of the distribution of the various lineages across the African landscape is critical for the implementation of effective control measures for this vector of river blindness (Morales-Hojas *et al.* 2002). Earlier work, leading to inferences of multiple lineages within the *S. damnosum* complex, rested on detailed analyses of chromosomal variation (see Krueger and Hennings 2006 for a review). More recent studies have utilized cytogenetic data, in concert with morphological and molecular characteristics, to further refine the geographic distribution and phylogenetic relationships of the different lineages (Morales-Hojas *et al.* 2002; Krueger and Hennings 2006; Krueger *et al.* 2006a, b; Mansiangi *et al.* 2007).

The availability of multiple data sets for *S. damnosum sensu lato* has not only afforded further resolution of the number of sibling species within this complex, but also facilitated the detection of numerous cases of putative introgressive hybridization. The following "species" were possibly affected by introgressive hybridization: (1) *Simulium sanctipauli*, *Simulium sirbanum*, *S. yahense*, *Simulium soubrense*, and *Simulium squamosum* (Morales-Hojas *et al.* 2002); (2) 'Kagera', 'Nkusi', 'Sanje' cytotypes (i.e., chromosomal types) along with *Simulium kilibanum*, *S. sanctipauli*, and *Simulium leonense* (Krueger and Hennings 2006); (3) *S. kilibanum* and the 'Sebwe' cytotype (Krueger *et al.* 2006a); (4) 'Linthipe' cytotype and *S. kilibanum* (Krueger *et al.* 2006b); and (5) *S. squamosum* and various "East and South African species"

(Mansiangi *et al.* 2007). Thus, a general pattern of reticulate evolutionary change marks a large proportion of the clade responsible for the transmission of river blindness.

8.2.3 Reticulate evolution and *Culex* species

Serious diseases of both humans (e.g., West Nile Virus, St. Louis encephalitis, and periodic lymphatic filariasis) and other animals (e.g., avian malaria) are caused by the transmission of pathogens by mosquitoes belonging to the *Culex pipiens* species assemblage (Fonseca *et al.* 2000; Hayes 2001; Smith and Fonseca 2004). Some of these diseases such as St. Louis encephalitis, though not affecting large numbers of persons relative to other diseases (~5,000 cases over a 40-year period in the United States), have mortality rates as high as 30% (Centers for Disease Control and Prevention 2007; http://www.cdc.gov/ncidod/dvbid/sle/Sle_FactSheet.html). Others, like the West Nile Virus, affect more people (in both the Old and New World), with the mortality rate still significant (~10%) among those infected (http://www.who.int/csr/don/2002_11_27/en/index.html). Finally, there are diseases such as filariasis for which it

> is estimated that there are about 600 million people living in the endemic areas in the South-East Asia Region constituting about 60% of the global burden, with about 60 million persons either harboring microfilaraemia or suffering from clinical manifestations which make up half of the global figure. (WHO 2001)

Members of the *Cx. pipiens* complex therefore account for the transmission of diseases that affect tens of millions of *H. sapiens* belonging to populations scattered around the world.

Several lineages within the *Cx. pipiens* clade have been implicated not only in disease transmission, but also in reticulate evolution. Of particular significance for the present discussion, *Cx. pipiens* (a vector of West Nile Virus; Hayes 2001) and *Culex quinquefasciatus* (vector of periodic lymphatic filariasis; Smith and Fonseca 2004) are two of the species commonly reported as participants in introgressive hybridization. Indeed, these two species overlap with one another geographically and are known to be able to form hybrids in the laboratory (Humeres *et al.* 1998; Smith and Fonseca 2004). Furthermore, numerous natural hybrid zones have been detected in various parts of their worldwide distribution—including North and South America and Asia (Humeres *et al.* 1998; Cornel *et al.* 2003; Smith and Fonseca 2004; Fonseca *et al.* 2004, 2006).

Introgression between lineages belonging to *Cx. pipiens* has also been inferred from genetic analyses. This introgression is hypothesized to have resulted in a unique set of feeding behaviors in the hybrids, relative to their parents, leading to unusual West Nile Virus outbreaks in North America (Fonseca *et al.* 2004). Unlike past epidemics in European cities, the North American episode persisted across years and spread from its center of origin (Hayes 2001). The role suggested for hybridization related to the fact that European and North American populations of the vector *Cx. pipiens* demonstrated behavioral differences. European mosquitoes are divided into two genetically differentiated forms that prefer either birds or mammals (including humans) as sources for their blood meals (Fonseca *et al.* 2004). In contrast, certain lineages of the North American *Cx. pipiens* species complex have been shown to feed from both humans and birds (Spielman 2001). Significantly, recent analyses have indeed confirmed that the presence of the North American hybrid lineage increases significantly the likelihood of exposure of humans to the West Nile Virus, due to the hybrid mosquitoes' unique feeding behavior (Kilpatrick *et al.* 2007).

8.2.4 Reticulate evolution and *Anopheles* species

Though mosquitoes of the genus *Anopheles* transmit numerous diseases worldwide (e.g., lymphatic filariasis in the Indonesian archipelago; WHO 2001), they are probably best known for their role as vectors of the causal agents of malaria—protozoans of the genus, *Plasmodium* (e.g., Prakash *et al.* 2006; Dusfour *et al.* 2007). Between 350 and 500 million cases of malaria occur each year, with ~1 million deaths attributed to the infections, mostly among young children in sub-Saharan African populations (http://www.cdc.gov/malaria). In developing countries in general, malaria has a profound impact on childhood mortality. For example, in 2002, malaria was ranked as the fourth most likely cause of death in children in these countries, with this disease accounting for ~11% of all cases of mortality (http://www.cdc.gov/malaria/facts.htm). A final statistic that helps highlight the profound effect of this disease on certain populations is that "In areas of Africa with high malaria transmission, an estimated 990,000 people died of malaria in 1995—over 2,700 deaths per day, or 2 deaths per minute" (http://www.cdc.gov/malaria/facts.htm). The hope for the future, for affected populations of humans, is that there are effective prevention and treatment procedures. Furthermore, curative measures are extremely inexpensive, with various options ranging from approximately US$0.13 to US$2.70 for treatment of an infected individual (http://www.cdc.gov/malaria/facts.htm).

Numerous species and species complexes of *Anopheles* are involved in the transmission of malaria into *H. sapiens* populations. Furthermore, many of these reflect the signature of reticulate evolution. To exemplify both the taxonomic diversity of the vector species and the extent of genetic-exchange-mediated evolution within this mosquito genus, I will discuss three assemblages. These include the *Anopheles dirus, Anopheles funestus,* and *Anopheles gambiae* species groups.

Anopheles dirus complex

The *An. dirus* species complex is a major contributor to malarial infections in both South and Southeast Asia. For example, species placed within this clade account for many of the 200,000–300,000 *Plasmodium falciparum* infections reported annually from the states of northeast India (Prakash *et al.* 2006). Thus, *An. dirus sensu lato* provides one of the primary vector bridges for 30% of all malarial cases worldwide, which occur each year in the South and Southeast Asia region (WHO 2004). Various analyses have detected multiple genetically divergent lineages within "*An. dirus*" (e.g., Manguin *et al.* 2002), reflected taxonomically by the recognition of seven named species: *Anopheles cracens, Anopheles scanloni, Anopheles baimaii, Anopheles nemophilous, Anopheles elegans, Anopheles takasagoensis,* and *An. dirus* (Sallum *et al.* 2007).

In addition to being an example of an evolutionary radiation, the *An. dirus* clade also reflects instances of putative introgression among the member species. For example, while inferring no contemporary genetic exchange among the various species in the complex, Walton *et al.* (2000) suggested "historical introgression" from *An. dirus* into other species to explain the pattern of mtDNA variation. The introgression hypothesis was subsequently supported by a study of nuclear variation that detected distinct nuclear lineages in the absence of mtDNA differentiation (Walton *et al.* 2001). Finally, Sallum *et al.* (2007) also found evidence consistent with the past occurrence of mtDNA introgression when they detected admixtures of *An. dirus* and *An. baimaii* lineages in their phylogenetic analyses.

Anopheles funestus complex

Of the nine species that belong to its complex, *An. funestus* is considered the most human-associated taxon, with individuals from the remaining lineages being mainly zoophilic (i.e., associated with animals other than *H. sapiens*; Temu *et al.* 2007). In terms of its role in the transmission of the malarial protozoan, this species is known to be among the most important vectors in sub-Saharan Africa (Koekemoer *et al.* 2006). Indeed, the prevalence of *An. funestus* in some regions results in it being the most important causal agent of transmission, even exceeding the impact of species belonging to the *An. gambiae* assemblage. For example, in the western highlands of Kenya, *An. gambiae* predominated at higher elevations, but *An. funestus* was numerically superior at lower elevations at which a much higher frequency of infectious bites occurred (Ndenga *et al.* 2006).

Evolution within the *An. funestus* clade has not been restricted to the origin of the nine species, but rather to the diversification of lineages within *An. funestus* as well. For example, Koekemoer *et al.* (2006) resolved three genetic "clusters" corresponding to West/central Africa, East Africa/Madagascar, and Angola/Mozambique/South Africa. These authors also detected additional, unique genotypes in individuals from Malawi, Angola, Ghana, and Zambia. The detection of genetic divergence within this single taxonomic species mirrors an earlier chromosomal study within Senegal that defined three lineages (Lochouarn *et al.* 1998). The genetic heterogeneity within *An. funestus* is likely partially due to introgressive hybridization between its own divergent lineages and possibly with other species of this complex as well. The following results provide support for this conclusion: (1) an analysis of mtDNA variation found Ugandan forms to be associated with three separate phylogenetic subdivisions (Michel *et al.* 2005); (2) divergent mtDNA lineages were found together in both Mozambique and Madagascar (Michel *et al.* 2005); (3) patterns of chromosomal rearrangement and nuclear loci variation indicate some level of gene flow between chromosomal forms in Cameroon (Cohuet *et al.* 2005); and similarly (4) a transect sampled across Burkina Faso, West Africa revealed differential genetic exchange depending on the position of nuclear loci relative to chromosomal inversions (Michel *et al.* 2006).

Anopheles gambiae complex

The impetus for the analysis of the *An. gambiae* assemblage (e.g., resulting in the sequencing of its genome; Holt *et al.* 2002) is due to the ranking of *An. gambiae* as the primary vector of *Plasmodium* in sub-Saharan Africa (White 1974). Thus, this species is the main cause of up to half a billion cases of malaria, and ~1 million deaths among children, annually (http://www.cdc.gov/malaria). Yet, the interest in understanding the *An. gambiae* species complex from a disease control standpoint has resulted in its development into a model system for defining evolutionary processes that lead to diversification and adaptation. For example, recent investigations have determined the evolutionary origin of certain paracentric rearrangement events on chromosome 2, a type of chromosomal change thought to be important for organismal evolution (Sharakhov *et al.* 2007). Similarly, reduced genetic recombination in regions proximal to the centromeres has been suggested as a catalyst for diversification, giving rise to divergent lineages within *An. gambiae* (Stump *et al.* 2005). Slotman *et al.* (2006) emphasized that such regions could result in the accumulation of genes causal in reproductive isolation, even as introgression was occurring in other portions of the genome.

The last point mentioned above—that certain portions of the genomes of the *An. gambiae* species are resistant to introgression, while others are not—is indicative of past and ongoing reticulate evolution in this assemblage. Numerous studies have detected both reduced and elevated frequencies of introgression depending on which regions of the genome were examined (e.g., Besansky *et al.* 1997; della Torre 1997; Lanzaro *et al.* 1998; Gentile *et al.* 2002; Turner *et al.* 2005; Boulesteix *et al.* 2007; see Arnold 2006, pp. 77–78 for additional references). Significantly, it has been shown that genomic isolation between two divergent *An. gambiae* lineages (the "M" and "S" forms) is restricted to a few (nuclear loci) "islands" with the remainder of the mtDNA and nuclear genomes relatively permeable to introgression (e.g., Turner *et al.* 2005; Turner and Hahn 2007). Furthermore, it appears that the genetic constitution of the M and S form mosquitoes is greatly affected by their ecological setting. For example, Yawson *et al.* (2007) found that genotypes were more similar if they occupied similar ecological settings. This similarity is likely due to a combination of both environmentally mediated selection as well as genetic exchange between M and S form individuals/populations (Yawson *et al.* 2007). Introgression may thus play a role in the adaptation of individuals to certain environments by the spread of beneficial alleles. Significantly, Besansky *et al.* (2003) hypothesized such an adaptive trait transfer event from the xeric-adapted *Anopheles arabiensis* into *An. gambiae*, thereby allowing the spread of the latter vector into additional geographic regions.

8.3 Reticulate evolution and human diseases

Anatomically modern humans have been intimately associated with, and thus under constant attack from, various pathogens even before their migration from Africa (e.g., Linz *et al.* 2007). Our own evolution as a host species for a range of viral, bacterial, protozoan, protist, and fungal lineages has therefore benefited many pathogens. In return, pathogens of one type or another infect all humans, accounting for the loss of large portions of *H. sapiens* populations, particularly in developing countries. The loss in human life is also reflected in severe social and economic repercussions. In the following sections, I will review information concerning the biology and impact on human societies of organisms as diverse as influenza virus and *Ascaris* roundworms. This level of diversity will be used to emphasize the negative pressure that such organisms exert on humans exposed to the various disease organisms. This information will also establish the platform for the discussion of the reticulate evolutionary trajectory of these widely divergent human pathogens. This in turn will indicate not only the breadth of biological examples, but also the causal role that genetic exchange between lineages might have played in the evolution of the organisms that contribute to the mortality of countless millions of our own species.

8.3.1 Reticulate evolution and viral pathogens

Viral infections cause millions of deaths annually among humans. For example, in the year 2003, the HIV progressed to AIDS leading to the death of >2 million people in Africa alone (Asiimwe-Okiror *et al.* 2005). Similarly, influenza kills an estimated 250,000–300,000 people annually (WHO 2003), with pandemics such as the "Spanish flu" accounting for 10 times that number (Gibbs *et al.* 2002). Furthermore, as will be seen in the following discussion, the origin and evolutionary change among many of the most deadly present-day viral pathogens has been relatively recent and extremely rapid. The recency and rapidity of evolutionary origin and change among these viruses—and indeed viruses in general—often results from their ability to incorporate many mutations quickly. Though many of these mutations are at the level of single base pairs, a high frequency of genomic modifications in these human pathogens result from recombination between divergent evolutionary lineages and thus reflect reticulate evolution.

Ebolavirus

The 1995 movie "Outbreak" starring Dustin Hoffman, Morgan Freeman, *et al.* begins with a July 1967 viral epidemic in the Motaba River Valley, Zaire. The so-called Motaba virus causes mortality in 100% of infected persons. Horrifyingly, the virus causes the liquefaction of internal organs, causing death within three days postinfection (http://www.imdb.com/title/tt0114069/plotsummary). The writers of the screenplay must have taken their cue from the origin of the Ebolavirus, first recognized in 1976 from an outbreak in the Sudan, near the border with Zaire. This virus causes Ebola hemorrhagic fever that, as its name implies, causes internal and external bleeding in at least some of those infected (Centers for Disease Control and Prevention 2002). Furthermore, the extraordinarily high frequency of mortality from Ebolavirus infection—~50–88% for the most virulent "Zaire ebolavirus" form (Centers for Disease Control and Prevention 2002)—is also reminiscent of the "Outbreak" plot.

The encapsulation of the real-world epidemic by the fictionalized story both over- and undershoots the mark. As mentioned above, the mortality rate is not 100%, and the horrific pathology of the movie-rendition of the epidemic is seen in only a proportion of infected humans. Furthermore, the major casualties of the spread of ebolavirus are actually nonhuman primates such as gorillas and chimpanzees as well as other mammals including duiker and bush pigs (Leroy *et al.* 2004; Rouquet *et al.* 2005; Lahm *et al.* 2007). Thus, in a single decade ~5,000 gorillas were killed by ebolavirus infections (Bermejo *et al.* 2006). Indeed, the population sizes of gorillas and chimpanzees in Gabon were reduced by 50% during the period 1983–2000 (Walsh *et al.* 2003). Though much of this reduction was due to bush-meat/market hunting, the spread of ebolavirus now ranks as an equal threat to the survival of the great apes (Walsh *et al.* 2003).

In total, several thousand people have died to date from the various epidemics of ebolavirus that began in 1976 (Pourrut *et al.* 2005). As mentioned above, the most virulent strain of the virus is known as the Zaire ebolavirus "species." Until recently, only a few viral genotypes, all belonging to a single lineage and all isolates derived from infected humans, had been identified (Wittmann *et al.* 2007). In their recent analysis, Wittmann *et al.* (2007) determined the genotypes at two viral loci from both human and nonhuman infections occurring from 1976 to the present. Their results indicated that the more recent outbreaks involved a unique set of genotypes relative to earlier incidences. Of most significance in the present context, is that the most recent epidemics in humans and nonhuman primates have involved a hybrid virus. Figure 8.2 illustrates the phylogenetic evidence for this inference. As expected from a hybrid origin hypothesis, the position of the "recombinant viruses" in the two trees (derived from sequences from two different genes) alternates between the two putative parental strains (i.e., "Group A" and "Group B" viruses; Figure 8.2). Wittmann *et al.* (2007) estimated that the origin of the A+B hybrid viruses occurred sometime between 1996 and 2001, resulting in viruses responsible for the highly virulent epidemics in 2001–2003. Critically, Wittmann *et al.* (2007) pointed out that "The potential for recombination adds an additional level of complexity to unraveling and potentially controlling" the Zaire form of the ebolavirus.

Poliovirus

Poliovirus (PV), the causative agent of poliomyelitis…is a member of the genus *Enterovirus* of the Picornaviridae…Other members of the picornavirus family include the genera of *Rhinovirus* (common cold virus)…*Hepatovirus* including human hepatitis A virus, and…*Apthovirus* (foot and mouth disease virus). (Mueller *et al.* 2005)

The fight to eradicate the occurrence of poliomyelitis through vaccinations has been very successful as reflected by the reduction in cases from 350,000 in 1988 to ~2,000 in 2005 (Arita *et al.* 2006).

Though the WHO set the year 2000 as the target date for worldwide eradication, outbreaks of poliomyelitis in various parts of the world are still occurring (e.g., 46 cases in the "Eastern Mediterranean Region" in 2007; http://www.emro.who.int/polio). In this regard, much concern continues to be voiced over the evolution of new forms of poliovirus (PV). In particular, there have been recent discussions of data indicating the origin of highly divergent "vaccine-derived poliovirus" (i.e., "VDPV") lineages (see Kew *et al.* 2005 and Agol 2006 for reviews). The origin of such viruses is doubly significant in that (1) the vaccination-based

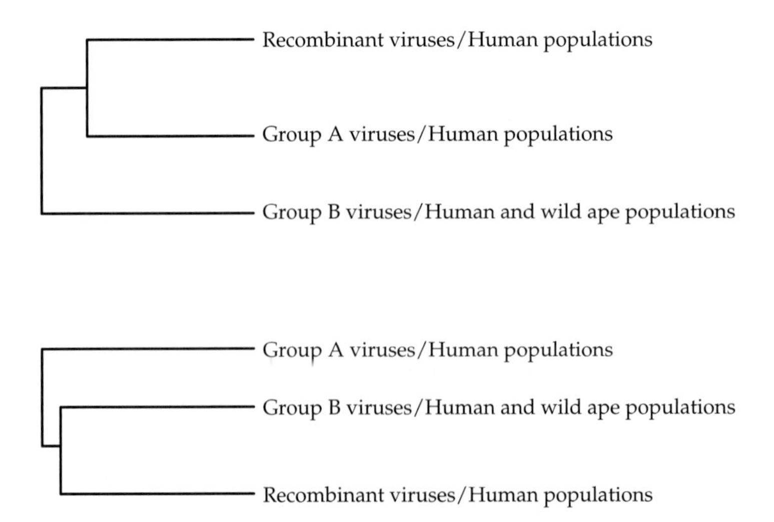

Figure 8.2 Phylogenetic trees for genotypes of the Zaire ebolavirus. The upper and lower phylogenies were derived from sequences at the glycoprotein and the nucleoprotein loci, respectively. Note that the position of the "recombinant viruses," isolated from recent human infections, changes depending on which gene is used. This is characteristic of hybrid lineages and indicates that this group of viruses originated from recombination between Group A and Group B lineages (Wittmann *et al.* 2007).

extinction of PVs was thought possible because there was no zoonotic reservoir in which the virus could be maintained and thus, elimination in the human host would drive the virus to extinction, but, instead, (2) the tool used to eradicate the virus is leading to new epidemics (Kew *et al.* 2005; Agol *et al.* 2006; Arita *et al.* 2006).

Jiang *et al.* (2007) recently inferred a reticulate origin for the diverse VDPVs. In particular, these workers detected phylogenetic and phenotypic (i.e., in terms of pathology) signatures indicative of their genesis from recombination between the attenuated viruses contained in the vaccines and related coxsackie A viral ("CAV") lineages. These viruses co-occur and utilize the gastrointestinal (GI) tract as their primary region for replication (Mueller *et al.* 2005; Jiang *et al.* 2007), with the CAV class considered to be the evolutionary ancestors of PVs. Kew *et al.* (2005) and Jiang *et al.* (2007) have presented evidence supporting a hybrid derivation for new PVs from recombination between these ancestral lineages and the viruses contained in vaccines. Their co-occurrence in the GI tract is considered a possible catalyst for this genetic exchange (Jiang *et al.* 2007).

The evidence for recombination-driven evolution of unique PVs resulting in new outbreaks of poliomyelitis is particularly strong from both the sequence data and experimental manipulations carried out by Jiang *et al.* (2007). These workers first detected discordance between phylogenies derived from sequence information for either the "P1" or "P2+P3" regions of the genome. The P1 sequences resolved the CAV and PV viruses into separate clades. In contrast, the phylogeny derived from the P2+P3 region sequence information reflected an admixture of the CAV and PV lineages. The reticulate origin of these viruses was supported by their analyses of experimentally derived CAV × PV hybrids. These latter viruses were similar to PVs in their replication and patterns of neurovirulence in mice (Jiang *et al.* 2007). The sobering conclusions drawn from their results included: (1) highly virulent, recombinant lineages will likely replace the original PVs; and, paradoxically, (2) the origin of such hybrid lineages will have even more serious consequences as vaccination frequencies drop due to the [mistaken] perception that poliomyelitis has been eradicated (Jiang *et al.* 2007).

Influenza virus

I have previously discussed some of the evidence for the importance of reticulate evolution in the origin and diversification of both the influenza and human immunodeficiency (i.e., HIV) viral clades (Arnold 2006 pp. 18–22 and pp. 62–63, 67, 69–71, respectively). However, the critical role of these organisms as major human pathogens requires that I discuss them further in this, and the following, section. I will expand my earlier description of the reticulate nature of their evolution by highlighting data derived mainly from the most recent analyses. I will omit a large number of findings, by numerous authors, and thus direct the reader to Arnold (2006) for additional references.

As mentioned at the outset of this section concerning viral pathogens, influenza in its various iterations has accounted for hundreds of millions of deaths during the twentieth and twenty-first centuries alone. About 10–20% of the world's population of *H. sapiens* is affected by epidemics each year, with an estimated 3–5 million severe cases of influenza occurring annually (WHO 2003); occasional pandemics lead to human mortality in the tens of millions (Gibbs *et al.* 2002). However, the effect of influenza is not limited to human suffering and death. For example, it has been estimated that the annual impact of epidemics on the US economy—in terms of health care costs and lost productivity at work—is ~US$5 billion (WHO 2003). A further (major) cost accrues from the loss of poultry and livestock that also become infected with various influenza lineages. The sacrifice of the human-associated animals is a control measure for limiting both further spread among the animals, and the transfer of variants from the domesticated animals into human populations (Normile and Enserink 2004).

Reticulate evolution among influenza viruses comes in the form of genetic exchange via "reassortment" (i.e., "the exchange of gene segments between two different influenza viruses"; Ma *et al.* 2007). This definition given by Ma *et al.* (2007) was in the context of a study of viral reassortants isolated from infected swine populations in the United States. These viral lineages had a combined genetic constitution derived from both avian and swine influenza subtypes. The concern generated by the isolation of this particular hybrid type was that it was similar to the influenza lineage leading to the human pandemic of 1957 (Ma *et al.* 2007). Human epidemics and pandemics normally result from viruses with a combination of genes from avian and human viral sources. In this regard, the possibility of cocirculation of human- and avian-type viruses in pigs is seen as a major reservoir for new viral lineages. Specifically, "Pigs are purported to be a mixing vessel for avian and human influenza viruses because their tracheal epithelial cells carry receptors for both human and avian influenza viruses... In this light, pigs have often been implicated in the emergence of human pandemic strains" (Ma *et al.* 2007).

Regardless of whether or not there are intermediate hosts in which influenza viruses with mosaic genomes arise, as mentioned above, reassortant lineages have been the basis for both domesticated animal and human epidemics/pandemics. Thus, the 1957 and 1968 pandemics, along with the present-day, drug-resistant lineages and the "bird flu" viruses are all products of reassortment (Obenauer *et al.* 2006; Stevens *et al.* 2006; Simonsen *et al.* 2007). With regard to the heightened fears concerning the evolution of the bird flu variants into a form that will cause a human pandemic, it is significant that recent genetic analyses have revealed extensive reassortment, with newly isolated variants possessing novel gene combinations (Chen *et al.* 2006; Obenauer *et al.* 2006). Of particular concern is that genetic exchange of this type will facilitate some recombinants the opportunity to "acquire many of the key adaptive mutations" (Kuiken *et al.* 2006) for increased infective capabilities resulting in another devastating pandemic.

Human immunodeficiency virus

The organization UNAIDS estimated that 33 million people worldwide were living with HIV infection during 2007, with ~2.5 million new infections and 2 million deaths occurring during this same period (UNAIDS 2007). The African region remains at the top of the list of most affected areas, with infection frequencies exceeding 25% in some southern African countries (WHO 2005). However, human populations in other regions of the world are also profoundly affected. For example, an estimated 2,500,000 people are infected with HIV in India; the disease is being spread primarily by unprotected sexual contact between sex workers and their clients (UNAIDS 2007). Similarly, in Vietnam the AIDS epidemic

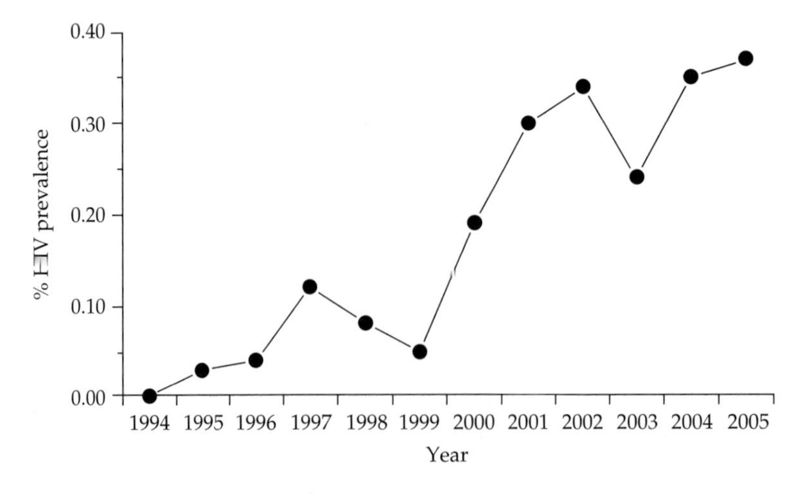

Figure 8.3 Frequency of HIV infections within the population of pregnant women in Vietnam who attended antenatal clinics in 1994–2005 (UNAIDS 2007).

is driven by unprotected sex involving sex workers, and the use of contaminated injection systems for drug application. This has led to a doubling in the number of infections from 120,000 in the year 2000 to 260,000 in the year 2005 (UNAIDS 2007). Figure 8.3 illustrates one of the most tragic outcomes with the frequency of pregnant women increasing from 0% in 1994 to over 0.35% in 2005. The introduction of the virus into this portion of the population is due largely to sexual contact with "males who were infected during unsafe paid sex and injecting drug use" (UNAIDS 2007).

The reticulate evolutionary trajectory of HIV involves recombination between divergent viral lineages in multiply infected humans (Heeney *et al.* 2006). The likelihood that such recombination-based changes act as catalysts for HIV evolution is reflected by the ~0.5 ratio of recombinational to point mutations (Edwards *et al.* 2006). Thus, half of all changes are due to genetic exchange between divergent viruses, with a proportion of these events expected to lead to novel adaptations and thus increased fitness in the recombinant viruses (Heeney *et al.* 2006; but see Bretscher *et al.* 2004). Levy *et al.* (2004) suggested that increased viral infection rates (with multiple lineages) would increase the frequency of recombination resulting in the "escape" of viruses from attacks by the host immune system. Dang *et al.* (2004) determined experimentally that multiple infections occurred more often than expected under a random model. This latter observation suggests that the foundation (i.e., multiple infections) for recombination is present in infected humans.

The predicted recombinant HIV lineages do indeed exist within human lineages, with up to 40% and 30% of HIV infections in Africa and Asia, respectively, involving hybrid viruses (Chin *et al.* 2005). Significantly, it is now understood that the pattern of rampant genetic exchange leading to these mosaic viral types is facilitated by regions of the genome that are hotspots for recombination (Minin *et al.* 2007). The prevalence of recombination in HIV is less surprising when one realizes that even the direct ancestral lineage was also a hybrid. HIV is derived from simian immunodeficiency viral lineages. The ancestor of HIV-1 (the cause of the worldwide human pandemic) was a virus termed, SIVcpz. The origin of the SIVcpz virus involved a reticulate event in which SIV lineages from red-capped mangabeys and *Cercopithecus* monkeys recombined to form the progenitor of HIV-1 (Keele *et al.* 2006). Thus, reticulate evolution has resulted in both the formation of the original viral lineage as well as the basis for the continued evolution of this viral plague.

8.3.2 Reticulate evolution and bacterial pathogens

There is a wealth of scientific literature, supported by historical treatises, that demonstrates the profound, negative impact of prokaryotic (i.e., bacterial) pathogens on human populations. For example, plague (caused by the bacterial species, *Yersinia pestis*) is estimated to have killed 200 million people during the period of human recorded history (Perry and Fetherston 1997). Although pandemics caused by *Y. pestis* are unlikely at present (due to effective antibiotic controls), other bacterial species cause severe illnesses in millions of *H. sapiens* annually. One such bacterial assemblage is the so-called Group A streptococci (i.e., hemolytic *Streptococcus pyogenes*). These bacteria cause a wide array of diseases, ranging from relatively mild skin infections to fatal invasive illnesses. There are currently an estimated 18 million infections that lead to pathologies such as rheumatic heart disease in worldwide human populations, causing >500,000 deaths per year (http://www.who.int/vaccine_research/diseases/soa_bacterial/en/print.html). In addition,

> Before the introduction of antimicrobials in the 1940s, the mortality rate of *S. aureus* invasive infection was about 90%. The initial success of antibiotherapy was rapidly countered by the successive emergence of penicillin-resistant, then methicillin-resistant *S. aureus* (MRSA) strains and, since 2002, by that of vancomycin-resistant strains…Case fatality rates in some *S. aureus* infections today still can reach 30%. (http://www.who.int/vaccine_research/diseases/soa_bacterial/en/index4.html)

It is well established that bacterial evolution is greatly affected by their ability to exchange genetic material with closely or distantly related lineages (e.g., see Ochman *et al.* 2000; Arnold 2006; Chapters 1 and 6 of this book). One indication of the pervasive effects from such horizontal gene transfers is reflected by the debate over the concept of bacterial "species." Ward *et al.* (2008) recently reviewed data from evolutionary studies of prokaryotes to illustrate the need for correlating genomic and ecological information to identify units definable as prokaryotic species. One problem with classical taxonomic designations of prokaryotic species, as highlighted by Ward *et al.* (2008), has been the use of phenotype to group bacteria into a species. As pointed out by Ochman *et al.* (2000), the weakness of such a systematic approach resides in the fact that many of the phenotypes used (e.g., "utilize lactose") are due to the lateral exchange of genes, rather than being a reflection of descent from a common ancestor. In the following sections, I will present examples of human bacterial pathogens for which reticulate evolution has been detected. I have omitted numerous important lineages and entire (e.g., *Staphylococcus* and *Streptococcus*) clades due to space limitations. I would direct the reader to Arnold (2006) for discussions of additional human bacterial pathogens.

Rickettsia

When a pathogen affects human populations, world affairs can be altered. HIV is a present-day example of an organism with such an effect. In particular, sub-Saharan African countries have lost up to one-fourth of their populations with all of the expected dire effects seen from a social, economic, and health standpoint. Similarly, when >200 million humans were lost from the series of plague pandemics, the same types of detrimental effects occurred. One of the associated effects inferred for many epidemics/pandemics is the loss of wars owing to the disease-mediated death of soldiers. In this regard, Raoult *et al.* (2006) tested the hypothesis that the defeat of Napoleon's army during their Russian campaign might have been catalyzed by diseases caused by bacterial pathogens. In particular, they tested the remains of Napoleonic soldiers from a mass grave in Vilnius, Lithuania for the presence of (1) lice known to act as a vector for numerous bacteria, including *Rickettsia prowazekii*, the causative agent of epidemic typhus (Raoult and Roux 1999) and (2) pathogenic bacteria including *R. prowazekii*. Their findings demonstrated the presence of both the disease vectors and the bacteria. Specifically, they found samples of lice associated with fragments of the soldiers' clothing in the gravesite. Genotyping of dental material from the remains of 35 soldiers identified 10 (i.e., 29%) that were infected with bacterial pathogens, 3 (i.e., 10%) of whom were infected with *R. prowazekii* (Raoult *et al.* 2006). Of the 25,000 French soldiers who reached Vilnius, only 3,000 are thought to have survived. Given that one-third appear to have been infected with louse-borne pathogens, Raoult *et al.* (2006) argued that *R. prowazekii* and other disease-causing bacteria may "have been a major factor in the French retreat from Russia."

The genus *Rickettsia* is divided phylogenetically into three subgroups—the "typhus," "spotted fever," and "*Rickettsia bellii*" groups (Roux *et al.* 1997; Blanc *et al.* 2007). Analyses of the genomes of two members of the spotted fever group (*Rickettsia massiliae* and *Rickettsia felis*) detected sequence variation consistent with reticulate evolutionary processes. First, in a study of the genomes of *R. felis*, Ogata *et al.* (2005) detected the unusual presence (for this class of bacterial pathogen) of a conjugative plasmid (i.e., capable of horizontal transfer between cells). Furthermore, they detected the lateral exchange of phage-related genes into the *R. felis* chromosome. Finally, though no nonambiguous candidates for lateral gene acquisition from other *Rickettsia* species were identified, it was inferred that many of the "*R. felis*-specific genes" had been obtained from lateral gene transfer from distantly related lineages (Ogata *et al.* 2005).

A second study, involving phylogenetic comparisons of the pathogen, *R. massiliae* with other members of the *Rickettsia* species complex also detected genomic variation indicative of lateral transfer. Figure 8.4 illustrates the phylogenetic relationships of various *Rickettsia* species based on either ribosomal or *tra* gene sequences. The products from the *tra* gene family have been suggested to function in conjugal DNA transfer (Blanc *et al.* 2007). Of most significance is that all of the *tra* loci apparently reflect genetic exchange (Blanc *et al.* 2007). The basis of this inference is partially

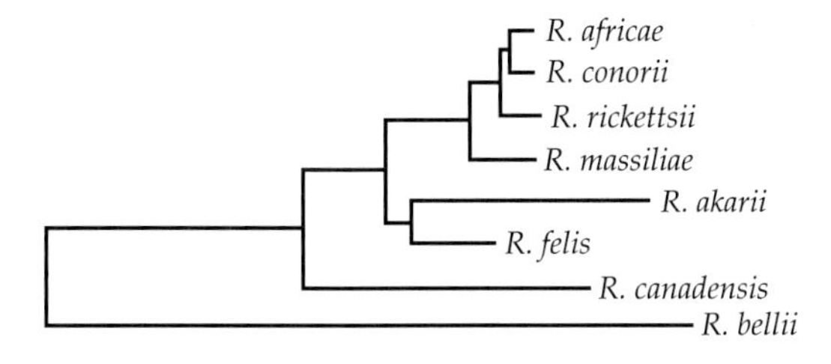

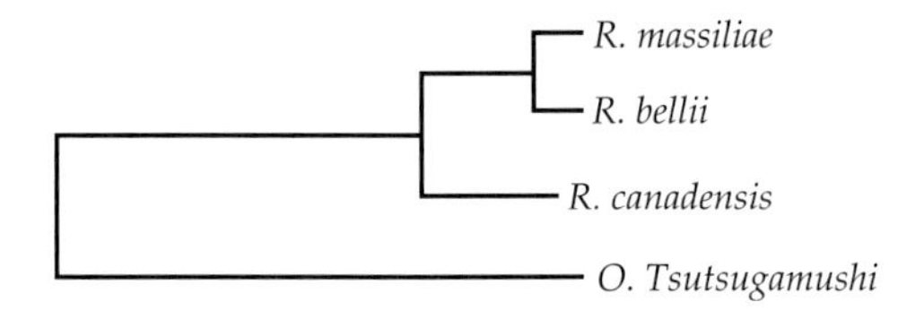

Figure 8.4 Phylogenetic trees based on ribosomal DNA (top) or *tra* gene (bottom) sequences. The top phylogeny is considered to reflect accurately species relationships. The different topology in the *tra* sequence phylogeny reflects the affect of lateral transfer events of these genes among various *Rickettsia* bacterial species (Blanc *et al.* 2007).

illustrated by the phylogenies in Figure 8.4. The ribosomal gene tree reflects the accepted species relationships (i.e., those supported by most gene sequences) while the *tra* phylogeny is discordant from the species tree. The conclusion of Blanc *et al.* (2007) was that horizontal gene transfer of this region among the various *Rickettsia* species had resulted in this discordant phylogenetic pattern. Although the molecular mechanisms are quite different, the incorporation of the conjugative plasmid into *R. felis* and the lateral transfer of the *tra* gene region into *R. massiliae* both reflect the reticulate evolutionary history of human pathogens.

Vibrio cholerae

During the 19th century, cholera spread repeatedly from its original reservoir or source in the Ganges delta in India to the rest of the world, before receding to South Asia. Six pandemics were recorded that killed millions of people across Europe, Africa, and the Americas. The seventh pandemic, which is still ongoing, started in 1961 in South Asia, reached Africa in 1971 and the Americas in 1991...A total of 236,896 cases were notified [in 2006] from 52 countries, including 6,311 deaths, an overall increase of 79% compared with the number of cases reported in 2005 (WHO 2007).

The causative agent of cholera, the bacteria *Vibrio cholerae,* is mainly transmitted through contaminated food and water. The people most at risk from cholera are those living under the most difficult of conditions—those in urban slums, refugee camps, and areas disrupted by natural disasters (WHO 2007). Furthermore, the virulence of *V. cholerae* is extreme in that, unlike other diarrhea-causing infections, otherwise healthy adults may die within hours of disease onset (WHO 2007).

Genetic exchange-mediated evolution of *V. cholerae* has been documented through a variety of analyses. Initially, Waldor and Mekalanos (1996) and Rubin *et al.* (1998) demonstrated that the toxigenic strains of this bacterial lineage differed from the nonpathogenic strains by the presence of genes encoded by a bacteriophage that had infected *V. cholerae*. This led these authors to infer several horizontal transfer events resulting in the transition from nonpathogenic bacteria to the pathogenic cholera-producing strain. These events included at least the acquisition of the "TCP pathogenicity island" genomic region, a bacteriophage receptor allowing the transfer of genes from the infection by the bacteriophage CTXϕ (Waldor and Mekalanos 1996; Rubin *et al.* 1998).

More recent analyses have revealed further the degree to which reticulate evolutionary processes have contributed to the evolution of the *V. cholerae* pathogen. Dziejman *et al.* (2002) defined genes characteristic to all pathogenic strains (relative to nonpathogenic lineages) and a set of genes unique to the strain of *V. cholerae* responsible for the seventh (i.e., current) cholera pandemic. These unique genes were gained by horizontal transfer—resulting in the evolution of the seventh pandemic pathogen—and were hypothesized to have provided this strain with the ability to displace the preexisting pathogenic forms and to become established in formerly cholera-free geographical regions (Dziejman *et al.* 2002). Faruque *et al.* (2007) extended the definition of horizontal transfer events that led to the evolution of the latest pandemic-causing strain. In particular, they produced data consistent with the model of

Rubin *et al.* (1998) in which an additional horizontal transfer event was necessary for the infection by the toxin gene donor bacteriophage. It thus appears that the evolutionary trajectory and the pathogenicity of *V. cholerae* have largely resulted from the incorporation of genetic material from nonrelated (i.e., bacteriophage) sources.

Helicobacter pylori

Human infection with the Gram-negative bacterium, *Helicobacter pylori*, exceeds 80% frequency in middle-aged adults in developing countries and 20–50% in the same demographic group in industrialized countries (Suerbaum and Michetti 2002). However, these infections are not a recent phenomenon. In fact, as determined from the geographic patterning of its genetic variation, *H. pylori* was a pathogen of *H. sapiens* before the latter's exodus from Africa (Linz *et al.* 2007). The consequences from infections by this bacterium include peptic ulcers, other upper GI pathologies, and gastric cancers (Suerbaum and Michetti 2002). Similar to the frequency of infections, the modes of transmission of *H. pylori* appear to differ between developing and developed countries. In the former, waterborne transmission is prevalent due to humans swimming in rivers and streams and because they often utilize streams as a source of drinking water (Goodman *et al.* 1996). In contrast, there is evidence that transmission in countries such as the United States is facilitated by bouts of GI illness (Parsonnet *et al.* 1999). However, whether in developed or developing societies, transmission is inversely correlated with socioeconomic classification—that is, children in the poorest and wealthiest situations demonstrated infection rates of 85% and 11%, respectively (Malaty and Graham 1994). Furthermore, those children living in crowded households had a 4.5 times greater chance of infection (Malaty and Graham 1994).

Xia and Palidwor (2005) tested the mechanism of gain of novel gene function or "adaptability" in *H. pylori*. They inferred this species to be highly adaptable. Xia and Palidwor (2005) assigned causality for this adaptability to several factors, including a high frequency of interstrain recombination and lateral gene transfer due to *H. pylori*'s natural competence for DNA transformation (i.e., the uptake and incorporation of DNA from the environment; Suerbaum *et al.* 1998). In this regard, two types of observations point to the role of genetic exchange in the evolutionary history and adaptive progression of *H. pylori*. The first class of data reflects recombination between different strains that have coinfected the same human host (Suerbaum and Achtman 2004). This type of recombination is apparently so common that "different loci and polymorphisms within each locus are all at linkage equilibrium" (Suerbaum *et al.* 1998). This frequency of recombination was unique relative to other bacterial lineages as well as the insect, *Drosophila melanogaster*, and resulted in *H. pylori* phylogenies that were unresolved (Suerbaum *et al.* 1998). The second category of observation comes from genome sequence analysis and has led to the inference of lateral gene transfer from other lineages. Specifically, Oh *et al.* (2006) detected several "restriction-modification" loci that were most similar to genes from other bacterial species. Their data also suggested a role for plasmid-mediated lateral gene transfers into *H. pylori* (Oh *et al.* 2006).

Orientia tsutsugamushi

As with the species that cause epidemic typhus and spotted fever, *Orientia tsutsugamushi*, the causative agent of "scrub typhus" or "tsutsugamushi disease," is an obligate intracellular rickettsia bacterium (Seong *et al.* 2001). Infection with this pathogen is caused by bites of trombiculid mites, resulting in a disease pathology that can include fever, rash, eschar (i.e., dead tissue that sloughs off from healthy skin), pneumonia, myocarditis, and, in untreated cases, multiple organ failure leading to death (Philip 1948; Seong *et al.* 2001). During World War II the extent to which tsutsugamushi disease affected regiments equaled or exceeded the results of prolonged combat (Philip 1948). The severity of the effect from this disease, like pathogens encountered during other wars, was compounded by a lack of control and/or medical facilities, with mortality rates for some years exceeding 20% (Philip 1948). Epidemic outbreaks of *O. tsutsugamushi* are, however, not limited to periods of war. For example, during the summer of 2002, an outbreak of scrub typhus in the Republic of Maldives claimed the lives of about 6% of all those infected (Lewis *et al.* 2003).

The sequencing of this species' genome has recently facilitated a description of processes associated with the evolution of *O. tsutsugamushi*. For example, Cho *et al.* (2007) detected the "massive proliferation of conjugative type IV secretion system" genes. This large-scale duplication has resulted in hundreds of copies of gene sequences scattered throughout this genome. Though Cho *et al.* (2007) argued for the importance of *intragenomic* recombination as a source for the multitude of gene duplicates, they also recognized that the genomic patterns are explainable by the widely seen process of plasmid invasions. Indeed, Cho *et al.* (2007) pointed to the "host-switching" life history of the *O. tsutsugamushi* pathogen as a catalyst for such horizontal transfers. This lifestyle may have thus led to conjugation between divergent lineages of this pathogen and/or with *Rickettsia* species with which it coinfects its hosts, leading to genetic exchange (Cho *et al.* 2007).

Pseudomonas aeruginosa

The bacterial species *Pseudomonas aeruginosa* fulfills roles as both a common component of soil and water as well as a human pathogen (see Liang *et al.* 2001 for a review). In the latter niche, this species causes diseases ranging from mild skin infections to acute and invasive pathologies of damaged tissues such as eyes, burned areas, and other types of wounds. *P. aeruginosa* also causes urinary and respiratory tract infections (Liang *et al.* 2001). Often infections with this bacterium are associated with hospitalization. Furthermore, those admitted to Intensive Care Units and/or for surgery may be at particular risk of contracting *P. aeruginosa* (Tacconelli *et al.* 2002). This is not surprising given the bacterium's propensity for invading damaged tissue. Of most serious concern is the difficulty encountered in ridding patients of these infections. It appears that the major impediment for eradicating *P. aeruginosa* from those infected has been its ability to gain multidrug resistance to antibiotics (Tacconelli *et al.* 2002). The difficulty in curing *P. aeruginosa* infections is thus reflected in relatively high mortality rates (Tacconelli *et al.* 2002).

The adaptive breadth, both in terms of resistance to antibiotics and habitats occupied, of *P. aeruginosa* is of major public health concern and also reflects the role of genetic exchange in the origin and continued evolution of this species. One of the primary mechanisms by which bacterial lineages may adapt to their environment is through the incorporation of DNA from outside sources in the form of "genomic islands." Hacker and Carniel (2001) coined this term to reflect the incorporation of "blocks of DNA with signatures of mobile genetic elements." These authors also hypothesized that many genomic islands might have increased the recipient bacterium's fitness and thus have been selected for (Hacker and Carniel 2001). Figure 8.5 illustrates one such genomic island identified in 85% of the strains of *P. aeruginosa* isolated from human infections (Liang *et al.* 2001). In many of these strains, there

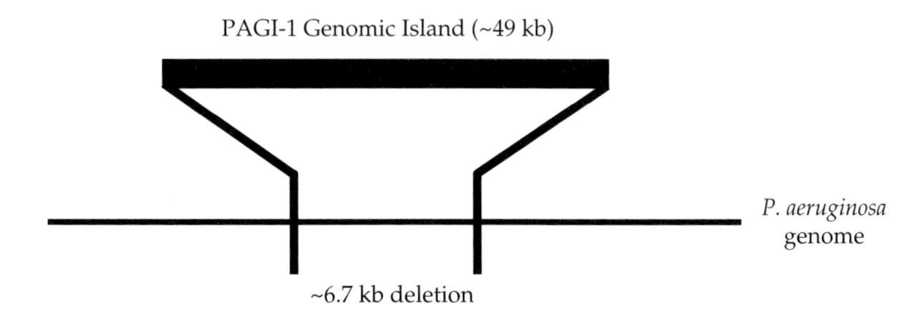

Figure 8.5 Diagram illustrating the insertion of the "PAGI-1 Genomic Island" into the genome of the bacterial pathogen, *P. aeruginosa*. This genomic island is associated with about 85% of all clinical isolates and, in approximately half of the isolates, is associated with a 6.7 kb deletion (Liang *et al.* 2001).

has been a concomitant deletion of ~6.7 kb of DNA at the insertion site (Figure 8.5). Of particular significance is the hypothesis that some of the >50 genes encoded by this genomic island allow *P. aeruginosa* to survive under harsh (for this bacterial species) environmental conditions (Liang *et al.* 2001). Similarly, Larbig *et al.* (2002) detected a second genomic island within this bacterial genome consisting of as many as 111 genes. As with the genomic island identified by Liang *et al.* (2001), this genomic insert was also hypothesized to have provided recipient *P. aeruginosa* with a selective advantage (Larbig *et al.* 2002).

Numerous studies have revealed both the lack of recombination within large portions of the *P. aeruginosa* genome, and at the same time the plasticity of regions associated with genomic islands (e.g., Hacker and Carniel 2001; Qiu *et al.* 2006; Wiehlmann *et al.* 2007). The plastic nature of the genomic islands, and in particular those regions identified as harboring virulence genes (i.e., "pathogenicity islands"; Hacker and Carniel 2001), is due to their freely recombining nature (Wiehlmann *et al.* 2007). As with genomic islands in general, pathogenicity islands are seen as potential targets of selection (He *et al.* 2004). In particular, these regions have likely facilitated evolutionary changes in *P. aeruginosa* resulting in an increase in the number of niches it could invade thus leading to its ecological transition from a soil/water inhabitant to a major human pathogen (He *et al.* 2004; Qiu *et al.* 2006).

Minibacterium massiliensis

Water-borne pathogens are of major concern for human populations. However, from the pathogen's perspective, many of their potential habitats are extremely hostile due to measures taken by *H. sapiens*. For example, water sources for semiconductor, pharmaceutical, and food-industrial complexes are routinely treated to affect the eradication of water-associated bacteria (Kulakov *et al.* 2002). Although purification steps may include filtration, ultraviolet light, heat, and ozonation, the "complete removal of contaminating microorganisms is considered to be nearly impossible" (Kulakov *et al.* 2002). I will use a newly named bacterium (inferred to be a human pathogen) isolated from purified hospital water supplies as an example of such a species. Though the literature on this species is limited by its recent discovery to a single reference (Audic *et al.* 2007), it is a wonderful example of the evolutionary processes (including reticulation) associated with the origin and adaptation of a bacterial pathogen. It also illustrates our limited knowledge of the diversity of pathogens that surround us, even in the relatively rarefied (at least in terms of bacterial populations) environments of hospital water supplies.

As alluded to above, microorganisms that are associated with water systems utilized by humans for industrial or health care applications often demonstrate adaptations that allow them to escape from biocontrol measures. For example, the causative agent of Legionnaires disease, *Legionella pneumophila*, has the ability to maintain its populations in the incredibly demanding environment of plumbing systems treated with biocides (Chien *et al.* 2004). The repeated isolation of the species, *Minibacterium massiliensis*, from hemodialysis units over a 7.5-month period reflects another bacterial species capable of resisting the extreme purification measures taken to produce ultrapure water supplies. In particular, this organism had the ability to sustain its populations in the 0.22-µm filtered water used for the dialysis system (Audic *et al.* 2007). The obvious danger of a pathogen in direct contact with patients' blood led to a genome sequencing exercise to define the genetic components of this bacterial lineage. This analysis not only indicated the presence of a new species, but also illustrated the degree to which horizontal gene transfer had helped shape its genomic composition and likely its evolutionary and ecological trajectory (Audic *et al.* 2007).

Several classes of genomic information from the *M. massiliensis* sequence analysis indicated the pervasive effects from reticulate evolution. These included: (1) detection of phage insertion into two genomic regions; (2) discordance between phylogenies constructed from either ribosomal DNA genes or the sequences of genes associated with the water-borne lifestyle of this bacterium; and (3) the identification of >600 potential gene sequences that are more similar to bacteria from clades other than that within which *M. massiliensis* was thought to belong (Audic *et al.* 2007). Once again, the lateral transfer of so many genes from divergent lineages includes events that likely resulted in the transfer of adaptations necessary for survivorship in the environments now inhabited by *M. massiliensis*. For example, Audic *et al.* (2007) identified a genomic island encoding genes that would provide this bacterium with resistance to metalloids and heavy-metal ions regularly used for purification. In addition, this organism possessed antibiotic resistance loci that protected it against both penicillin and streptomycin (Audic *et al.* 2007). Finally, numerous genes associated with virulence traits in other pathogens were also detected. Taken together, these findings led Audic *et al.* (2007) to recognize *M. massiliensis* as a new health threat for humans, particularly because of its "capacity to acquire and promote the exchange of virulence factors and resistance genes."

8.3.3 Reticulate evolution and protozoan pathogens

Infections of protozoan parasites affect a large proportion of *H. sapiens* individuals. For example, >1 million people are infected with *Toxoplasma gondii* (Boyle *et al.* 2006), with Toxoplasmosis (the disease caused by *T. gondii* infections) accounting for the third highest number of food-related fatalities in the United States (http://www.cdc.gov/toxoplasmosis). Furthermore, members of the protozoan genus *Cryptosporidium* cause the diarrheal disease, Cryptosporidiosis. Over the past 20 years, *Cryptosporidium* infections have become one of the most frequent causes of waterborne disease (both from recreational and drinking water supplies) in the United States. Furthermore, *Cryptosporidium* and people burdened by its infections are present throughout the world. Many protozoan infections have a limited mortality rate among otherwise healthy humans, because their immune system is capable of ridding their bodies of the pathogen (Fricker *et al.* 1998). However, with the growing percentage of the human species affected by immunocompromising diseases such as HIV/AIDS, infections by the numerous protozoan pathogens are becoming of increasing concern (Fricker *et al.* 1998).

Toxoplasma gondii

As cited above, an estimated one-fourth of the human population has chronic infections of the protozoan parasite, *T. gondii* (Boyle *et al.* 2006; Khan *et al.* 2007). A large proportion of these infections, like those of other protozoans, will remain unnoticed by those infected (http://www.cdc.gov/toxoplasmosis). However, a significant number of the *T. gondii* infections will result in serious harm and even death. For example, of the population of congenitally infected European newborns, 1–2% will exhibit learning disabilities and/or die from the parasite load (Cook *et al.* 2000). In this same sample, 4–27% of the babies develop lesions resulting in permanent impairment of their vision (Cook *et al.* 2000). The percentage of pregnant women infected with *T. gondii* has ranged from a low of 10% in the United Kingdom and Norway to >50% in France and Greece (Cook *et al.* 2000). Although there are signs that the overall frequency of infection in pregnant women may have temporarily declined, it appears that pregnant women are now at greater risk than in the past. This is due to a shift in eating habits in the European population away from (1) food animals raised indoors, to those that are raised outdoors (i.e., "organic meat"), leading to a higher potential that the animals will contract *T. gondii* and (2) beef to the more-likely-to-be-infected lamb and pork products (Cook *et al.* 2000).

Comparisons of lineages within *T. gondii* have detected significant divergence. In particular, comparisons of the pathogen assemblages found in South America with those from North America and Europe indicated a divergence from a common ancestor ~10^6 ybp (Khan *et al.* 2007). These genomic comparisons also indicated that the divergent lineages were recently brought into contact, probably through the inadvertent transport of nonhuman hosts of *T. gondii* (i.e. rats, mice, and cats) on transatlantic slave ships (Lehmann *et al.* 2006; Khan *et al.* 2007). The evolutionary history of this pathogen clade has also involved hybridization between various lineages and, indeed, the formation of widespread hybrid pathogens (Boyle *et al.* 2006). Although the frequency of hybridization events (i.e., through sexual recombination) between divergent members of the *T. gondii* assemblage has been estimated at less than 5%, it has resulted in novel evolutionary lineages (Figure 8.6; Boyle *et al.* 2006; Lehmann *et al.* 2006). Boyle *et al.* (2006) used genomic comparisons to infer a hybrid derivation of so-called types "I" and "III" from a cross between a type "II" strain and one of two common ancestral lineages (Figure 8.6). Specifically, these workers determined that strains III and I were first- and second-generation hybrid offspring, respectively (Figure 8.6; Boyle *et al.* 2006). In spite of the relative rarity of sexual recombination and thus hybridization, reticulate evolution has obviously played a significant role in both the origin and genetic structuring of the globally distributed lineages of *T. gondii* (Boyle *et al.* 2006; Khan *et al.* 2007).

Consistent with the above conclusion—that hybridization between different lineages of *T. gondii* can have evolutionary as well as epidemiological significance—were findings from an experimental analysis by Grigg *et al.* (2001). These authors reported the outcome of introducing an F_1 generation hybrid (between two relatively avirulent, lineage II and III strain isolates) into mice. The result of the recombination between these two benign forms was a hybrid that demonstrated high lethality for the mice, even when they were inoculated with only one parasite (Grigg *et al.* 2001). This resulted in the conclusion that "through random genetic reassortment, two avirulent strains of *Toxoplasma gondii* can give rise to highly virulent progeny" (Grigg *et al.* 2001). This suggests that the reticulate evolutionary history of the present-day *T. gondii* strains may have provided the genetic basis for virulence.

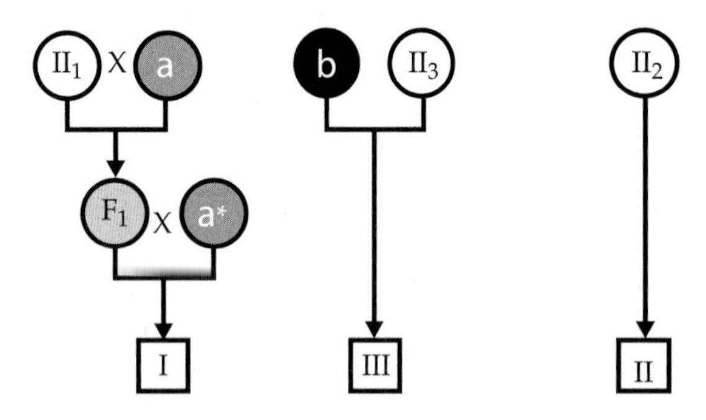

Figure 8.6 The evolutionary derivation of the three present-day *T. gondii* lineages ("I", "II," and "III"—at the bottom of diagram) inferred by Boyle *et al.* (2006) from their genomic analyses. Strain II was proposed to have derived directly from the ancestral strain II_2, while strains I and III were inferred to be a first generation backcross and an F_1 hybrid lineage, respectively.

Leishmania

Leishmaniasis is a disease complex caused by various species of *Leishmania* (Dujardin 2006). Vectors differ between the Old and New Worlds. In its Old World distribution, the protozoan is usually transmitted to humans by the bite of female sandflies (genus *Phlebotomus*) that are initially infected by feeding from blood sources (either humans or other animals) laden with *Leishmania* (Desjeux 2001, 2004). The disease epidemiology takes several forms, including those that affect the skin or the internal organs. The former can result in disfigurement due to tissue destruction, while the latter is fatal if left untreated because of damage to such organs as the liver and spleen (Desjeux 2004; http://www.cdc.gov/healthypets/diseases/leishmania.htm).

Leishmaniasis occurs in 88 countries, threatens some 350 million persons, with a total of ~12 million infections, and 1.5–2 million new infections annually (Desjeux 2001; Dujardin 2006). Approximately 90% of all infections occur within the disease foci of the Indian subcontinent and Sudan (Dujardin 2006). Owing to the poor monitoring of cases of Leishmaniasis, accurate estimates of the annual death toll are lacking (Desjeux 2004). However, what has to be considered as potentially a significant underestimate (Desjeux 2004) has been set at ~60,000 persons (Dujardin 2006). Recently, this disease complex has reemerged as a major threat to human health. This renewed, and increasing, danger for *H. sapiens* populations has been caused by three significant factors—human-mediated environmental changes, the increase in immunocompromised individuals, and treatment failures/drug resistance (Dujardin 2006).

Recently, both genomic sequencing and extensive genotyping of numerous globally distributed isolates have highlighted the effect of reticulate evolutionary change within the *Leishmania* assemblage (Ivens *et al.* 2005; Ravel *et al.* 2006; Lukes *et al.* 2007). First, the description of the complete genome sequence for *Leishmania major* included findings consistent with lateral gene transfers into this lineage. Specifically, a number of genes coding for peptidases (seven metallopeptidases, one peptidase T, and a serine peptidase) were inferred to have arisen through genetic exchange into *L. major* from prokaryotic lineages (Ivens *et al.* 2005). Reticulate evolution affecting *Leishmania* has also occurred through introgressive hybridization between various taxa. Ravel *et al.* (2006) identified hybrid progeny from crosses between the two species *Leishmania infantum* and *L. major*. These hybrids strains were detected in samples taken from immunocompromised (i.e., HIV/AIDS) individuals from Portugal. Their hybrid identity was inferred from the admixture of DNA sequences from the two parental species (Ravel *et al.* 2006).

The most extensive analysis to date of *Leishmania* lineages involved the description of evolutionary patterns and processes for members of the *Leishmania donovani* complex (Lukes *et al.* 2007). Of the many inferences drawn from the data, the most germane for the present discussion were those that identified hybrid lineages. It is possible to summarize these inferences with the following points (Lukes *et al.* 2007): (1) the uncertainties in grouping taxa phylogenetically is likely due to hybridization between divergent lineages; and (2) the highly heterozygous genomes of strains from Sudan and East Africa, in general, reflect the establishment of numerous hybrid genotypes. Lukes *et al.* (2007) suggested that the high frequency of hybrid genotypes in Sudan might be causal in the observed emergence of strains with elevated virulence responsible for increased frequencies of human mortality.

Trypanosoma

In Section 8.2.1, I discussed the evolutionary processes associated with the insect vector of the causal organism of Chagas disease. Though I have outlined elsewhere (see Arnold 2006, pp. 149–150) some of the evidence for the reticulate nature of the evolutionary trajectory of this causal agent, that is, *T. cruzi*, I will review these data again here. There are three reasons for this. First, *T. cruzi* is of major health concern for millions of humans. Second, this pathogen reflects well the processes of genetic-exchange-mediated evolution. Third, this is an excellent example of how both the vector of a human pathogen and the pathogen itself can reflect reticulate evolution.

As discussed above, an estimated 125 million Latin Americans live in danger of contracting Chagas disease, with ~18 million living with *T. cruzi* infections (Scientific Working Group on Chagas Disease 2005). Sadly, *T. cruzi* "acute" infections often remain undetected, and one-third of those with such infections will succumb to fatal damage to their heart, intestines, and/or nervous systems (Teixeira *et al.* 2006). One reason that *T. cruzi* infections can be so damaging is due to the occurrence of the integration of their mitochondrial (i.e., the "kinetoplast"; Hartwell *et al.* 2004, p. 529) genome into that of the host. Teixeira *et al.* (2006) have provided a review of studies that have detected this integration, and, most importantly, they have also reviewed the findings that suggest that this integration is causal in the pathology of Chagas disease. In particular, it has been suggested that the correlation between the integration events and the disease effects is likely due to the interruption of host gene sequences by the insertion of the parasite's genome (Teixeira *et al.* 2006).

Given that the manifestation of Chagas disease pathologies depend at least partly on insertion of pathogen DNA into the host genome, it can be said that continual genetic exchange is associated with *Trypanosoma* infections. However, the evolutionary history of this clade is also based on genetic exchange, in the form of both horizontal transfer and introgressive hybridization. I will illustrate this with only two studies, but I direct the reader to Arnold (2006, pp. 149–150) for a discussion of numerous references not cited here. The first analysis by Zamudio *et al.* (2007) described some of the regulatory mechanisms for three loci associated with methylation (i.e., "TbMTr1–3"). These functional analyses, along with previous studies suggested the inference of the horizontal acquisition of these loci from viral lineages (Zamudio *et al.* 2007).

In addition to lateral exchanges, the present-day *Trypanosoma* assemblage also possesses genetic

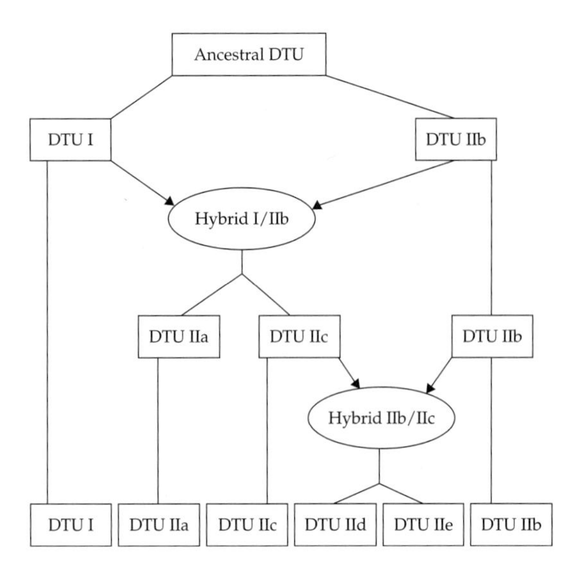

Figure 8.7 Model of the evolutionary processes that have given rise to present-day *T. cruzi* strains or "discrete typing units" (i.e., "DTU I, DTU IIa, DTU IIc, DTU IId, DTU IIe, and DTU IIb"). Two hybridization events are proposed leading to the formation of "Hybrid I/IIb" and "Hybrid IIb/IIc," which then formed the basis for further evolutionary diversification (Westenberger *et al.* 2005).

signatures reflecting the formation of new lineages via hybridization (Machado and Ayala 2001). For example, Figure 8.7 illustrates a model describing the evolutionary diversification of the pathogen, *T. cruzi*. Westenberger *et al.* (2005) described this hypothesis to account for the pattern of genetic variation found within the current isolates of this species. Thus, the ancestral "DTU" or "discrete typing unit" is proposed to have given rise to two lineages, both of which contributed to the formation of the DTU I/IIb hybrid lineage (Figure 8.7). DTU IIb also hybridized later in the evolutionary history of this clade with DTU IIc (i.e., a derivative of the original hybrid) to form a second recombinant lineage that then gave rise to DTU IId and DTU IIe (Figure 8.7; Westenberger *et al.* 2005). As with all the other examples of protozoan pathogens, the evolutionary history of the *Trypanosoma* clade is better typified as reticulate—from both lateral exchanges and introgressive hybridization events rather than simply divergent.

8.3.4 Reticulate evolution and protist pathogens

Hundreds of millions of humans are infected by protist pathogens. As one example (not discussed below, but see Arnold 2006, pp. 87–89), *Entamoeba histolytica*, the causative agent of amoebiasis (1) infects >1 million citizens of Mexico, (2) accounts for nearly 40% of all Egyptians treated as outpatients for acute diarrhea, and (3) is the second leading cause of deaths owing to parasitic infections, with an annual mortality of 40,000–100,000 persons worldwide (Stanley 2003). The two examples given below of protist-based diseases in humans—trichomoniasis and giardiasis—are also of major health concern, with the former present at frequencies of 2–13% of those belonging to at-risk populations and the latter present in about 2 out of every 10 persons in such groups (Fricker *et al.* 1998; Sutton *et al.* 2007). Of most significance for the present discussion, the evolution of parasitic protists has involved significant bouts of genetic exchange events. For example, Nosenko and Bhattacharya (2007) gave the following numbers of potential horizontal gene transfer events for the anaerobic parasitic protists: *E. histolytica*—96; *Trichomonas vaginalis*—152; *Cryptosporidium parvum*—24.

Trichomonas vaginalis

Infections with *Trichomonas vaginalis*, a sexually transmitted protist, are associated with the pathologies of preterm births, low-birth weight newborns, invasive cervical cancer, prostate cancer, and an increased probability of HIV transmission through sexual contact (Hirt *et al.* 2007; Sutton *et al.* 2007). As a part of the tremendous health burden of sexually transmitted diseases, trichomoniasis affects women and children most seriously (Sihavong *et al.* 2007). The 170 million cases worldwide (Carlton *et al.* 2007) reflect the severity of the overall risk to women and the young. However, there are factors that increase the risk of infection within these groups. For example, women in the United States were found to have a greater likelihood of being infected with *T. vaginalis* if they had a lower educational level, lived in poverty, were older, and/or had a greater number of sexual partners (Sutton *et al.* 2007). This disease does not just infect women, urethritis and prostatitis are the outcomes of infection in males (Carlton *et al.* 2007).

As mentioned above, Nosenko and Bhattacharya (2007) reported over 150 putative lateral exchange events involving the *T. vaginalis* genome. This estimate was based on the whole-genome sequence reported by Carlton *et al.* (2007). It is instructive to consider specific examples to highlight the pervasive impact of these genetic exchange events on entire gene complexes. Carlton *et al.* (2007) illustrated one such complex, involving those genes responsible for amino acid metabolism. These authors estimated that 15% of the total number of those genes inferred to have been transferred into the *T. vaginalis* genome were involved in amino acid metabolism. A similarly impressive number of putative surface protein encoding genes also demonstrated sequences indicating their lateral transfer from prokaryotes into the *T. vaginalis* lineage. In this latter case, the so-called *TpLRR* loci numbered >650 and were the largest of this type of gene-encoding family (Carlton *et al.* 2007; Hirt *et al.* 2007). The large numbers of these genes reflects not only their initial introduction via genetic exchange from divergent organisms, but also duplications during the evolution of the trichomoniasis-causing pathogen (Carlton *et al.* 2007).

Giardia lamblia

Giardiasis is a GI disease that affects hundreds of millions of people annually in both developing and developed countries. In the former, the frequency of *Giardia* infections can exceed 40%, while in the latter maximum infection rates are nearer 5% or less (Fricker *et al.* 1998). Reflecting its major impact on the infrastructure of "industrialized" populations, *Giardia* is the most common intestinal parasite identified by US public health laboratories (Yoder and Beach 2007). The presence of giardiasis in both industrialized and developing countries reflects the worldwide distribution of this water-borne pathogen. Thus,

> cysts of *Giardia* occur in the aquatic environment throughout the world. They have been found in most surface waters, where their concentration is related to the level of fecal pollution or human use of the water...The environmentally robust cysts are very persistent in water...and extremely resistant to the disinfectants commonly used in drinking-water treatment. (Fricker *et al.* 1998)

Such characteristics "place these organisms among the most critical pathogens in the production of safe drinking-water from surface water" (Fricker *et al.* 1998).

Phylogenetic analyses of *G. lamblia* have resolved numerous discordances in regard to this species' placement relative to other lineages. For example, some studies have placed this pathogen near the base of the eukaryotic clade (Hashimoto *et al.* 1995), while others have indicated a nearly simultaneous divergence with lineages that gave rise to clades such as plants (Morrison *et al.* 2007). These discrepancies are likely due to the lineage of origin of the different genes chosen to construct the phylogenies. This conclusion is supported by the analysis of the entire genome sequence of *G. lamblia*. In particular, Morrison *et al.* (2007) defined a major impact from lateral gene transfers from archaebacterial and bacterial lineages during the evolution of this pathogen's genome. Morrison *et al.* (2007) identified ~100 putative, laterally transferred genes. Many of these loci were a part of major metabolic pathways. For example, nearly 50% of the genes involved in the glucose, pentose-phosphate, and arginine metabolism pathways (for which the lineage of origin was identifiable) were present in the *G. lamblia* genome due to lateral gene transfer (Morrison *et al.* 2007). This result reflects the degree to which genetic exchange has, like for the other human pathogens, affected the evolutionary trajectory of the causal agent of giardiasis.

8.3.5 Reticulate evolution and fungal pathogens

Martin *et al.* (2003) and Shao *et al.* (2007) listed the following sobering statistics reflecting the impact of fungal infections on humans:

1. There has been a significant increase in the frequency of mortality owing to these organisms,

largely due to the difficulty for prompt diagnosis and the fragility of many of the patients who become infected.
2. With regard to "1," there is a significant correlation between the frequency of life-threatening fungal infections and the recent increase in the number of immunosuppressed patients.
3. The frequency of sepsis cases owing to fungal infections rose by 207% between 1979 and 2000—the largest increase for any pathogen category.
4. The most commonly encountered pathogens causing opportunistic invasive fungal infections have been *Candida albicans, Cryptococcus neoformans,* and *Aspergillus fumigatus.*
5. Owing to the widespread application of antifungal prophylaxis, new fungal lineages have become important as pathogenic agents, leading to even higher mortality rates.

Thus, although the ratio of mortality from fungal infections to the frequency of infections may be dropping, the overall number of deaths continues to increase (Martin *et al.* 2003).

Cryptococcus neoformans
Over a decade ago Mitchell and Perfect (1995) outlined the seriousness of *C. neoformans* infections on those persons with compromised immune systems. Specifically, they discussed the role that infections with this fungal pathogen played in mortality rates among those suffering with AIDS. First, they noted that the syndrome associated with these infections—that is, cryptococcosis—was often the first evidence that a patient had AIDS (Mitchell and Perfect 1995). Second, they reviewed data concerning the frequency of fatal *C. neoformans* infections in AIDS patients in various geographical regions. For example, 6–10% and 15–30% frequencies were encountered in the United States/western Europe/Australia and sub-Saharan Africa, respectively. Furthermore, it was found that AIDS patients from the southeastern United States and equatorial Africa had the highest mortality rates: 10–25% died during the initial therapy and 30–60% died within 12 months (Mitchell and Perfect 1995). As mentioned above, the number of deaths among immunocompromised patients from fungal infections, including those from *C. neoformans,* have continued to increase over the last two decades. The statistics reported by Mitchell and Perfect (1995) are thus dated in the sense that they are likely underestimates of the number of deaths, if not the overall frequency of infections (Martin *et al.* 2003).

There is a wealth of data that illuminate the genetic structure of present-day populations and the evolutionary history of the *C. neoformans* clade. These data illustrate specifically the extent to which reticulate events have helped shape the genetic variation and pathogenicity of this fungal species. Five distinct strains have been identified within this assemblage. Four of these are the "A", "B", "C," and "D" lineages. Genetic exchange between divergent subtypes within the various major strains has been identified (e.g., within the A and D lineages; Xu and Mitchell 2003; Litvintseva *et al.* 2006). In addition, interstrain introgressive hybridization—also involving the A and D lineages—has been inferred. In particular, Kavanaugh *et al.* (2006) detected the transfer of an ~40 kilobase stretch of DNA from the A lineage into the genome of the D strain. The fifth strain of *C. neoformans* is a hybrid lineage (termed "AD") that has arisen repeatedly in nature (Boekhout *et al.* 2001; Xu *et al.* 2002; Lin *et al.* 2007); there is evidence that this hybrid strain has originated independently at least three times (Xu *et al.* 2002). The formation of the AD lineage is significant for several reasons. First, it is a major infective agent of humans and has been shown to be virulent under experimental conditions (Cogliati *et al.* 2001; Lengeler *et al.* 2001). Second, its prevalence in nature (in spite of its known deficiencies in sporulation) and its elevated fitness detected by experimental analyses suggest that the hybrid origin of the AD lineage was causal in its evolutionary and pathogenic success (Lin *et al.* 2007; Litvintseva *et al.* 2007).

8.3.6 Reticulate evolution and parasitic roundworms

The final example of how reticulate evolution has impacted human parasites/pathogens involves the parasitic roundworms of the genus *Ascaris.* O'Lorcain and Holland (2000) reviewed the public health implications from infections by the widely distributed *Ascaris lumbricoides*. At the time of their publication, it was estimated that one-fourth of all humans were infected with this parasite (i.e.,~1.3 billion people; O'Lorcain and Holland 2000). The estimated geographic distribution of infection frequencies of *A. lumbricoides* was 73% in Asia, 12% in Africa, and 8% in Latin America (O'Lorcain and Holland 2000). These infections are more common in those people living in poor socioeconomic conditions. Furthermore, children infected with *A. lumbricoides* are likely to suffer from malnourishment and restricted mental abilities (O'Lorcain and Holland 2000). The large proportion of the human population affected by this parasite, and its significant toll on the young, reflects its negative impact on not only human health, but also the economic stability of countries whose populations are most affected. One positive outlook, however, is the finding that treatment programs that focus their attention on the young—and particularly those children most heavily infected—are most beneficial in terms of their cost effectiveness (Guyatt *et al.* 1995).

Evidence for genetic admixture between divergent lineages of *Ascaris* has come from studies of the variation in populations of these parasites found either in pigs or humans. For example, Peng *et al.* (1998, 2005) used both nuclear and mitochondrial gene markers to test for introgression (due to mixed infections of pig- and human-related *Ascaris*) between these strains. Although both studies inferred limited gene flow (Peng *et al.* 1998, 2005), it was unlikely that introgression was nonexistent. This was indeed shown to be the case by a recent analysis that utilized many nuclear markers thereby allowing a previously unavailable level of sensitivity for detecting introgression. Criscione *et al.* (2007) utilized 23 loci to decipher the genetic variation in both pig and human lineages of *Ascaris*. From this detailed survey, they were able to infer that 4% and 7% of the parasites from Guatemala and China, respectively, were recombinants between the two lineages (Criscione *et al.* 2007). This indicated that there was "contemporary cross-transmission between populations of human and pig *Ascaris*" (Criscione *et al.* 2007). Such a finding reflects again the importance of reticulate evolution in the origin and genetic structure of human pathogens. It also poses another problem for control programs given that humans can become infected with divergent lineages thus facilitating the origin of novel (hybrid) pathogens (Criscione *et al.* 2007).

8.4 Summary and conclusions

One goal of this chapter, shared with all the other parts of this book, was to emphasize the ubiquity of reticulate evolution as an agent of genetic and evolutionary change. Another goal, unique to the topic in this chapter, was to address the hypothesis that genetic exchange events have impacted greatly the evolutionary trajectory of organisms that parasitize and kill humans. The first of these goals, I hope, is accomplished as this chapter is considered in the context of the previous seven. The second objective, I believe, is obvious from the examples. Thus, the organisms that make up the lion-share of the pathogens that maim and kill humans are represented in the examples discussed in the various instances. Furthermore, the degree to which many of these organisms have impacted *H. sapiens* has been inferred to be due to their hybrid constitution. In some cases—for example, the bacterial pathogens—the mechanism of genetic exchange has been the uptake of DNA from other, very divergent, lineages. In other organisms such as the insect vectors of many of our diseases, the genetic transfer has been through sexual reproduction and introgression. Yet, regardless of the mechanism and even the ultimate effect of the genetic transfers on the recipient's fitness, it is clear that the organisms that cause so much pain and suffering for humans are mosaics of divergent genomes.

CHAPTER 9

Epilogue

I will review the diverse and very scattered literature on population differentiation in the absence of barriers to gene flow, and point out some of the largely unexplored, but important and fascinating, problems in genetic biogeography.

(Endler 1977, p. 3)

As far as animals are concerned, there can be no doubt that even in certain groups where premating isolating mechanisms...have played a major part in speciation, occasional hybridization occurs and may lead to a significant level of genic "introgression" from one species into another.

(White 1978, p. 346)

Hybridization, introgression, and speciation are examples of natural and dynamic evolutionary processes that exert great influence on how genetic diversity is organized.

(Avise 1994, p. 361)

More challenging is evidence that most archaeal and bacterial genomes (and the inferred ancestral eukaryotic nuclear genome) contain genes from multiple sources. If "chimerism" or "lateral gene transfer" cannot be dismissed as trivial in extent or limited to special categories of genes, then no hierarchical universal classification can be taken as natural.

(Doolittle 1999)

Bacteria reproduce asexually, yet they are also capable of obtaining genes from other organisms, even those of different kingdoms.

(Ochman *et al.* 2005)

9.1 Reticulate evolution and humans: what have we learned?

My goals for this book are relatively limited. First and foremost, I desire to reflect some of the complexity of evolutionary patterns and processes, specifically in the context of how both have been enriched by genetic exchange between divergent evolutionary lineages. Because of this primary goal, I have used interchangeably such terms as "reticulate evolution," "introgressive hybridization," "lateral (or horizontal) gene transfer," "the web-of-life metaphor," and indeed even "genetic exchange." This synonymy of words that are usually applied in a much more restrictive manner has been intentional, even at the risk of appearing sloppy. I have also approached this primary goal by exploring data sets for organisms as different as viruses and mammals. Both the terminological and organismal diversity has, I hope, enlivened the discussions and brought a better understanding of the mosaic nature of the genomes of organisms and thus the organisms themselves.

Obviously, as indicated by everything from the title of this book to the vast majority of examples I have chosen to discuss, this text is intentionally human-centric. This was not necessary for illustrating the central thesis—that the majority (if not all) of organisms have an evolutionary history, whether distant or more recent, that includes genetic exchange events. I know such an approach/limitation is unnecessary because I published, in 2006, a book that outlined the evidence for the web-of-life metaphor using mostly organisms that do not impact our own species. So what can we learn from the topics covered in this book that could not be discovered by this earlier text? I believe there are several lessons from the present text that make it unique. The first of these is that our own lineage is pockmarked with genetic material from lineages other than those who were our immediate, evolutionary ancestors. For example, pre-*Homo* individuals appear to have participated in introgressive hybridization with primates belonging to lineages that would eventually give rise to what we now recognize as chimpanzees and gorillas. However, this introgression did not end when our genus appeared. Archaic members of the *Homo* clade introgressed as did our own species when it came into contact with its archaic cousins. This statement will raise the hackles of those who hold to a conceptual framework—possibly based more on philosophy than science—that denies such exchanges between *H. sapiens* and other lineages. Yet, the data seem compelling, and will likely become more so as new data and analyses come to light.

Another message that should be plain from this book is that not only our evolutionary trajectory, but also the trajectories of the vast majority of organisms with which we interact favorably, and not so favorably, have been changed by the introduction of "foreign" genomic material. This observation has a twofold importance. First, it again reflects the explanatory and predictive nature of the web-of-life metaphor. Second, and most important for the specific topic of this book, is that it demonstrates how pervasively genetic exchange has affected the daily life and existence of *H. sapiens*. We cannot escape the conclusion that we are constantly fed, entertained, sheltered, attacked, and killed by organisms that possess mosaic genomes reflective of introgression and/or lateral gene transfer. This realization provides the context for how we might move forward in developing a deeper understanding of the mechanisms underlying genetic exchange, thus, allowing a more informed set of predictions concerning the outcome (both from a basic and applied standpoint) of such events.

Both of the above "lessons" derived from the focus of this book reflect clearly my belief that most people are inherently more interested in discussions concerning humans and organisms with which they interact, than they are in less human-centered topics. In particular, this hypothesis has been repeatedly supported by my interactions with hundreds of undergraduate students in various evolutionary biology courses. Thus, I believe that the message that genetic exchange is widespread and pervasive may be best understood when presented in the context of what we humans see, hear, touch, and contract.

9.2 Reticulate evolution and humans: how do we apply what we have learned?

Multiple applications—both in regard to future studies and in our appreciation of the benefits and perils of reticulate evolution for our species—can be made from what has been presented in this text. The first is that many lineages, whether our own or those of organisms with which we interact, have not diverged in isolation. Instead, our and their evolutionary divergence has been intimately tied to the genomes of other organisms by the bridge of genetic exchange. This has profound implications for our understanding of the role played by geographic distribution, mutation, gene flow, genetic drift, and natural selection on the process of evolutionary diversification. Specifically, we assume at our scientific peril that the separation of populations into nonoverlapping (i.e., allopatric) geographic distributions is necessary for speciation. In contrast, the data reviewed in the previous eight chapters illustrates the pervasiveness of divergence in the face of gene flow. Given parapatric or sympatric divergence as a major mechanism by which new lineages evolve, it is necessary to ask how much more of a role natural selection must play in evolutionary radiations. This reflects a major transition from previous models constructed during the Neo-Darwinian synthesis (e.g., Mayr 1942, 1963), but is encompassed by more recently embraced models (e.g., Schluter 2000).

Another application that arises from the studies reviewed above is that humans must respect the degree to which genetic exchange has, and continues to, structure our world. Two examples can be used to illustrate this point. First, we must continue to fight against the urge to relegate "hybrid" lineages to a category of less value than that occupied by "pure" lineages. Thus, a lack of appreciation for the fact that the majority of organisms have a checkered past with regard to their genetic makeup has led to the conclusion that any population of hybrid origin should not be protected from extirpation/extinction from human-mediated habitat disturbances. Such conservation issues continue to roil as reflected by papers in scientific journals such as *Conservation Genetics.* A second reason our species should respect the processes leading to the web of life is that it could facilitate attempts to control disease vectors and pathogens. This was highlighted in Chapter 8, but it is worthwhile to emphasize again that a lack of appreciation for the degree to which genetic exchange continues to impact those organisms that we have the most reason to avoid will limit greatly our ability to make predictions concerning disease spread and evolution. A naïve (in regard to the pervasive effects from introgression and lateral gene transfer) outlook on the evolution of disease vectors and pathogens may also greatly hinder control measures (e.g., the application of vaccines that can lead to new viral lineages through recombination) that could relieve the suffering of hundreds of millions of our own species.

Given the above, it is now apparent that genetic exchange cannot be accurately placed into categories typified by the terms "anomalous" or "rare" or "unimportant." Thus, I hope that a consideration of the concepts and examples reviewed in the preceding chapters will yield a wealth of studies that approach evolutionary diversification as a dynamic process that is likely to include the exchange of genetic material between organisms on different strands of the web of life.

Glossary

adaptive radiation The evolution of ecological and phenotypic diversity within a rapidly multiplying lineage (Seehausen 2004).

adaptive trait transfer (or adaptive trait introgression) The transfer of genes and thus the phenotype of an adaptive trait through viral recombination, lateral gene transfer, or introgressive hybridization (Arnold 2006).

allopolyploid species A species formed through hybridization between members of evolutionary lineages with "strongly differentiated genomes" followed by chromosomal doubling, trebling, and so on (Stebbins 1947).

anagenesis Transformation from one taxonomic form into another involving a single evolutionary lineage. The ancestral taxon thus goes to extinction in the process of the origin of the new form.

autopolyploid speciation A type of polyploid speciation in which the derivative lineage arises from crosses within a single evolutionary lineage (Stebbins 1947).

conjugation The transfer of DNA through physical contact between donor and recipient cells. This process can mediate the transfer of genetic material between such divergent evolutionary lineages as bacteria and plants (Ochman *et al.* 2000).

hybrid speciation Process in which *natural hybridization* (see below) results in the production of an evolutionary lineage (i.e., a *hybrid species*; see below) that is at least partially reproductively isolated from both parental lineages, and which demonstrates a distinct evolutionary and ecological trajectory (Arnold and Burke 2006).

hybrid species At least partially reproductively isolated lineages arising from natural hybridization, which demonstrate distinct evolutionary and ecological trajectories as defined by distinguishable (and heritable) morphological, ecological, and/or reproductive differences relative to their progenitors (Arnold and Burke 2006).

hybrid swarm model of adaptive radiation "Because hybridization is common when populations invade new environments and potentially elevates rates of response to selection, it predisposes colonizing populations to rapid adaptive diversification under disruptive or divergent selection" (Seehausen 2004).

incomplete lineage sorting "Particularly for species or populations that have separated recently (relative to their effective population sizes), the probabilities are high that a state of reciprocal monophyly has not yet been achieved with respect to the true genealogical ancestry of haplotypes at particular loci" (Avise *et al.* 1990).

introgressive hybridization or introgression The transfer of DNA between individuals from two populations, or groups of populations, that are distinguishable on the basis of one or more heritable characters via hybridization followed by repeated backcrossing between hybrid and parental individuals (modified from Anderson and Hubricht 1938).

lateral or horizontal gene transfer The transfer of genetic material between individuals from two populations, or groups of populations, that are distinguishable on the basis of one or more heritable characters through the processes of *transformation, transduction, conjugation,* or *vector-mediated transfer* (see below; Arnold 2006).

levant A term associated with a region in the Middle East south of the Taurus Mountains, bounded by the Mediterranean Sea on the west, and by the northern Arabian Desert and Upper Mesopotamia to the east.

natural hybrid Offspring resulting from a cross in nature between individuals from two populations, or groups of populations, that are distinguishable on the basis of one or more heritable characters (Arnold 1997 as adapted from Harrison 1990).

natural hybridization Successful matings in nature between individuals from two populations, or groups of populations, that are distinguishable on the basis of one or more heritable characters, resulting in the formation of *natural hybrid* (see above) individuals (Arnold 1997 as adapted from Harrison 1990).

prezygotic reproductive barriers Processes that limit the formation of hybrid zygotes (e.g., mate selection in animals that favors pairings between members of the same evolutionary lineage).

postzygotic reproductive barriers Processes that result in lower fitness of hybrid genotypes, relative to progeny formed between individuals from the same evolutionary lineage (e.g., lower fertility in F_1 hybrids due to chromosomal rearrangement differences in the parental lineages that formed the hybrid).

reticulate evolution Web-like phylogenetic relationships reflecting genetic exchange (through lateral transfer, viral recombination, introgressive hybridization, etc.) between divergent lineages (Arnold 2006).

syngameon "an habitually interbreeding community" (Lotsy 1931) or "the most inclusive unit of interbreeding in a hybridizing species group" (Grant 1981).

transduction The transfer of DNA through a bacteriophage intermediate (Ochman *et al.* 2000).

transformation The uptake of naked DNA from the environment (Ochman *et al.* 2000).

vector-mediated transfer The transfer of DNA from a donor to a recipient by a vector intermediate (e.g., between insect species through a shared parasite; Arnold 2006).

References

Abbott RJ (1992). Plant invasions, interspecific hybridization and the evolution of new plant taxa. *Trends in Ecology and Evolution* **7**, 401–5.

Abzhanov A, Kuo WP, Hartmann C, Grant BR, Grant PR, and Tabin CJ (2006). The calmodulin pathway and evolution of elongated beak morphology in Darwin's finches. *Nature* **442**, 563–7.

Ackermann RR, Rogers J, and Cheverud JM (2006). Identifying the morphological signatures of hybridization in primate and human evolution. *Journal of Human Evolution* **51**, 632–45.

Adams JR, Leonard JA, and Waits LP (2003). Widespread occurrence of a domestic dog mitochondrial DNA haplotype in southeastern US coyotes. *Molecular Ecology* **12**, 541–6.

Adams JR, Lucash C, Schutte L, and Waits LP (2007). Locating hybrid individuals in the red wolf (*Canis rufus*) experimental population area using a spatially targeted sampling strategy and faecal DNA genotyping. *Molecular Ecology* **16**, 1823–34.

Addison G and Tavares R (1952). Hybridization and grafting in species of *Theobroma* which occur in Amazonia. *Evolution* **6**, 380–6.

Agol VI (2006). Vaccine-derived polioviruses. *Biologicals* **34**, 103–8.

Aguiar LM, Mellek DM, Abreu KC, *et al.* (2007). Sympatry between *Alouatta caraya* and *Alouatta clamitans* and the rediscovery of free-ranging potential hybrids in Southern Brazil. *Primates* **48**, 245–8.

Aguiar LM, Pie MR, and Passos FC (2008). Wild mixed groups of howler species (*Alouatta caraya* and *Alouatta clamitans*) and new evidence for their hybridization. *Primates* **49**, 149–52.

Albert V, Jónsson B, and Bernatchez L (2006). Natural hybrids in Atlantic eels (*Anguilla anguilla, A. rostrata*): Evidence for successful reproduction and fluctuating abundance in space and time. *Molecular Ecology* **15**, 1903–16.

Alberts SC and Altmann J (2001). Immigration and hybridization patterns of yellow and anubis baboons in and around Amboseli, Kenya. *American Journal of Primatology* **53**, 139–54.

Allendorf FW, Leary RF, Spruell P, and Wenburg JK (2001). The problems with hybrids: Setting conservation guidelines. *Trends in Ecology and Evolution* **16**, 613–22.

Almodóvar A, Nicola GG, Elvira B, and García-Marín JL (2006). Introgression variability among Iberian brown trout Evolutionary Significant Units: The influence of local management and environmental features. *Freshwater Biology* **51**, 1175–87.

Alvarado Bremer JR, Naseri I, and Ely B (1997). Orthodox and unorthodox phylogenetic relationships among tunas revealed by the nucleotide sequence analysis of the mitochondrial DNA control region. *Journal of Fish Biology* **50**, 540–54.

Alvarado Bremer JR, Viñas J, Mejuto J, Ely B, and Pla C (2005). Comparative phylogeography of Atlantic bluefin tuna and swordfish: The combined effects of vicariance, secondary contact, introgression, and population expansion on the regional phylogenies of two highly migratory pelagic fishes. *Molecular Phylogenetics and Evolution* **36**, 169–87.

Alves PC, Ferrand N, Suchentrunk F, and Harris DJ (2003). Ancient introgression of *Lepus timidus* mtDNA into *L. granatensis* and *L. europaeus* in the Iberian Peninsula. *Molecular Phylogenetics and Evolution* **27**, 70–80.

Alves PC, Harris DJ, Melo-Ferreira J, Branco M, and Ferrand N (2006). Hares on thin ice: Introgression of mitochondrial DNA in hares and its implications for recent phylogenetic analyses. *Molecular Phylogenetics and Evolution* **40**, 640–1.

Alves PC, Melo-Ferreira J, Branco M, Suchentrunk F, Ferrand N, and Harris DJ (2008). Evidence for genetic similarity of two allopatric European hares (*Lepus corsicanus* and *L. castroviejoi*) inferred from nuclear DNA sequences. *Molecular Phylogenetics and Evolution* **46**, 1191–7.

Ambrose SE (1996). *Crazy Horse and Custer—The Parallel Lives of Two American Warriors*. Random House Inc, New York.

American Honey Producers Association (2007). Statement for the United States Senate Committee on Agriculture, Nutrition and Forestry, pp. 1–15.

Ames M and Spooner DM (2008). DNA from herbarium specimens settles a controversy about origins of the European potato. *American Journal of Botany* **95,** 252–7.

Amoukou AI and Marchais L (1993). Evidence of a partial reproductive barrier between wild and cultivated pearl millets (*Pennisetum glaucum*). *Euphytica* **67,** 19–26.

Anderson E (1936). An experimental study of hybridization in the genus *Apocynum*. *Annals of the Missouri Botanical Garden* **23,** 159–168.

Anderson E (1948). Hybridization of the habitat. *Evolution* **2,** 1–9.

Anderson E (1949). *Introgressive Hybridization*. John Wiley and Sons, Inc, New York.

Anderson E and Hubricht L (1938). Hybridization in *Tradescantia*. III. The evidence for introgressive hybridization. *American Journal of Botany* **25,** 396–402.

Anderson E and Stebbins GL, Jr (1954). Hybridization as an evolutionary stimulus. *Evolution* **8,** 378–88.

Anderson GJ (1975). The variation and evolution of selected species of *Solanum* section *Basarthrum*. *Brittonia* **27,** 209–22.

Andersone Z, Lucchini V, Randi E, and Ozolins J (2002). Hybridisation between wolves and dogs in Latvia as documented using mitochondrial and microsatellite DNA markers. *Mammalian Biology* **67,** 79–90.

Anthony F, Bertrand B, Quiros O, *et al.* (2001). Genetic diversity of wild coffee (*Coffea arabica* L.) using molecular markers. *Euphytica* **118,** 53–65.

Anthony F, Combes MC, Astorga C, Bertrand B, Graziosi G, and Lashermes P (2002). The origin of cultivated *Coffea arabica* L. varieties revealed by AFLP and SSR markers. *Theoretical and Applied Genetics* **104,** 894–900.

Anthony NM, Johnson-Bawe M, Jeffery K, *et al.* (2007). The role of Pleistocene refugia and rivers in shaping gorilla genetic diversity in central Africa. *Proceedings of the National Academy of Sciences USA* **104,** 20432–6.

Antón SC and Swisher CC III (2004). Early dispersals of *Homo* from Africa. *Annual Review of Anthropology* **33,** 271–96.

Aradhya MK, Dangl GS, Prins BH, *et al.* (2003). Genetic structure and differentiation in cultivated grape, *Vitis vinifera* L. *Genetical Research* **81,** 179–92.

Arias MC and Sheppard WS (1996). Molecular phylogenetics of honey bee subspecies (*Apis mellifera* L.) inferred from mitochondrial DNA sequence. *Molecular Phylogenetics and Evolution* **5,** 557–66.

Ariefdjohan MW and Savaiano DA (2005). Chocolate and cardiovascular health: Is it too good to be true? *Nutrition Reviews* **63,** 427–30.

Arita I, Nakane M, and Fenner F (2006). Is polio eradication realistic? *Science* **312,** 852–4.

Arnold ML (1992). Natural hybridization as an evolutionary process. *Annual Review of Ecology and Systematics* **23,** 237–61.

Arnold ML (1993). *Iris nelsonii*: Origin and genetic composition of a homoploid hybrid species. *American Journal of Botany* **80,** 577–83.

Arnold ML (1997). *Natural Hybridization and Evolution*. Oxford University Press, Oxford.

Arnold ML (2006). *Evolution Through Genetic Exchange*. Oxford University Press, Oxford.

Arnold ML, Buckner CM, and Robinson JJ (1991). Pollen mediated introgression and hybrid speciation in Louisiana irises. *Proceedings of the National Academy of Sciences, USA* **88,** 1398–1402.

Arnold ML and Burke JM (2006). Natural Hybridization. In CW Fox and JB Wolf, eds. *Evolutionary Genetics: Concepts and Case Studies,* pp. 399–413. Oxford University Press, Oxford.

Arnold ML, Hamrick JL, and Bennett BD (1990). Allozyme variation in Louisiana Irises: A test for introgression and hybrid speciation. *Heredity* **65,** 297–306.

Arnold ML and Hodges SA (1995). Are natural hybrids fit or unfit relative to their parents? *Trends in Ecology and Evolution* **10,** 67–71.

Arnold ML and Larson EJ (2004). Evolution's new look. *The Wilson Quarterly* (Autumn) 60–72.

Arnold ML and Meyer A (2006). Natural hybridization in primates: One evolutionary mechanism. *Zoology* **109,** 261–76.

Arroyo-García R, Ruiz-García L, Bolling L, *et al.* (2006). Multiple origins of cultivated grapevine (*Vitis vinifera* L. ssp. *sativa*) based on chloroplast DNA polymorphisms. *Molecular Ecology* **15,** 3707–14.

Asiimwe-Okiror G, Ntabangana S, Asamoah-Odei E, and Calleja JMG (2005). HIV/AIDS epidemiological surveillance report for the WHO African region—2005 update. World Health Organization. 74 pp.

Audic S, Robert C, Campagna B, *et al.* (2007). Genome analysis of *Minibacterium massiliensis* highlights the convergent evolution of water-living bacteria. *PLoS Genetics* **3,** 1454–63.

Australian Bureau of Statistics (2007). Agricultural production. *Year Book Australia*. pp 1–20. Canberra, Australia.

Avey JT, Ballard WB, Wallace MC, *et al.* (2003). Habitat relationships between sympatric mule deer and white-tailed deer in Texas. *Southwestern Naturalist* **48,** 644–53.

Avise JC (1994). *Molecular Markers, Natural History and Evolution*. Chapman & Hall Inc, New York.

Avise JC, Ankney CD, and Nelson WS (1990). Mitochondrial gene trees and the evolutionary relationship of mallard and black ducks. *Evolution* **44,** 1109–19.

Avise JC, Helfman GS, Saunders NC, and Hales LS (1986). Mitochondrial DNA differentiation in North Atlantic eels: Population genetic consequences of an unusual life history pattern. *Proceedings of the National Academy of Sciences USA* **83,** 4350–4.

Avise JC, Nelson WS, Arnold J, Koehn RK, Williams GC, and Thorsteinsson V (1990). The evolutionary genetic status of Icelandic eels. *Evolution* **44,** 1254–62.

Awadalla P, Eyre-Walker A, and Maynard Smith J (1999). Linkage disequilibrium and recombination in hominid mitochondrial DNA. *Science* **286,** 2524–5.

Ba FS, Pasquet RS, and Gepts P (2004). Genetic diversity in cowpea [*Vigna unguiculata* (L.) Walp.] as revealed by RAPD markers. *Genetic Resources and Crop Evolution* **51,** 539–50.

Balashov SP, Imasheva ES, Boichenko VA, Antón J, Wang JM, and Lanyi JK (2005). Xanthorhodopsin: A proton pump with a light-harvesting carotenoid antenna. *Science* **309,** 2061–4.

Baltazar BM, de Jesús Sánchez-Gonzalez J, de la Cruz-Larios L, and Schoper JB (2005). Pollination between maize and Teosinte: An important determinant of gene flow in Mexico. *Theoretical and Applied Genetics* **110,** 519–26.

Baratti M, Ammannati M, Magnelli C, and Dessì-Fulgheri F (2004). Introgression of *chukar* genes into a reintroduced red-legged partridge (*Alectoris rufa*) population in central Italy. *Animal Genetics* **36,** 29–35.

Barbanera F, Negro JJ, Di Giuseppe G, Bertoncini F, Cappelli F, and Dini F (2005). Analysis of the genetic structure of red-legged partridge (*Alectoris rufa*, Galliformes) populations by means of mitochondrial DNA and RAPD markers: A study from central Italy. *Biological Conservation* **122,** 275–87.

Barilani M, Sfougaris A, Giannakopoulos A, Mucci N, Tabarroni C, and Randi E (2007). Detecting introgressive hybridisation in rock partridge populations (*Alectoris graeca*) in Greece through Bayesian admixture analyses of multilocus genotypes. *Conservation Genetics* **8,** 343–54.

Barkley NA, Dean RE, Pittman RN, Wang ML, Holbrook CC, and Pederson GA (2007). Genetic diversity of cultivated and wild-type peanuts evaluated with M13-tailed SSR markers and sequencing. *Genetical Research, Cambridge* **89,** 93–106.

Barnes NG (2004). A market analysis of the US pet food industry to determine new opportunities for the cranberry industry. pp. 1–192. Report for the Center for Business Research, University of Massachusetts, Dartmouth, Massachusetts.

Barrett HC and Rhodes AM (1976). A numerical taxonomic study of affinity relationships in cultivated *Citrus* and its close relatives. *Systematic Botany* **1,** 105–36.

Barton NH (2006). Evolutionary biology: How did the human species form? *Current Biology* **16,** R647–50.

Barton NH and Hewitt GM (1985). Analysis of hybrid zones. *Annual Review of Ecology and Systematics* **16,** 113–48.

Baudouin L, Lebrun P, Konan JL, Ritter E, Berger A, and Billotte N (2006). QTL analysis of fruit components in the progeny of a Rennell Island Tall coconut (*Cocos nucifera* L.) individual. *Theoretical and Applied Genetics* **112,** 258–68.

Baumann M, Babotai C, and Schibler J (2005). Native or naturalized? Validating alpine chamois habitat models with archaeozoological data. *Ecological Applications* **15,** 1096–110.

Beaumont M, Barratt EM, Gottelli D, *et al.* (2001). Genetic diversity and introgression in the Scottish wildcat. *Molecular Ecology* **10,** 319–36.

Becquet C, Patterson N, Stone AC, Przeworski M, and Reich D (2007). Genetic structure of chimpanzee populations. *PLoS Genetics* **3,** 0617–26.

Beebe S, Rengifo J, Gaitan E, Duque MC, and Tohme J (2001). Diversity and origin of Andean landraces of common bean. *Crop Science* **41,** 854–62.

Beebe S, Toro Ch O, González AV, Chacón MI, and Debouck DG (1997). Wild-weed-crop complexes of common bean (*Phaseolus vulgaris* L., Fabaceae) in the Andes of Peru and Colombia, and their implications for conservation and breeding. *Genetic Resources and Crop Evolution* **44,** 73–91.

Béjà O, Aravind L, Koonin EV, *et al.* (2000). Bacterial rhodopsin: Evidence for a new type of phototrophy in the sea. *Science* **289,** 1902–6.

Béjà O, Spudich EN, Spudich JL, Leclerc M, and DeLong EF (2001). Proteorhodopsin phototrophy in the ocean. *Nature* **411,** 786–9.

Beja-Pereira A, Caramelli D, Lalueza-Fox C, *et al.* (2006). The origin of European cattle: Evidence from modern and ancient DNA. *Proceedings of the National Academy of Sciences USA* **103,** 8113–8.

Beja-Pereira A, England PR, Ferrand N, *et al.* (2004). African origins of the domestic donkey. *Science* **304,** 1781

Belahbib N, Pemonge M-H, Ouassou A, Sbay H, Kremer A, and Petit RJ (2001). Frequent cytoplasmic exchanges between oak species that are not closely related: *Quercus suber* and *Q. ilex* in Morocco. *Molecular Ecology* **10,** 2003–12.

Bergman TJ and Beehner JC (2004). Social system of a hybrid baboon group (*Papio anubis* × *P. hamadryas*). *International Journal of Primatology* **25,** 1313–30.

Bergman TJ, Phillips-Conroy JE, and Jolly CJ (2008). Behavioral variation and reproductive success of male baboons (*Papio anubis* × *Papio hamadryas*) in a hybrid social group. *American Journal of Primatology* **70,** 136–47.

Bergthorsson U, Adams KL, Thomason B, and Palmer JD (2003). Widespread horizontal transfer of mitochondrial genes in flowering plants. *Nature* **424,** 197–201.

Bermejo M, Rodríguez-Teijeiro JD, Illera G, Barroso A, Vilà C and Walsh PD (2006). Ebola outbreak killed 5000 gorillas. *Science* **314,** 1564.

Bermúdez de Castro JM, Arsuaga JL, Carbonell E, Rosas A, Martinez I, and Mosquera M (1997). A hominid from the lower Pleistocene of Atapuerca, Spain: Possible ancestor to Neandertals and modern humans. *Science* **276,** 1392–5.

Besansky NJ, Krzywinski J, Lehmann T, *et al.* (2003). Semipermeable species boundaries between *Anopheles gambiae* and *Anopheles arabiensis*: Evidence from multilocus DNA sequence variation. *Proceedings of the National Academy of Sciences USA* **100,** 10818–23.

Besansky NJ, Lehmann T, Fahey GT, *et al.* (1997). Patterns of mitochondrial variation within and between African malaria vectors, *Anopheles gambiae* and *An. arabiensis,* suggest extensive gene flow. *Genetics* **147,** 1817–28.

Bierne N, Bonhomme F, Boudry P, Szulkin M, and David P (2006). Fitness landscapes support the dominance theory of post-zygotic isolation in the mussels *Mytilus edulis* and *M. galloprovincialis. Proceedings of the Royal Society of London B* **273,** 1253–60.

Bierne N, Borsa P, Daguin C, *et al.* (2003). Introgression patterns in the mosaic hybrid zone between *Mytilus edulis* and *M. galloprovincialis. Molecular Ecology* **12,** 447–61.

Birungi J and Arctander P (2000). Large sequence divergence of mitochondrial DNA genotypes of the control region within populations of the African antelope, kob (*Kobus kob*). *Molecular Ecology* **9,** 1997–2008.

Blair MW, Díaz JM, Hidalgo R, Díaz LM, and Duque MC (2007). Microsatellite characterization of Andean races of common bean (*Phaseolus vulgaris* L.). *Theoretical and Applied Genetics* **116,** 29–43.

Blanc G, Ogata H, Robert C, Audic S, Claverie J-M, and Raoult D (2007). Lateral gene transfer between obligate intracellular bacteria: Evidence from the *Rickettsia massiliae* genome. *Genome Research* **17,** 1657–64.

Blanca JM, Prohens J, Anderson GJ, Zuriaga E, Cañizares J, and Nuez F (2007). AFLP and DNA sequence variation in an Andean domesticate, pepino (*Solanum muricatum,* Solanaceae): Implications for evolution and domestication. *American Journal of Botany* **94,** 1219–29.

Boddington CT (1997). *Where Lions Roar.* Safari Press Inc., Long Beach, California.

Boekhout T, Theelen B, Diaz M, *et al.* (2001). Hybrid genotypes in the pathogenic yeast *Cryptococcus neoformans. Microbiology* **147,** 891–907.

Bolling C and Suarez NR (2001). The Brazilian sugar industry: Recent developments. pp. 14–18. Economic Research Service, United States Department of Agriculture.

Bollongino R, Edwards CJ, Alt KW, Burger J, and Bradley DG (2006). Early history of European domestic cattle as revealed by ancient DNA. *Biology Letters* **2,** 155–9.

Bonen L and Calixte S (2006). Comparative analysis of bacterial-origin genes for plant mitochondrial ribosomal proteins. *Molecular Biology and Evolution* **23,** 701–12.

Borges EC, Dujardin J-P, Schofield CJ, Romanha AJ, and Diotaiuti L (2005). Dynamics between sylvatic, peridomestic and domestic populations of *Triatoma brasiliensis* (Hemiptera: Reduviidae) in Ceará State, northeastern Brazil. *Acta Tropica* **93,** 119–26.

Borsa P, Daguin C, and Bierne N (2007). Genomic reticulation indicates mixed ancestry in southern-hemisphere *Mytilus* spp. mussels. *Biological Journal of the Linnean Society* **92,** 747–54.

Bouck AC, Peeler R, Arnold ML, and Wessler SR (2005). Genetic mapping of species boundaries in Louisiana Irises using *IRRE* retrotransposon display markers. *Genetics* **171,** 1289–1303.

Boulesteix M, Simard F, Antonio-Nkondjio C, Awono-Ambene HP, Fontenille D, and Biémont C (2007). Insertion polymorphism of transposable elements and population structure of *Anopheles gambiae* M and S molecular forms in Cameroon. *Molecular Ecology* **16,** 441–52.

Bourdain A (2004). *Anthony Bourdain's Les Halles Cookbook—Strategies, Recipes, and Techniques of Classic Bistro Cooking.* Bloomsbury, New York.

Bouza C, Vilas R, Castro J and Martínez P (2007). Mitochondrial haplotype variability of brown trout populations from northwestern Iberian Peninsula, a secondary contact area between lineages. *Conservation Genetics* **in press.**

Bowers J, Boursiquot J-M, This P, Chu K, Johansson H, and Meredith C (1999). Historical genetics: The parentage of Chardonnay, Gamay, and other wine grapes of northeastern France. *Science* **285,** 1562–5.

Bowers JE, Arias MA, Asher R, *et al.* (2005). Comparative physical mapping links conservation of microsynteny to chromosome structure and recombination in grasses. *Proceedings of the National Academy of Sciences USA* **102,** 13206–11.

Boyle JP, Rajasekar B, Saeij JPJ, *et al.* (2006). Just one cross appears capable of dramatically altering the population biology of a eukaryotic pathogen like *Toxoplasma gondii*. *Proceedings of the National Academy of Sciences USA* **103,** 10514–19.

Bradley RD, Bryant FC, Bradley LC, Haynie ML, and Baker RJ (2003). Implications of hybridization between white-tailed deer and mule deer. *Southwestern Naturalist* **48**, 654–60.

Branco M, Ferrand N, and Monnerot M (2000). Phylogeography of the European rabbit (*Oryctolagus cuniculus*) in the Iberian Peninsula inferred from RFLP analysis of the cytochrome *b* gene. *Heredity* **85,** 307–17.

Branco M, Monnerot M, Ferrand N, and Templeton AR (2002). Postglacial dispersal of the European rabbit (*Oryctolagus cuniculus*) on the Iberian Peninsula reconstructed from nested clade and mismatch analyses of the mitochondrial DNA genetic variation. *Evolution* **56,** 792–803.

Brandes EW and Sartoris GB (1936). Sugarcane: Its origin and improvement, pp. 561–623. Yearbook of the United States Department of Agriculture.

Bretscher MT, Althaus CL, Müller V, and Bonhoeffer S (2004). Recombination in HIV and the evolution of drug resistance: For better or for worse? *BioEssays* **26,** 180–8.

Britten RJ (2002). Divergence between samples of chimpanzee and human DNA sequences is 5%, counting indels. *Proceedings of the National Academy of Sciences USA* **99,** 13633–5.

Broughton WJ, Hernández G, Blair M, Beebe S, Gepts P, and Vanderleyden J (2003). Beans (*Phaseolus* spp.)—model food legumes. *Plant and Soil* **252,** 55–128.

Brown HP, Panshin AJ, and Forsaith CC (1949). *Textbook of Wood Technology—Volume I—Structure, Identification, Defects, and Uses of the Commercial Woods of the United States*. McGraw-Hill Book Company Inc, New York.

Brown P, Sutikna T, Morwood MJ, *et al.* (2004). A new small-bodied hominin from the Late Pleistocene of Flores, Indonesia. *Nature* **431,** 1055–61.

Brubaker CL and Wendel JF (1994). Reevaluating the origin of domesticated cotton (*Gossypium hirsutum*; Malvaceae) using nuclear restriction fragment length polymorphisms (RFLPs). *American Journal of Botany* **81,** 1309–26.

Bruman HJ (1945). Early coconut culture in western Mexico. *The Hispanic American Historical Review* **25,** 212–23.

Brunner PC, Douglas MR, Osinov A, Wilson CC, and Bernatchez L (2001). Holarctic phylogeography of arctic charr (*Salvelinus alpinus* L.) inferred from mitochondrial DNA sequences. *Evolution* **55,** 573–86.

Burke JM and Arnold ML (2001). Genetics and the fitness of hybrids. *Annual Review of Genetics* **35,** 31–52.

Burrows W and Ryder OA (1997). Y-chromosome variation in great apes. *Nature* **385,** 125–6.

Burzynski A, Zbawicka M, Skibinski DOF, and Wenne R (2006). Doubly uniparental inheritance is associated with high polymorphism for rearranged and recombinant control region haplotypes in Baltic *Mytilus trossulus*. *Genetics* **174,** 1081–94.

Busso CS, Devos KM, Ross G, *et al.* (2000). Genetic diversity within and among landraces of pearl millet (*Pennisetum glaucum*) under farmer management in West Africa. *Genetic Resources and Crop Evolution* **47,** 561–8.

Bynum EL, Bynum DZ, and Supriatna J (1997). Confirmation and location of the hybrid zone between wild populations of *Macaca tonkeana* and *Macaca hecki* in central Sulawesi, Indonesia. *American Journal of Primatology* **43,** 181–209.

Bynum N (2002). Morphological variation within a macaque hybrid zone. *American Journal of Physical Anthropology* **118,** 45–9.

Caccone A and Powell JR (1989). DNA divergence among hominoids. *Evolution* **43,** 925–42.

Calvete C, Angulo E, and Estrada R (2005). Conservation of European wild rabbit populations when hunting is age and sex selective. *Biological Conservation* **121,** 623–34.

Cann RL, Stoneking M, and Wilson AC (1987). Mitochondrial DNA and human evolution. *Nature* **325,** 31–6.

Capelli C, MacPhee RDE, Roca AL, *et al.* (2006). A nuclear DNA phylogeny of the wooly mammoth (*Mammuthus primigenius*). *Molecular Phylogenetics and Evolution* **40,** 620–7.

Caramelli D, Lalueza-Fox C, Vernesi C, *et al.* (2003). Evidence for a genetic discontinuity between Neandertals and 24,000-year-old anatomically modern Europeans. *Proceedings of the National Academy of Sciences USA* **100,** 6593–7.

Carlton JM, Hirt RP, Silva JC, *et al.* (2007). Draft genome sequence of the sexually transmitted pathogen *Trichomonas vaginalis*. *Science* **315,** 207–12.

Carr SM, Ballinger SW, Derr JN, Blankenship LH, and Bickham JW (1986). Mitochondrial DNA analysis of hybridization between sympatric white-tailed deer and mule deer in west Texas. *Proceedings of the National Academy of Sciences USA* **83,** 9576–80.

Castillo RO and Spooner DM (1997). Phylogenetic relationships of wild potatoes, *Solanum* series *Conicibaccata* (sect. *Petota*). *Systematic Botany* **22,** 45–83.

Cathey JC, Bickham JW, and Patton JC (1998). Introgressive hybridization and nonconcordant evolutionary history of maternal and paternal lineages in North American deer. *Evolution* **52**, 1224–9.

Celeira de Lima MM, Sampaio I, dos Santos Vieira R, and Schneider H (2007). Spider monkey, Muriqui and Woolly monkey relationships revisited. *Primates* **48**, 55–63.

Centers for Disease Control and Prevention (2002). Ebola hemorrhagic fever information packet.

Centers for Disease Control and Prevention (2004). Onchocerciasis/River Blindness.

Centers for Disease Control and Prevention (2007). St. Louis encephalitis fact sheet.

Chacón MI, Pickersgill SB, and Debouck DG (2005). Domestication patterns in common bean (*Phaseolus vulgaris* L.) and the origin of the Mesoamerican and Andean cultivated races. *Theoretical and Applied Genetics* **110**, 432–44.

Chakraborty D, Ramakrishnan U, Panor J, Mishra C, and Sinha A (2007). Phylogenetic relationships and morphometric affinities of the Arunachal macaque *Macaca munzala*, a newly described primate from Arunachal Pradesh, northeastern India. *Molecular Phylogenetics and Evolution* **44**, 838–49.

Champigneulle A and Cachera S (2003). Evaluation of large-scale stocking of early stages of brown trout, *Salmo trutta*, to angler catches in the French-Swiss part of the River Doubs. *Fisheries Management and Ecology* **10**, 79–85.

Chen H, Smith GJD, Li KS, *et al.* (2006). Establishment of multiple sublineages of H5N1 influenza virus in Asia: Implications for pandemic control. *Proceedings of the National Academy of Sciences USA* **103**, 2845–50.

Chen J-H, Pan D, Groves C, *et al.* (2006). Molecular phylogeny of *Nycticebus* inferred from mitochondrial genes. *International Journal of Primatology* **27**, 1187–200.

Chen S-Y, Duan Z-Y, Sha T, Xiangyu J, Wu S-F, and Zhang Y-P (2006). Origin, genetic diversity, and population structure of Chinese domestic sheep. *Gene* **376**, 216–23.

Cheng C, Motohashi R, Tsuchimoto S, Fukuta Y, Ohtsubo H, and Ohtsubo E (2003). Polyphyletic origin of cultivated rice: Based on the interspersion pattern of SINEs. *Molecular Biology and Evolution* **20**, 67–75.

Chin MPS, Rhodes TD, Chen J, Fu W, and Hu W-S (2005). Identification of a major restriction in HIV-1 intersubtype recombination. *Proceedings of the National Academy of Sciences USA* **102**, 9002–7.

Cho N-H, Kim H-R, Lee J-H, *et al.* (2007). The *Orientia tsutsugamushi* genome reveals massive proliferation of conjugative type IV secretion system and host-cell interaction genes. *Proceedings of the National Academy of Sciences USA* **104**, 7981–6.

Chow S and Kishino H (1995). Phylogenetic relationships between tuna species of the genus *Thunnus* (Scombridae: Teleostei): Inconsistent implications from morphology, nuclear and mitochondrial genomes. *Journal of Molecular Evolution* **41**, 741–8.

Chu J-H, Lin Y-S, and Wu H-Y (2007). Evolution and dispersal of three closely related macaque species, *Macaca mulatta, M. cyclopis,* and *M. fuscata,* in the eastern Asia. *Molecular Phylogenetics and Evolution* **43**, 418–29.

Clabaut C, Bunje PME, Salzburger W, and Meyer A (2007). Geometric morphometric analyses provide evidence for the adaptive character of the Tanganyikan cichlid fish radiations. *Evolution* **61**, 560–78.

Clark RM, Linton E, Messing J, and Doebley JF (2004). Pattern of diversity in the genomic region near the maize domestication gene *tb1*. *Proceedings of the National Academy of Sciences USA* **101**, 700–7.

Clarke AC, Burtenshaw MK, McLenachan PA, Erickson DL, and Penny D (2006). Reconstructing the origins and dispersal of the Polynesian bottle gourd (*Lagenaria siceraria*). *Molecular Biology and Evolution* **23**, 893–900.

Clarke KE, Rinderer TE, Franck P, Quezada-Euán JG, and Oldroyd BP (2002). The Africanization of honeybees (*Apis mellifera* L.) of the Yucatan: A study of a massive hybridization event across time. *Evolution* **56**, 1462–74.

Clarkson JJ, Knapp S, Garcia VF, Olmstead RG, Leitch AR, and Chase MW (2004). Phylogenetic relationships in *Nicotiana* (Solanaceae) inferred from multiple plastid DNA regions. *Molecular Phylogenetics and Evolution* **33**, 75–90.

Clarkson JJ, Lim KY, Kovarik A, Chase MW, Knapp S, and Leitch AR (2005). Long-term genome diploidization in allopolyploid *Nicotiana* section *Repandae* (Solanaceae). *New Phytologist* **168**, 241–52.

Clauss MJ and Koch MA (2006). Poorly known relatives of *Arabidopsis thaliana*. *Trends in Plant Science* **11**, 449–59.

Clement CR and Manshardt RM (2000). A review of the importance of spines for pejibaye heart-of-palm production. *Scientia Horticulturae* **83**, 11–23.

Clifford SL, Anthony NM, Bawe-Johnson M, *et al.* (2004). Mitochondrial DNA phylogeography of western lowland gorillas (*Gorilla gorilla gorilla*). *Molecular Ecology* **13**, 1551–65.

Coart E, Van Glabeke S, De Loose M, Larsen AS, and Roldán-Ruiz I (2006). Chloroplast diversity in the genus *Malus*: New insights into the relationship between the European wild apple (*Malus sylvestris* (L.) Mill.) and the domesticated apple (*Malus domestica* Borkh.). *Molecular Ecology* **15**, 2171–82.

Coart E, Vekemans X, Smulders MJM, *et al.* (2003). Genetic variation in the endangered wild apple (*Malus sylvestris* (L.) Mill.) in Belgium as revealed by amplified fragment length polymorphism and microsatellite markers. *Molecular Ecology* **12,** 845–57.

Cock JH (1982). Cassava: A basic energy source in the tropics. *Science* **218,** 755–62.

Coelho AC, Lima MB, Neves D, and Cravador A (2006). Genetic diversity of two evergreen oaks [*Quercus suber* (L.) and *Quercus ilex* subsp. *rotundifolia* (Lam.)] in Portugal using AFLP markers. *Silvae Genetica* **55,** 105–18.

Cogliati M, Esposto MC, Clarke DL, Wickes BL, and Viviani MA (2001). Origin of *Cryptococcus neoformans* var. *neoformans* diploid strains. *Journal of Clinical Microbiology* **39,** 3889–94.

Cohuet A, Dia I, Simard F, *et al.* (2005). Gene flow between chromosomal forms of the malaria vector *Anopheles funestus* in Cameroon, Central Africa, and its relevance in malaria fighting. *Genetics* **169,** 301–11.

Collet T, Ferreira KM, Arias MC, Soares AEE, and Del Lama MA (2006). Genetic structure of Africanized honeybee populations (*Apis mellifera* L.) from Brazil and Uruguay viewed through mitochondrial DNA COI-COII patterns. *Heredity* **97,** 329–35.

Collins AC (2004). Atelinae phylogenetic relationships: The trichotomy revived? *American Journal of Physical Anthropology* **124,** 285–96.

Collins AC and Dubach JM (2000). Biogeographic and ecological forces responsible for speciation in *Ateles*. *International Journal of Primatology* **21,** 421–44.

Collins AC and Dubach JM (2001). Nuclear DNA variation in Spider monkeys (*Ateles*). *Molecular Phylogenetics and Evolution* **19,** 67–75.

Comstock KE, Georgiadis N, Pecon-Slattery J, *et al.* (2002). Patterns of molecular genetic variation among African elephant populations. *Molecular Ecology* **11,** 2489–98.

Conroy GC, Weber GW, Seidler H, Tobias PV, Kane A, and Brunsden B (1998). Endocranial capacity in an early hominid cranium from Sterkfontein, South Africa. *Science* **280,** 1730–1.

Conte L, Cotti C and Cristofolini G (2007). Molecular evidence for hybrid origin of *Quercus crenata* Lam. (Fagaceae) from *Q. cerris* L and *Q. suber* L. *Plant Biosystems* **141,** 181–93.

Cook AJC, Gilbert RE, Buffolano W, *et al.* (2000). Sources of toxoplasma infection in pregnant women: European multicentre case-control study. *British Medical Journal* **321,** 142–7.

Cornel AJ, McAbee RD, Rasgon J, Stanich MA, Scott TW, and Coetzee M (2003). Differences in extent of genetic introgression between sympatric *Culex pipiens* and *Culex quinquefasciatus* (Diptera: Culicidae) in California and South Africa. *Journal of Medical Entomology* **40,** 36–51.

Cornman RS, Burke JM, Wesselingh RA, and Arnold ML (2004). Contrasting genetic structure of adults and progeny in a Louisiana Iris hybrid population. *Evolution* **58,** 2669–81.

Corpuz PG (2004). Philippines: Oilseeds and Products—GRP Promotes Biodiesel Use. pp. 1–9. Foreign Agricultural Service, United States Department of Agriculture.

Cortés-Ortiz L, Bermingham E, Rico C, Rodríquez-Luna E, Sampaio I, and Ruiz-García M (2003). Molecular systematics and biogeography of the Neotropical monkey genus, *Alouatta*. *Molecular Phylogenetics and Evolution* **26,** 64–81.

Cortés-Ortiz L, Duda TF Jr, Canales-Espinosa D, García-Orduña F, Rodríquez-Luna E, and Bermingham E (2007). Hybridization in large-bodied New World primates. *Genetics* **176,** 2421–5.

Costa A and Oliveira AC (2001). Variation in cork production of the cork oak between two consecutive cork harvests. *Forestry* **74,** 337–46.

Costa A, Pereira H, and Oliveira A (2004). The effect of cork-stripping damage on diameter growth of *Quercus suber* L. *Forestry* **77,** 1–8.

Costa J, Almeida CE, Dujardin JP, and Beard CB (2003). Crossing experiments detect genetic incompatibility among populations of *Triatoma brasiliensis* Neiva, 1911 (Heteroptera, Reduviidae, Triatominae). *Memórias do Instituto Oswaldo Cruz* **98,** 637–9.

Costa J, Peterson AT, and Beard CB (2002). Ecologic niche modeling and differentiation of populations of *Triatoma brasiliensis* Neiva, 1911, the most important Chagas' disease vector in northeastern Brazil (Hemiptera, Reduviidae, Triatominae). *American Journal of Tropical Medicine and Hygiene* **67,** 516–20.

Coulibaly S, Pasquet RS, Papa R, and Gepts P (2002). AFLP analysis of the phenetic organization and genetic diversity of *Vigna unguiculata* L. Walp. reveals extensive gene flow between wild and domesticated types. *Theoretical and Applied Genetics* **104,** 358–66.

Couvreur TLP, Billotte N, Risterucci A-M, *et al.* (2006). Close genetic proximity between cultivated and wild *Bactris gasipaes* Kunth revealed by microsatellite markers in western Ecuador. *Genetic Resources and Crop Evolution* **53,** 1361–73.

Couvreur TLP, Hahn WJ, de Granville J-J, Pham J-L, Ludeña B and Pintaud J-C (2007). Phylogenetic relationships of the cultivated neotropical palm *Bactris gasipaes* (Arecaceae) with its wild relatives inferred from chloroplast and nuclear DNA polymorphisms. *Systematic Botany* **32,** 519–30.

Cox MP, Mendez FL, Karafet TM, *et al.* (2008). Testing for archaic hominin admixture on the X chromosome: Model likelihoods for the modern human *RRM2P4* region from summaries of genealogical topology under the structured coalescent. *Genetics* **178,** 427–37.

Coyne JA and Orr HA (2004). *Speciation*. Sinauer Associates, Inc. Sunderland, Massachusetts.

Crespi BJ and Fulton MJ (2004). Molecular systematics of Salmonidae: Combined nuclear data yields a robust phylogeny. *Molecular Phylogenetics and Evolution* **31,** 658–79.

Criscione CD, Anderson JD, Sudimack D, *et al.* (2007). Disentangling hybridization and host colonization in parasitic roundworms of humans and pigs. *Proceedings of the Royal Society of London B* **274,** 2669–77.

Critchfield WB and Kinloch BB (1986). Sugar pine and its hybrids. *Silvae Genetica* **35,** 138–45.

Cronin MA, MacNeil MD, and Patton JC (2005). Variation in mitochondrial DNA and microsatellite DNA in caribou (*Rangifer tarandus*) in North America. *Journal of Mammalogy* **86,** 495–505.

Cronin MA, MacNeil MD, and Patton JC (2006). Mitochondrial DNA and microsatellite DNA variation in domestic reindeer (*Rangifer tarandus tarandus*) and relationships with wild caribou (*Rangifer tarandus granti, Rangifer tarandus groenlandicus,* and *Rangifer tarandus caribou*). *Journal of Heredity* **97,** 525–30.

Cronin MA, Patton JC, Balmysheva N, and MacNeil MD (2003). Genetic variation in caribou and reindeer (*Rangifer tarandus*). *Animal Genetics* **34,** 33–41.

Cropp SJ, Larson A, and Cheverud JM (1999). Historical biogeography of tamarins, genus *Saguinus*: The molecular phylogenetic evidence. *American Journal of Physical Anthropology* **108,** 65–89.

Cros J, Combes MC, Trouslot P, *et al.* (1998). Phylogenetic analysis of chloroplast DNA variation in *Coffea* L. *Molecular Phylogenetics and Evolution* **9,** 109–17.

Cruzan MB and Arnold ML (1993). Ecological and genetic associations in an *Iris* hybrid zone. *Evolution* **47,** 1432–45.

Currat M and Excoffier L (2004). Modern humans did not admix with Neanderthals during their range expansion into Europe. *PLoS Biology* **2,** e421

Dadejová M, Lim KY, Soucková-Skalická K, *et al.* (2007). Transcription activity of rRNA genes correlates with a tendency towards intergenomic homogenization in *Nicotiana* allotetraploids. *New Phytologist* **174,** 658–68.

Daguin C, Bonhomme F, and Borsa P (2001). The zone of sympatry and hybridization of *Mytilus edulis* and *M. galloprovincialis,* as described by intron length polymorphism at locus *mac-1*. *Heredity* **86,** 342–54.

Dang Q, Chen J, Unutmaz D, *et al.* (2004). Nonrandom HIV-1 infection and double infection via direct and cell-mediated pathways. *Proceedings of the National Academy of Sciences USA* **101,** 632–7.

Darwin C (1845). *The Voyage of the Beagle,* Second Edition. PF Collier & Son, New York.

Darwin C (1859). *On the Origin of Species by Means of Natural Selection or the Preservation of Favoured Races in the Struggle for Life.* John Murray, London.

Davis CC, Anderson WR, and Wurdack KJ (2005). Gene transfer from a parasitic flowering plant to a fern. *Proceedings of the Royal Society of London B* **272,** 2237–42.

Day JJ, Simona S, and Garcia-Moreno J (2007). Phylogenetic relationships of the Lake Tanganyika cichlid tribe Lamprologini: The story from mitochondrial DNA. *Molecular Phylogenetics and Evolution* **45,** 629–42.

de Barros Lopes M, Bellon JR, Shirley NJ, and Ganter PF (2002). Evidence for multiple interspecific hybridization in *Saccharomyces sensu stricto. FEMS Yeast Research* **1,** 323–31.

Debruyne R (2005). A case study of apparent conflict between molecular phylogenies: The interrelationships of African elephants. *Cladistics* **21,** 31–50.

Decker-Walters DS, Wilkins-Ellert M, Chung S-M, and Staub JE (2004). Discovery and genetic assessment of wild bottle gourd [*Lagenaria siceraria* (Mol.) Standley; Cucurbitaceae] from Zimbabwe. *Economic Botany* **58,** 501–8.

Dediu D and Ladd DR (2007). Linguistic tone is related to the population frequency of the adaptive haplogroups of two brain size genes, *ASPM* and *microcephalin*. *Proceedings of the National Academy of Sciences USA* **104,** 10944–9.

de Heinzelin J, Clark JD, White T, *et al.* (1999). Environment and behavior of 2.5-million-year-old Bouri hominids. *Science* **284,** 625–35.

Deinard A and Kidd K (1999). Evolution of a HOXB6 intergenic region within the great apes and humans. *Journal of Human Evolution* **36,** 687–703.

Deinard A and Kidd K (2000). Identifying conservation units within captive chimpanzee populations. *American Journal of Physical Anthropology* **111,** 25–44.

della Torre A, Merzagora L, Powell JR, and Coluzzi M (1997). Selective introgression of paracentric inversions between two sibling species of the *Anopheles gambiae* complex. *Genetics* **146,** 239–44.

DeLong EF (1992). Archaea in coastal marine environments. *Proceedings of the National Academy of Sciences USA* **89,** 5685–9.

DeMarais BD, Dowling TE, Douglas ME, Minckley WL, and Marsh PC (1992). Origin of *Gila seminuda* (Teleostei:

Cyprinidae) through introgressive hybridization: Implications for evolution and conservation. *Proceedings of the National Academy of Sciences USA* **89,** 2747–51.

de Meijer EPM and van Soest LJM (1992). The CPRO *Cannabis* germplasm collection. *Euphytica* **62,** 201–11.

de Moraes AP, dos Santos Soares Filho W, and Guerra M (2007). Karyotype diversity and the origin of grapefruit. *Chromosome Research* **15,** 115–21.

Desjeux P (2001). The increase in risk factors for leishmaniasis worldwide. *Transactions of the Royal Society of Tropical Medicine and Hygiene* **95,** 239–43.

Desjeux P (2004). Leishmaniasis: Current situation and new perspectives. *Comparative Immunology, Microbiology and Infectious Diseases* **27,** 305–18.

Detwiler KM, Burrell AS, and Jolly CJ (2005). Conservation implications of hybridization in African cercopithecine monkeys. *International Journal of Primatology* **26,** 661–84.

Dillinger TL, Barriga P, Escárcega S, Jimenez M, Lowe DS, and Grivetti LE (2000). Food of the gods: Cure for humanity? A cultural history of the medicinal and ritual use of chocolate. *Journal of Nutrition* **130 (supp),** 2057S–72S.

Dillon SL, Shapter FM, Henry RJ, Cordeiro G, Izquierdo L, and Lee LS (2007). Domestication to crop improvement: Genetic resources for *Sorghum* and *Saccharum* (Andropogoneae). *Annals of Botany* **100,** 975–89.

Dirks W, Reid DJ, Jolly CJ, Phillips-Conroy JE, and Brett FL (2002). Out of the mouths of baboons: Stress, life history, and dental development in the Awash National Park hybrid zone, Ethiopia. *American Journal of Physical Anthropology* **118,** 239–52.

Disotell TR (1994). Generic level relationships of the Papionini (Cercopithecoidea). *American Journal of Physical Anthropology* **94,** 47–57.

Dodd RS and Afzal-Rafii Z (2004). Selection and dispersal in a multispecies oak hybrid zone. *Evolution* 58, 261–9.

Dohlman E and Livezey J (2005). Peanut Backgrounder. pp. 1–30. Economic Research Service, United States Department of Agriculture.

Doolittle RF, Feng D-F, Tsang S, Cho G ,and Little E (1996). Determining divergence times of the major kingdoms of living organisms with a protein clock. *Science* **271**, 470–7.

Doolittle WF (1999). Phylogenetic classification and the universal tree. *Science* **284,** 2124–8.

Doolittle WF and Bapteste E (2007). Pattern pluralism and the Tree of Life hypothesis. *Proceedings of the National Academy of Sciences USA* **104,** 2043–2049.

Doolittle WF, Boucher Y, Nesbø CL, Douady CJ, Andersson JO, and Roger AJ (2003). How big is the iceberg of which organellar genes in nuclear genomes are but the tip? *Philosophical Transactions of the Royal Society of London, B* **358,** 39–58.

Dowling TE and DeMarais BD (1993). Evolutionary significance of introgressive hybridization in cyprinid fishes. *Nature* **362**, 444–6.

Dowling TE, Minckley WL, Douglas ME, Marsh PC, and DeMarais BD (1992). Response to Wayne, Nowak, and Phillips and Henry: Use of molecular characters in conservation biology. *Conservation Biology* **6,** 600–3.

Driscoll CA, Menotti-Raymond M, Roca AL, *et al.* (2007). The Near Eastern origin of cat domestication. *Science* **317,** 519–23.

Duarte C, Maurício J, Pettitt PB, *et al.* (1999). The early Upper Paleolithic human skeleton from the Abrigo do Lagar Velho (Portugal) and modern human emergence in Iberia. *Proceedings of the National Academy of Sciences USA* **96,** 7604–9.

Dubcovsky J and Dvorak J (2007). Genome plasticity a key factor in the success of polyploid wheat under domestication. *Science* **316,** 1862–6.

Dujardin J-C (2006). Risk factors in the spread of leishmaniasis: Towards integrated monitoring? *Trends in Parasitology* **22,** 4–6.

Duputié A, David P, Debain C, and McKey D (2007). Natural hybridization between a clonally propagated crop, cassava (*Manihot esculenta* Crantz) and a wild relative in French Guiana. *Molecular Ecology* **16,** 3025–38.

Dusfour I, Blondeau J, Harbach RE, *et al.* (2007). Polymerase chain reaction identification of three members of the *Anopheles sundaicus* (Diptera: Culicidae) complex, malaria vectors in Southeast Asia. *Journal of Medical Entomology* **44,** 723–31.

Dvorak J, Akhunov ED, Akhunov AR, Deal KR, and Luo M-C (2006). Molecular characterization of a diagnostic DNA marker for domesticated tetraploid wheat provides evidence for gene flow from wild tetraploid wheat to hexaploid wheat. *Molecular Biology and Evolution* **23,** 1386–96.

Dziejman M, Balon E, Boyd D, Fraser CM, Heidelberg JF, and Mekalanos JJ (2002). Comparative genomic analysis of *Vibrio cholerae*: Genes that correlate with cholera endemic and pandemic disease. *Proceedings of the National Academy of Sciences USA* **99,** 1556–61.

Ebersberger I, Galgoczy P, Taudien S, Taenzer S, Platzer M, and von Haeseler A (2007). Mapping human genetic ancestry. *Molecular Biology and Evolution* **24,** 2266–76.

Eckert AJ and Hall BD (2006). Phylogeny, historical biogeography, and patterns of diversification for *Pinus* (Pinaceae): Phylogenetic tests of fossil-based hypotheses. *Molecular Phylogenetics and Evolution* **40,** 166–82.

Edmond-Blanc F (1947). A contribution to the knowledge of the Cambodian wild ox or kouproh. *Journal of Mammalogy* **28,** 245–8.

Edwards CJ, Bollongino R, Scheu A, *et al.* (2007). Mitochondrial DNA analysis shows a Near Eastern Neolithic origin for domestic cattle and no indication of domestication of European aurochs. *Proceedings of the Royal Society of London* B 274, 1377–85.

Edwards CTT, Holmes EC, Pybus OG, *et al.* (2006). Evolution of the human immunodeficiency virus envelope gene is dominated by purifying selection. *Genetics* **174,** 1441–53.

Elias M, Mühlen GS, McKey D, Roa AC, and Tohme J (2004). Genetic diversity of traditional South American landraces of cassava (*Manihot esculenta* Crantz): An analysis using microsatellites. *Economic Botany* **58,** 242–56.

Elias M, Penet L, Vindry P, McKey D, Panaud O, and Robert T (2001). Unmanaged sexual reproduction and the dynamics of genetic diversity of a vegetatively propagated crop plant, cassava (*Manihot esculenta* Crantz), in a traditional farming system. *Molecular Ecology* **10,** 1895–907.

Ellstrand NC, Garner LC, Hegde S, Guadagnuolo R, and Blancas L (2007). Spontaneous hybridization between maize and Teosinte. *Journal of Heredity* **98,** 183–7.

Ely JJ, Dye B, Frels WI, *et al.* (2005). Subspecies composition and founder contribution of the captive U.S. chimpanzee (*Pan troglodytes*) population. *American Journal of Primatology* **67,** 223–41.

Enard W and Pääbo S (2004). Comparative primate genomics. *Annual Review of Genomics and Human Genetics* **5,** 351–78.

Endler JA (1977). *Geographic Variation, Speciation, and Clines.* Princeton University Press, Princeton.

Erickson DL, Smith BD, Clarke AC, Sandweiss DH, and Tuross N (2005). An Asian origin for a 10,000-year-old domesticated plant in the Americas. *Proceedings of the National Academy of Sciences USA* **102,** 18315–20.

Evans BJ, Morales JC, Supriatna J, and Melnick DJ (1999). Origin of the Sulawesi macaques (Cercopithecidae: *Macaca*) as suggested by mitochondrial DNA phylogeny. *Biological Journal of the Linnean Society* **66,** 539–60.

Evans BJ, Supriatna J, Andayani N, and Melnick DJ (2003). Diversification of Sulawesi macaque monkeys: Decoupled evolution of mitochondrial and autosomal DNA. *Evolution* **57,** 1931–46.

Evans BJ, Supriatna J, and Melnick DJ (2001). Hybridization and population genetics of two macaque species in Sulawesi, Indonesia. *Evolution* **55,** 1686–702.

Evans PD, Anderson JR, Vallender EJ, Choi SS, and Lahn BT (2004). Reconstructing the evolutionary history of *microcephalin*, a gene controlling human brain size. *Human Molecular Genetics* **13,** 1139–45.

Evans PD, Gilbert SL, Mekel-Bobrov N, *et al.* (2005). *Microcephalin*, a gene regulating brain size, continues to evolve adaptively in humans. *Science* **309,** 1717–20.

Evans PD, Mekel-Bobrov N, Vallender EJ, Hudson RR, and Lahn BT (2006). Evidence that the adaptive allele of the brain size gene *microcephalin* introgressed into *Homo sapiens* from an archaic *Homo* lineage. *Proceedings of the National Academy of Sciences USA* **103,** 18178–83.

Fabbri E, Miquel C, Lucchini V, *et al.* (2007). From the Apennines to the Alps: Colonization genetics of the naturally expanding Italian wolf (*Canis lupus*) population. *Molecular Ecology* **16,** 1661–71.

Fagundes NJR, Ray N, Beaumont M, *et al.* (2007). Statistical evaluation of alternative models of human evolution. *Proceedings of the National Academy of Sciences USA* **104,** 17614–9.

Falk D (1998). Hominid brain evolution: Looks can be deceiving. *Science* **280,** 1714.

Fang J, Chao C-CT, Roberts PA, and Ehlers JD (2007). Genetic diversity of cowpea [*Vigna unguiculata* (L.) Walp.] in four West African and USA breeding programs as determined by AFLP analysis. *Genetic Resources and Crop Evolution* **54,** 1197–209.

Farias IP, Ortí G, Sampaio I, Schneider H, and Meyer A (2001). The cytochrome *b* gene as a phylogenetic marker: The limits of resolution for analyzing relationships among cichlid fishes. *Journal of Molecular Evolution* **53,** 89–103.

Faruque SM, Tam VC, and Chowdhury N, *et al.* (2007). Genomic analysis of the Mozambique strain of *Vibrio cholerae* O1 reveals the origin of El Tor strains carrying classical CTX prophage. *Proceedings of the National Academy of Sciences USA* **104,** 5151–6.

Fauquet CM and Tohme J (2004). The global cassava partnership for genetic improvement. *Plant Molecular Biology* **56,** v–x.

Fávero AP, Simpson CE, Valls JFM, and Vello NA (2006). Study of the evolution of cultivated peanut through crossability studies among *Arachis ipaënsis, A. duranensis,* and *A. hypogaea. Crop Science* **46,** 1546–52.

Feleke Y, Pasquet RS, and Gepts P (2006). Development of PCR-based chloroplast DNA markers that characterize domesticated cowpea (*Vigna unguiculata* ssp. *unguiculata* var. *unguiculata*) and highlight its crop-weed complex. *Plant Systematics and Evolution* **262,** 75–87.

Fenchel T (2003). Biogeography for bacteria. *Science* 301, 925–6.

Filée J, Siguier P, and Chandler M (2007). I am what I eat and I eat what I am: Acquisition of bacterial genes by giant viruses. *Trends in Genetics* **23,** 10–15.

Filho HDC, Machado MA, Targon MLPN, Moreira MCPQDG, and Pompeu J (1998). Analysis of the genetic diversity among mandarins (*Citrus* spp.) using RAPD markers. *Euphytica* **102,** 133–9.

Finlay BJ and Fenchel T (2004). Cosmopolitan metapopulations of free-living microbial eukaryotes. *Protist* **155,** 237–44.

Finlayson C (2005). Biogeography and evolution of the genus *Homo. Trends in Ecology and Evolution* **20,** 457–63.

Fischer A, Pollack J, Thalmann O, Nickel B, and Pääbo S (2006). Demographic history and genetic differentiation in apes. *Current Biology* **16,** 1133–8.

Fischer A, Wiebe V, Pääbo S, and Przeworski M (2004). Evidence for a complex demographic history of chimpanzees. *Molecular Biology and Evolution* **21,** 799–808.

Fitzpatrick BM (2004). Rates of evolution of hybrid inviability in birds and mammals. *Evolution* **58,** 1865–70.

Flagstad O and Røed KH (2003). Refugial origins of reindeer (*Rangifer tarandus* L.) inferred from mitochondrial DNA sequences. *Evolution* **57,** 658–70.

Fonseca DM, Keyghobadi N, Malcolm CA, *et al.* (2004). Emerging vectors in the *Culex pipiens* complex. *Science* **303,** 1535–8.

Fonseca DM, LaPointe DA, and Fleischer RC (2000). Bottlenecks and multiple introductions: Population genetics of the vector of avian malaria in Hawaii. *Molecular Ecology* **9,** 1803–14.

Fonseca DM, Smith JL, Wilkerson RC, and Fleischer RC (2006). Pathways of expansion and multiple introductions illustrated by large genetic differentiation among worldwide populations of the southern house mosquito. *American Journal of Tropical Medicine and Hygiene* **74,** 284–9.

Fosberg FR (1960). Introgression in *Artocarpus* (Moraceae) in Micronesia. *Brittonia* **12,** 101–13.

Franck P, Garnery L, Celebrano G, Solignac M, and Cornuet J-M (2000). Hybrid origins of honeybees from Italy (*Apis mellifera ligustica*) and Sicily (*A. m. sicula*). *Molecular Ecology* **9,** 907–21.

Fredrickson RJ and Hedrick PW (2006). Dynamics of hybridization and introgression in red wolves. *Conservation Biology* **20,** 1272–83.

Freeland JR and Boag PT (1999). The mitochondrial and nuclear genetic homogeneity of the phenotypically diverse Darwin's ground finches. *Evolution* **53,** 1553–63.

Freeman AR, Hoggart CJ, Hanotte O, and Bradley DG (2006). Assessing the relative ages of admixture in the bovine hybrid zones of Africa and the Near East using X chromosome haplotype mosaicism. *Genetics* **173,** 1503–10.

Fricker, CR, Medema GD, and Smith HV (1998). Protozoan parasites (*Cryptosporidium, Giardia, Cyclospora*). pp. 70–118. World Health Organization.

Frigaard N-U, Martinez A, Mincer TJ, and DeLong EF (2006). Proteorhodopsin lateral gene transfer between marine planktonic Bacteria and Archaea. *Nature* **439,** 847–50.

Fujita K, Watanabe K, Widarto TH, and Suryobroto B (1997). Discrimination of macaques by macaques: The case of Sulawesi species. *Primates* **38,** 233–45.

Fuller DQ (2007). Contrasting patterns in crop domestication and domestication rates: Recent archaeobotanical insights from the Old World. *Annals of Botany* **100,** 903–24.

Fulnecek J, Lim KY, Leitch AR, Kovarík A, and Matyásek R (2002). Evolution and structure of 5S rDNA loci in allotetraploid *Nicotiana tabacum* and its putative parental species. *Heredity* **88,** 19–25.

Futuyma DJ (2005). *Evolution.* Sinauer Associates, Inc. Sunderland, Massachusetts.

Gaffrey A (2003). Statement for the United States House of Representatives Subcommittee on Criminal Justice, Drug Policy and Human Resources. pp. 1–7.

Gagneux P (2004). A *Pan*-oramic view: Insights into hominoid evolution through the chimpanzee genome. *Trends in Ecology and Evolution* **19,** 571–6.

Galbreath GJ, Mordacq JC, and Weiler FH (2006). Genetically solving a zoological mystery: Was the kouprey (*Bos sauveli*) a feral hybrid? *Journal of Zoology* **270,** 561–4.

Garner KJ and Ryder OA (1996). Mitochondrial DNA diversity in gorillas. *Molecular Phylogenetics and Evolution* **6,** 39–48.

Garrigan D, Mobasher Z, Severson T, Wilder JA, and Hammer MF (2005). Evidence for archaic Asian ancestry on the human X chromosome. *Molecular Biology and Evolution* **22,** 189–92.

Gaubert P and Begg CM (2007). Re-assessed molecular phylogeny and evolutionary scenario within genets (Carnivora, Viverridae, Genettinae). *Molecular Phylogenetics and Evolution* **44,** 920–7.

Gaubert P, Fernandes CA, Bruford MW, and Veron G (2004a). Genets (Carnivora, Viverridae) in Africa: An evolutionary synthesis based on cytochrome *b* sequences and morphological characters. *Biological Journal of the Linnean Society* **81,** 589–610.

Gaubert P, Papes M, and Peterson AT (2006). Natural history collections and the conservation of poorly known taxa: Ecological niche modeling in central African rainforest genets (*Genetta* spp.). *Biological Conservation* **130,** 106–17.

Gaubert P, Taylor PJ, Fernandes CA, Bruford MW, and Veron G (2005). Patterns of cryptic hybridization revealed using an integrative approach: A case study on genets (Carnivora, Viverridae, *Genetta* spp.) from the southern African subregion. *Biological Journal of the Linnean Society* **86,** 11–33.

Gaubert P, Trainer M, Delmas A-S, Colyn M, and Veron G (2004b). First molecular evidence for reassessing phylogenetic affinities between genets (*Genetta*) and the enigmatic genet-like taxa *Osbornictis, Poiana* and *Prionodon* (Carnivora, Viverridae). *Zoologica Scripta* **33,** 117–29.

Gaubert P, Veron G, and Trainer M (2002). Genets and 'genet-like' taxa (Carnivora, Viverrinae): Phylogenetic analysis, systematics and biogeographic implications. *Zoological Journal of the Linnean Society* **134,** 317–34.

Gaut BS and Doebley JF (1997). DNA sequence evidence for the segmental allotetraploid origin of maize. *Proceedings of the National Academy of Sciences USA* **94,** 6809–14.

Ge S, Sang T, Lu B-R, and Hong D-Y (1999). Phylogeny of rice genomes with emphasis on origins of allotetraploid species. *Proceedings of the National Academy of Sciences USA* **96,** 14400–5.

Genner MJ, Nichols P, Carvalho GR, Robinson RL, Shaw PW, and Turner GF (2007a). Reproductive isolation among deep-water cichlid fishes of Lake Malawi differing in monochromatic male breeding dress. *Molecular Ecology* **16,** 651–62.

Genner MJ, Seehausen O, Lunt DH, *et al.* (2007b). Age of cichlids: New dates for ancient lake fish radiations. *Molecular Biology and Evolution* **24,** 1269–82.

Gentile G, della Torre A, Maegga B, Powell JR, and Caccone A (2002). Genetic differentiation in the African malaria vector, *Anopheles gambiae* s.s., and the problem of taxonomic status. *Genetics* **161,** 1561–78.

Geraldes A, Ferrand N, and Nachman MW (2006). Contrasting patterns of introgression at X-linked loci across the hybrid zone between subspecies of the European rabbit (*Oryctolagus cuniculus*). *Genetics* **173,** 919–33.

Geraldes A, Rogel-Gaillard C, and Ferrand N (2005). High levels of nucleotide diversity in the European rabbit (*Oryctolagus cuniculus*) *SRY* gene. *Animal Genetics* **36,** 349–51.

Gernandt DS, López GG, Garcia SO, and Liston A (2005). Phylogeny and classification of *Pinus. Taxon* **54,** 29–42.

Gerstel DU and Sisson VA (1995). Tobacco—*Nicotiana tabacum* (Solanaceae). In J Smartt and NW Simmonds, eds. *Evolution of Crop Plants,* Second Edition, pp. 458–63. Longman Scientific & Technical, Essex, England.

Gibbs MJ, Armstrong JS, and Gibbs AJ (2002). Questioning the evidence for genetic recombination in the 1918 "Spanish flu" virus. *Science* **296,** 211a.

Gilmore S, Peakall R, and Robertson J (2003). Short tandem repeat (STR) DNA markers are hypervariable and informative in *Cannabis sativa*: Implications for forensic investigations. *Forensic Science International* **131,** 65–74.

Gilmore S, Peakall R, and Robertson J (2007). Organelle DNA haplotypes reflect crop–use characteristics and geographic origins of *Cannabis sativa. Forensic Science International* **172,** 179–90.

Giuffra E, Kijas JMH, Amarger V, Carlborg O, Jeon J-T, and Andersson L (2000). The origin of the domestic pig: Independent domestication and subsequent introgression. *Genetics* **154,** 1785–91.

Glémet H, Blier P, and Bernatchez L (1998). Geographical extent of arctic char (*Salvelinus alpinus*) mtDNA introgression in brook char populations (*S. fontinalis*) from eastern Québec, Canada. *Molecular Ecology* **7,** 1655–62.

Gobert V, Moja S, Colson M, and Taberlet P (2002). Hybridization in the section *Mentha* (Lamiaceae) inferred from AFLP markers. *American Journal of Botany* **89,** 2017–23.

Godwin P (2000). Bushmen. *National Geographic Magazine Online Extra.* National Geographic.com.

Gómez C (2003). Cowpea: Post-harvest operations. Chapter XXXII In Compendium on Post-harvest Operations. Mejía D and Parrucci E, eds. FAO, Rome.

Gomez-Alpizar L, Carbone I, and Ristaino JB (2007). An Andean origin of *Phytophthora infestans* inferred from mitochondrial and nuclear gene genealogies. *Proceedings of the National Academy of Sciences USA* **104,** 3306–11.

Gonder MK, Oates JF, Disotell TR, Forstner MRJ, Morales JC, and Melnick DJ (1997). A new west African chimpanzee subspecies? *Nature* **388,** 337.

Goodman KJ, Correa P, Aux HJT, *et al.* (1996). *Helicobacter pylori* infection in the Colombian Andes: A population-based study of transmission pathways. *American Journal of Epidemiology* **144,** 290–9.

Götherström A, Anderung C, Hellborg L, *et al.* (2005). Cattle domestication in the Near East was followed by hybridization with aurochs bulls in Europe. *Proceedings of the Royal Society of London B* **272,** 2345–50.

Gotoh S, Takenaka O, Watanabe K, *et al.* (2001). Hematological values and parasite fauna in free-ranging *Macaca hecki* and the *M. hecki/M. tonkeana* hybrid group of Sulawesi Island, Indonesia. *Primates* **42,** 27–34.

Gottelli D, Sillero-Zubiri C, Applebaum GD, *et al.* (1994). Molecular genetics of the most endangered canid: The Ethiopian wolf *Canis simensis. Molecular Ecology* **3,** 301–12.

Grant BR and Grant PR (1993). Evolution of Darwin's finches caused by a rare climatic event. *Proceedings of the Royal Society of London* B **251**, 111–7.

Grant BR and Grant PR (1996). High survival of Darwin's finch hybrids: Effects of beak morphology and diets. *Ecology* **77**, 500–9.

Grant PR (1993). Hybridization of Darwin's finches on Isla Daphne Major, Galápagos. *Philosophical Transactions of the Royal Society of London* B. **340**, 127–39.

Grant PR and Grant BR (1992). Hybridization of bird species. *Science* **256**, 193–7.

Grant PR and Grant BR (2006). Evolution of character displacement in Darwin's finches. *Science* **313**, 224–6.

Grant PR, Grant BR, Keller LF, Markert JA, and Petren K (2003). Inbreeding and interbreeding in Darwin's finches. *Evolution* **57**, 2911–6.

Grant PR, Grant BR, Markert JA, Keller LF, and Petren K (2004). Convergent evolution of Darwin's finches caused by introgressive hybridization and selection. *Evolution* **58**, 1588–99.

Grant PR, Grant BR, and Petren K (2005). Hybridization in the recent past. *American Naturalist* **166**, 56–67.

Grant V (1981). *Plant Speciation*. Columbia University Press, New York.

Gravina B, Mellars P, and Ramsey B (2005). Radiocarbon dating of interstratified Neanderthal and early modern human occupations at the Chatelperronian type-site. *Nature* **438**, 51–6.

Gravlund P, Meldgaard M, Pääbo S, and Arctander P (1998). Polyphyletic origin of the small-bodied, high-arctic subspecies of tundra reindeer (*Rangifer tarandus*). *Molecular Phylogenetics and Evolution* **10**, 151–9.

Green RE, Krause J, Ptak SE, *et al.* (2006). Analysis of one million base pairs of Neanderthal DNA. *Nature* **444,** 330–6.

Grigg ME, Bonnefoy S, Hehl AB, Suzuki Y, and Boothroyd JC (2001). Success and virulence in *Toxoplasma* as the result of sexual recombination between two distinct ancestries. *Science* **294,** 161–5.

Grivet L, D'Hont A, Roques D, Feldmann P, Lanaud C, and Glaszmann JC (1996). RFLP mapping in cultivated sugarcane (*Saccharum* spp.): Genome organization in a highly polyploid and aneuploid interspecific hybrid. *Genetics* **142,** 987–1000.

Groth C, Hansen J, and Piskur J (1999). A natural chimeric yeast containing genetic material from three species. *International Journal of Systematic Bacteriology* **49**, 1933–8.

Groves CP (1997). Taxonomy of wild pigs (*Sus*) of the Philippines. *Zoological Journal of the Linnean Society* **120,** 163–91.

Gunnell K, Tada MK, Hawthorne FA, Keeley ER, and Ptacek MB (2008). Geographic patterns of introgressive hybridization between native Yellowstone cutthroat trout (*Oncorhynchus clarkii bouvieri*) and introduced rainbow trout (*O. mykiss*) in the South Fork of the Snake River watershed, Idaho. *Conservation Genetics* **9,** 49–64.

Gürtler RE, Kitron U, Cecere MC, Segura EL, and Cohen JE (2007). Sustainable vector control and management of Chagas disease in the Gran Chaco, Argentina. *Proceedings of the National Academy of Sciences USA* **104,** 16194–9.

Guyatt HL, Chan MS, Medley GF, and Bundy DAP (1995). Control of *Ascaris* infection by chemotherapy: Which is the most cost-effective option? *Transactions of the Royal Society of Tropical Medicine and Hygiene* **89,** 16–20.

Hacker J and Carniel E (2001). Ecological fitness, genomic islands and bacterial pathogenicity. *EMBO Reports* **2,** 376–81.

Halbert ND and Derr JN (2007). A comprehensive evaluation of cattle introgression into US federal bison herds. *Journal of Heredity* **98,** 1–12.

Hall C and Dietrich FS (2007). The reacquisition of biotin prototrophy in *Saccharomyces cerevisiae* involved horizontal gene transfer, gene duplication and gene clustering. *Genetics* **177,** 2293–307.

Hamada Y, Urasopon N, Hadi I, and Malaivijitnond S (2006). Body size and proportions and pelage color of free-ranging *Macaca mulatta* from a zone of hybridization in northeastern Thailand. *International Journal of Primatology* **27,** 497–513.

Hammer MF (1995). A recent common ancestry for human Y chromosomes. *Nature* **378,** 376–8.

Hammer MF and Zegura SL (2002). The human Y chromosome haplogroup tree: Nomenclature and phylogeography of its major divisions. *Annual Review of Anthropology* **31,** 303–21.

Hapke A, Zinner D, and Zischler H (2001). Mitochondrial DNA variation in Eritrean hamadryas baboons (*Papio hamadryas hamadryas*): Life history influences population genetic structure. *Behavioral Ecology and Sociobiology* **50,** 483–92.

Hardin JW (1975). Hybridization and introgression in *Quercus alba*. *Journal of the Arnold Arboretum* **56,** 336–63.

Hardin JW (1979). Atlas of foliar surface features in woody plants, I. Vestiture and trichome types of eastern North American *Quercus*. *Bulletin of the Torrey Botanical Club* **106,** 313–25.

Harris EE and Disotell TR (1998). Nuclear gene trees and the phylogenetic relationships of the mangabeys

(Primates: Papionini). *Molecular Biology and Evolution* **15**, 892–900.

Harris SA, Robinson JP, and Juniper BE (2002). Genetic clues to the origin of the apple. *Trends in Genetics* **18**, 426–30.

Harrison RG (1986). Pattern and process in a narrow hybrid zone. *Heredity* **56**, 337–49.

Harrison RG (1990). Hybrid zones: Windows on evolutionary process. *Oxford Surveys in Evolutionary Biology* **7**, 69–128.

Hartwell LH, Hood L, Goldberg ML, Reynolds AE, Silver LM, and Veres RC (2004). *Genetics—From Genes to Genomes*, Second Edition. McGraw-Hill, New York.

Hashimoto T, Nakamura Y, Kamaishi T, *et al.* (1995). Phylogenetic place of mitochondrion-lacking protozoan, *Giardia lamblia*, inferred from amino acid sequences of elongation factor 2. *Molecular Biology and Evolution* **12**, 782–93.

Hassanin A and Ropiquet A (2004). Molecular phylogeny of the tribe Bovini (Bovidae, Bovinae) and the taxonomic status of the kouprey, *Bos sauveli* Urbain 1937. *Molecular Phylogenetics and Evolution* **33**, 896–907.

Hassanin A and Ropiquet A (2007a). Resolving a zoological mystery: The kouprey is a real species. *Proceedings of the Royal Society of London B* **274**, 2849–55.

Hassanin A and Ropiquet A (2007b). What is the taxonomic status of the Cambodian banteng and does it have close genetic links with the kouprey? *Journal of Zoology* **271**, 246–52.

Hassanin A, Ropiquet A, Cornette R, *et al.* (2006). Has the kouprey (*Bos sauveli* Urbain, 1937) been domesticated in Cambodia. *Comptes Rendus Biologies* **329**, 124–35.

Hawks J, Cochran G, Harpending HC, and Lahn BT (2008). A genetic legacy from archaic *Homo*. *Trends in Genetics* **24**, 19–23.

Hawks JD and Wolpoff MH (2001). The Accretion model of Neandertal evolution. *Evolution* **55**, 1474–85.

Hayakawa T, Aki I, Varki A, Satta Y, and Takahata N (2006). Fixation of the human-specific CMP-*N*-Acetylneuraminic Acid Hydroxylase pseudogene and implications of haplotype diversity for human evolution. *Genetics* **172**, 1109–16.

Hayakawa T, Satta Y, Gagneux P, Varki A, and Takahata N (2001). *Alu*-mediated inactivation of the human CMP-*N*-acetylneuraminic acid hydroxylase gene. *Proceedings of the National Academy of Sciences USA* **98**, 11399–404.

Hayashi S, Hayasaka K, Takenaka O, and Horai S (1995). Molecular phylogeny of gibbons inferred from mitochondrial DNA sequences: Preliminary report. *Journal of Molecular Evolution* **41**, 359–65.

Hayes CG (2001). West Nile virus: Uganda, 1937, to New York City, 1999. *Annals of the New York Academy of Sciences* **951**, 25–37.

He J, Baldini RL, Déziel E, *et al.* (2004). The broad host range pathogen *Pseudomonas aeruginosa* strain PA14 carries two pathogenicity islands harboring plant and animal virulence genes. *Proceedings of the National Academy of Sciences USA* **101**, 2530–5.

Heckman KL, Mariani CL, Rasoloarison R, and Yoder AD (2007). Multiple nuclear loci reveal patterns of incomplete lineage sorting and complex species history within western mouse lemurs (*Microcebus*). *Molecular Phylogenetics and Evolution* **43**, 353–67.

Hedges S (2000). *Bos javanicus*. In: IUCN 2007. IUCN Red List of Threatened Species. International Union for Conservation of Nature and Natural Resources

Heeney JL, Dalgleish AG, and Weiss RA (2006). Origins of HIV and the evolution of resistance to AIDS. *Science* **313**, 462–466.

Heiser CB Jr (1951). Hybridization in the annual sunflowers: *Helianthus annuus X H. debilis* var *cucumerifolius*. *Evolution* **5**, 42–51.

Helfgott DM and Mason-Gamer RJ (2004). The evolution of North American *Elymus* (Triticeae, Poaceae) allotetraploids: Evidence from phosphoenolpyruvate carboxylase gene sequences. *Systematic Botany* **29**, 850–61.

Henderson E (2005). Economic impact of waterfowl hunting in the United States—Addendum to the 2001 National Survey of Fishing, Hunting, and Wildlife-Associated Recreation. United States Fish & Wildlife Service. Arlington, Virginia.

Henshilwood CS, d'Errico F, Yates R, *et al.* (2002). Emergence of modern human behavior: Middle Stone Age engravings from South Africa. *Science* **295**, 1278–80.

Hernandez RD, Hubisz MJ, Wheeler DA, *et al.* (2007). Demographic histories and patterns of linkage disequilibrium in Chinese and Indian Rhesus macaques. *Science* **316**, 240–3.

Hey J (2003). Speciation and inversions: Chimps and humans. *BioEssays* **25**, 825–8.

Hey J and Nielsen R (2004). Multilocus methods for estimating population sizes, migration rates and divergence time, with applications to the divergence of *Drosophila pseudoobscura* and *D. persimilis*. *Genetics* **167**, 747–60.

Hey J, Won Y-J, Sivasundar AS, Nielsen R, and Markert JA (2004). Using nuclear haplotypes with microsatellites to study gene flow between recently separated cichlid species. *Molecular Ecology* **13**, 909–19.

Hibben FC (1992). *Indian Hunts and Indian Hunters of the Old West*. Safari Press Inc, Long Beach.

Hiendleder S, Kaupe B, Wassmuth R, and Janke A (2002). Molecular analysis of wild and domestic sheep questions current nomenclature and provides evidence for domestication from two different subspecies. *Proceedings of the Royal Society of London B* **269,** 893–904.

Hird H, Chisholm J, and Brown J (2005). The detection of commercial duck species in food using a single probe-multiple species-specific primer real-time PCR assay. *European Food Research and Technology* **221,** 559–63.

Hirt RP, Noel CJ, Sicheritz-Ponten T, Tachezy J, and Fiori P-L (2007). *Trichomonas vaginalis* surface proteins: A view from the genome. *Trends in Parasitology* **23,** 540–7.

Hockett BS and Bicho NF (2000). The rabbits of Picareiro Cave: Small mammal hunting during the Late Upper Palaeolithic in the Portuguese Estremadura. *Journal of Archaeological Science* **27,** 715–23.

Hoeh WR, Blakley KH, and Brown WM (1991). Heteroplasmy suggests limited biparental inheritance of *Mytilus* mitochondrial DNA. *Science* **251,** 1488–90.

Hofbauer P, Bauer F, and Paulsen P (2006). Meat of chamois—a note on quality traits of the m. longissimus of chamois (*Rupicapra rupicapra* L.) in Austrian sub-alpine regions. *Fleischwirtschaft* **86,** 100–2.

Hofreiter M, Siedel H, Van Neer W, and Vigilant L (2003). Mitochondrial DNA sequence from and enigmatic gorilla population (*Gorilla gorilla uellensis*). *American Journal of Physical Anthropology* **121,** 361–8.

Hogarth B (2007). National Marine Fisheries Service. October 2007 Statement from the Director. http://www.nmfs.noaa.gov.

Hokanson SC, Lamboy WF, Szewc-McFadden AK, and McFerson JR (2001). Microsatellite (SSR) variation in a collection of *Malus* (apple) species and hybrids. *Euphytica* **118,** 281–94.

Holliday TW (2003). Species concepts, reticulation, and human evolution. *Current Anthropology* **44,** 653–60.

Holt RA, Subramanian GM, Halpern A, *et al.* (2002). The genome sequence of the malaria mosquito *Anopheles gambiae*. *Science* **298,** 129–149.

Holzman RS (1998). The legacy of Atropos, the fate who cut the thread of life. *Anesthesiology* **89,** 241–9.

Horvath JE and Willard HF (2007). Primate comparative genomics: Lemur biology and evolution. *Trends in Genetics* **23,** 173–82.

Houck MA, Clark JB, Peterson KR, and Kidwell MG (1991). Possible horizontal transfer of *Drosophila* genes by the mite *Proctolaelaps regalis*. *Science* **253,** 1125–9.

Howard DJ (1982). Speciation and coexistence in a group of closely related ground crickets. PhD Dissertation, Yale University, New Haven, Connecticut.

Howard DJ (1986). A zone of overlap and hybridization between two ground cricket species. *Evolution* **40,** 34–43.

Howard DJ, Preszler RW, Williams J, Fenchel S, and Boecklen WJ (1997). How discrete are oak species? Insights from a hybrid zone between *Quercus grisea* and *Quercus gambelii*. *Evolution* **51,** 747–55.

Huamán Z and Spooner DM (2002). Reclassification of landrace populations of cultivated potatoes (*Solanum* sect. *Petota*). *American Journal of Botany* **89,** 947–65.

Hubartt D (1994). *Dolly Varden,* pp. 1–2. Alaska Department of Fish and Game.

Huber SK, De León LF, Hendry AP, Bermingham E, and Podos J (2007). Reproductive isolation of sympatric morphs in a population of Darwin's finches. *Proceedings of the Royal Society of London B* **274,** 1709–14.

Hughes CE, Govindarajulu R, Robertson A, Filer DL, Harris SA, and Bailey CD (2007). Serendipitous backyard hybridization and the origin of crops. *Proceedings of the National Academy of Sciences USA* **104,** 14389–94.

Humeres SG, Almirón WR, Sabattini MS, and Gardenal CN (1998). Estimation of genetic divergence and gene flow between *Culex pipiens* and *Culex quinquefasciatus* (Diptera: Culicidae) in Argentina. *Memórias do Instituto Oswaldo Cruz* **93,** 57–62.

International Commission for the Conservation of Atlantic Tunas (2007). Report of the standing committee on research and statistics (SCRS). pp. 216. Madrid, Spain.

Ingman M, Kaessmann H, Pääbo S, and Gyllensten U (2000). Mitochondrial genome variation and the origin of modern humans. *Nature* **408,** 708–13.

Ingram AL and Doyle JJ (2003). The origin and evolution of *Eragrostis tef* (Poaceae) and related polyploids: Evidence from nuclear *waxy* and plastid *rps*16. *American Journal of Botany* **90,** 116–22.

Ingram AL and Doyle JJ (2004). Is *Eragrostis* (Poaceae) monophyletic? Insights from nuclear and plastid sequence data. *Systematic Botany* **29,** 545–52.

Innan H and Watanabe H (2006). The effect of gene flow on the coalescent time in the human-chimpanzee ancestral population. *Molecular Biology and Evolution* **23,** 1040–7.

Irion DN, Schaffer AL, Famula TR, Eggleston ML, Hughes SS, and Pedersen NC (2003). Analysis of genetic variation in 28 dog breed populations with 100 microsatellite markers. *Journal of Heredity* **94,** 81–7.

Islam FMA, Beebe S, Muñoz M, Tohme J, Redden RJ, and Basford KE (2004). Using molecular markers to assess the effect of introgression on quantitative attributes of common bean in the Andean gene pool. *Theoretical and Applied Genetics* **108,** 243–52.

Ivens AC, Peacock CS, Worthey EA, *et al.* (2005). The genome of the kinetoplastid parasite, *Leishmania major*. *Science* **309,** 436–41.

Jackson AP, Eastwood H, Bell SM, *et al.* (2002). Identification of microcephalin, a protein implicated in determining the size of the human brain. *American Journal of Human Genetics* **71,** 136–42.

Jacob T, Indriati E, Soejono RP, *et al.* (2006). Pygmoid Australomelanesian *Homo sapiens* skeletal remains from Liang Bua, Flores: Population affinities and pathological abnormalities. *Proceedings of the National Academy of Sciences USA* **103,** 13421–6.

Jaillon O, Aury J-M, Noel B, *et al.* (2007). The grapevine genome sequence suggests ancestral hexaploidization in major angiosperm phyla. *Nature* **449,** 463–8.

Jakobsson M, Hagenblad J, Tavaré S, *et al.* (2006). A unique recent origin of the allotetraploid species *Arabidopsis suecica*: Evidence from nuclear DNA markers. *Molecular Biology and Evolution* **23,** 1217–31.

Jannoo N, Grivet L, Chantret N, *et al.* (2007). Orthologous comparison in a gene-rich region among grasses reveals stability in the sugarcane polyploid genome. *The Plant Journal* **50,** 574–85.

Jarvis DI and Hodgkin T (1999). Wild relatives and crop cultivars: Detecting natural introgression and farmer selection of new genetic combinations in agroecosystems. *Molecular Ecology* **8,** S159–73.

Jensen-Seaman MI, Deinard AS, and Kidd KK (2001). Modern African ape populations as genetic and demographic models of the last common ancestor of humans, chimpanzees, and gorillas. *Journal of Heredity* **92,** 475–80.

Jensen-Seaman MI and Kidd KK (2001). Mitochondrial DNA variation and biogeography of eastern gorillas. *Molecular Ecology* **10,** 2241–7.

Jensen-Seaman MI, Sarmiento EE, Deinard AS, and Kidd KK (2004). Nuclear integrations of mitochondrial DNA in gorillas. *American Journal of Primatology* **63,** 139–47.

Jepsen BI, Siegismund HR, and Fredholm M (2002). Population genetics of the native caribou (*Rangifer tarandus groenlandicus*) and the semi-domestic reindeer (*Rangifer tarandus tarandus*) in Southwestern Greenland: Evidence of introgression. *Conservation Genetics* **3,** 401–9.

Jiang C-X, Chee PW, Draye X, Morrell PL, Smith CW, and Paterson AH (2000). Multilocus interactions restrict gene introgression in interspecific populations of polyploid *Gossypium* (Cotton). *Evolution* **54,** 798–814.

Jiang C-X, Wright RJ, El-Zik KM, and Paterson AH (1998). Polyploid formation created unique avenues for response to selection in *Gossypium* (cotton). *Proceedings of the National Academy of Sciences USA* **95,** 4419–24.

Jiang P, Faase JAJ, Toyoda H, Paul A, Wimmer E, and Gorbalenya AE (2007). Evidence for emergence of diverse polioviruses from C-cluster coxsackie A viruses and implications for global poliovirus eradication. *Proceedings of the National Academy of Sciences USA* **104,** 9457–62.

Jiménez P, López de Heredia U, Collada C, Lorenzo Z, and Gil L (2004). High variability of chloroplast DNA in three Mediterranean evergreen oaks indicates complex evolutionary history. *Heredity* **93,** 510–5.

Jing R, Knox MR, Lee JM, *et al.* (2005). Insertional polymorphism and antiquity of *PDR1* retrotransposon insertions in *Pisum* species. *Genetics* **171,** 741–52.

Johnsgard PA (1960). Hybridization in the Anatidae and its taxonomic implications. *Condor* **62,** 25–33.

Johnson SE, Gordon AD, Stumpf RM, Overdorff DJ, and Wright PC (2005). Morphological variation in populations of *Eulemur albocollaris* and *E. fulvus rufus*. *International Journal of Primatology* **26,** 1399–416.

Johnson WE and O'Brien SJ (1997). Phylogenetic reconstruction of the Felidae using 16S rRNA and NADH-5 mitochondrial genes. *Journal of Molecular Evolution* **44** (Supplement 1), S98-S116.

Johnston JA, Wesselingh RA, Bouck AC, Donovan LA, and Arnold ML (2001). Intimately linked or hardly speaking? The relationship between genotypic variation and environmental gradients in a Louisiana Iris hybrid population. *Molecular Ecology* **10,** 673–81.

Jolly CJ (2001). A proper study for mankind: Analogies from the papionin monkeys and their implications for human evolution. *Yearbook of Physical Anthropology* **44,** 177–204.

Jolly CJ, Wooley-Barker T, Beyene S, Disotell TR, and Phillips-Conroy JE (1997). Intergeneric hybrid baboons. *International Journal of Primatology* **18,** 597–627.

Joshi MB, Rout PK, Mandal AK, Tyler-Smith C, Singh L, and Thangaraj K (2004). Phylogeography and origin of Indian domestic goats. *Molecular Biology and Evolution* **21,** 454–62.

Joyce DA, Lunt DH, Bills R, *et al.* (2005). An extant cichlid fish radiation emerged in an extinct Pleistocene lake. *Nature* **435,** 90–5.

Jung S, Tate PL, Horn R, Kochert G, Moore K, and Abbott AG (2003). The phylogenetic relationship of possible progenitors of the cultivated peanut. *Journal of Heredity* **94,** 334–40.

Kaessmann H and Pääbo S (2002). The genetical history of humans and the great apes. *Journal of Internal Medicine* **251,** 1–18.

Kaessmann H, Wiebe V, and Pääbo S (1999). Extensive nuclear DNA sequence diversity among chimpanzees. *Science* **286,** 1159–62.

Kaessmann H, Wiebe V, Weiss G, and Pääbo S (2001). Great ape DNA sequences reveal a reduced diversity and an expansion in humans. *Nature Genetics* **27,** 155–6.

Kanthaswamy S, Kurushima JD, and Smith DG (2006). Inferring *Pongo* conservation units: A perspective based on microsatellite and mitochondrial DNA analyses. *Primates* **47,** 310–21.

Kanthaswamy S and Smith DG (2002). Population subdivision and gene flow among wild orangutans. *Primates* **43,** 315–27.

Kavanaugh LA, Fraser JA, and Dietrich FS (2006). Recent evolution of the human pathogen *Cryptococcus neoformans* by intervarietal transfer of a 14-gene fragment. *Molecular Biology and Evolution* **23,** 1879–90.

Kawamoto Y (2005). NRAMP1 polymorphism in a hybrid population between Japanese and Taiwanese macaques in Wakayama, Japan. *Primates* **46,** 203–6.

Keele BF, Van Heuverswyn F, Li Y, *et al.* (2006). Chimpanzee reservoirs of pandemic and nonpandemic HIV-1. *Science* **313,** 523–6.

Kellogg EA, Appels R, and Mason-Gamer RJ (1996). When genes tell different stories: The diploid genera of Triticeae (Gramineae). *Systematic Botany* **21,** 321–47.

Kennedy AJ (1995). Cacao—*Theobroma cacao* (Sterculiaceae). In J Smartt and NW Simmonds, eds. *Evolution of Crop Plants,* Second Edition, pp. 472–5. Longman Scientific & Technical, Essex, England.

Kentner EK, Arnold ML, and Wessler SR (2003). Characterization of high copy number retrotransposons from the large genomes of the Louisiana Iris species and their use as molecular markers. *Genetics* **164,** 685–97.

Kew OM, Sutter RW, de Gourville EM, Dowdle WR, and Pallansch MA (2005). Vaccine-derived polioviruses and the endgame strategy for global polio eradication. *Annual Review of Microbiology* **59,** 587–635.

Key KHL (1968). The concept of stasipatric speciation. *Systematic Zoology* **17**, 14–22.

Khan A, Fux B, Su C, *et al.* (2007). Recent transcontinental sweep of *Toxoplasma gondii* driven by a single monomorphic chromosome. *Proceedings of the National Academy of Sciences USA* **104,** 14872–7.

Khanuja SPS, Shasany AK, Srivastava A, and Kumar S (2000). Assessment of genetic relationships in *Mentha* species. *Euphytica* **111,** 121–5.

Khush GS (1997). Origin, dispersal, cultivation and variation of rice. *Plant Molecular Biology* **35,** 25–34.

Kidwell MG (1993). Lateral transfer in natural populations of eukaryotes. *Annual Review of Genetics,* **27,** 235–56.

Kierstein G, Vallinoto M, Silva A, Schneider MP, Iannuzzi L, and Brenig B (2004). Analysis of mitochondrial D-loop region casts new light on domestic water buffalo (*Bubalus bubalis*) phylogeny. *Molecular Phylogenetics and Evolution* **30,** 308–24.

Kijas JMH and Andersson L (2001). A phylogenetic study of the origin of the domestic pig estimated from the near-complete mtDNA genome. *Journal of Molecular Evolution* **52,** 302–8.

Kikkawa Y, Takada T, Sutopo, *et al.* (2003). Phylogenies using mtDNA and *SRY* provide evidence for male-mediated introgression in Asian domestic cattle. *Animal Genetics* **34,** 96–101.

Kilian B, Özkan H, Deusch O, *et al.* (2006). Independent wheat B and G genome origins in outcrossing *Aegilops* progenitor haplotypes. *Molecular Biology and Evolution* **24,** 217–27.

Kilpatrick AM, Kramer LD, Jones MJ, Marra PP, Daszak P, and Fonseca DM (2007). Genetic influences on mosquito feeding behavior and the emergence of zoonotic pathogens. *American Journal of Tropical Medicine and Hygiene* **77,** 667–71.

Kim KS, Tanabe Y, Park CK, and Ha JH (2001). Genetic variability in East Asian dogs using microsatellite loci analysis. *Journal of Heredity* **92,** 398–403.

Kim S-C and Rieseberg LH (1999). Genetic architecture of species differences in annual sunflowers: Implications for adaptive trait introgression. *Genetics* **153**, 965–77.

Kim S-C and Rieseberg LH (2001). The contribution of epistasis to species differences in annual sunflowers. *Molecular Ecology* **10**, 683–90.

King M-C and Wilson AC (1975). Evolution at two levels in humans and chimpanzees. *Science* **188,** 107–16.

Kingswood SC, Kumamoto AT, Charter SJ, Aman RA, and Ryder OA (1998). Centric fusion polymorphisms in waterbuck (*Kobus ellipsiprymnus*). *Journal of Heredity* **89,** 96–100.

Kitamura S, Inoue M, Shikazono N, and Tanaka A (2001). Relationships among *Nicotiana* species revealed by the 5S rDNA spacer sequence and fluorescence in situ hybridization. *Theoretical and Applied Genetics* **103,** 678–86.

Kobasa D, Jones SM, Shinya K, *et al.* (2007). Aberrant innate immune response in lethal infection of macaques with the 1918 influenza virus. *Nature* **445,** 319–23.

Koblmüller S, Duftner N, Sefc K, *et al.* (2007a). Reticulate phylogeny of gastropod-shell-breeding cichlids from Lake Tanganyika—the result of repeated introgressive hybridization. *BMC Evolutionary Biology* **7,** 7.

Koblmüller S, Egger B, Sturmbauer C, and Sefc KM (2007b). Evolutionary history of Lake Tanganyika's scale-eating cichlid fishes. *Molecular Phylogenetics and Evolution.*

Koch MA and Matschinger M (2007). Evolution and genetic differentiation among relatives of *Arabidopsis thaliana*. *Proceedings of the National Academy of Sciences, USA* **104**, 6272–7.

Kochert G, Stalker HT, Gimenes M, Galgaro L, Lopes CR, and Moore K (1996). RFLP and cytogenetic evidence on the origin and evolution of allotetraploid domesticated peanut, *Arachis hypogaea* (Leguminosae). *American Journal of Botany* **83,** 1282–91.

Koekemoer LL, Kamau L, Garros C, Manguin S, Hunt RH, and Coetzee M (2006). Impact of the Rift Valley on restriction fragment length polymorphism typing of the major African malaria vector *Anopheles funestus* (Diptera: Culicidae). *Journal of Medical Entomology* **43,** 1178–84.

Kovach MJ, Sweeney MT, and McCouch SR (2007). New insights into the history of rice domestication. *Trends in Genetics* **23,** 578–87.

Kraus FB, Franck P and Vandame R (2007). Asymmetric introgression of African genes in honeybee populations (*Apis mellifera* L.) in central Mexico. *Heredity* **99,** 233–40.

Krause J, Dear PH, Pollack JL, *et al.* (2006). Multiplex amplification of the mammoth mitochondrial genome and the evolution of Elephantidae. *Nature* **439,** 724–7.

Krings M, Stone A, Schmitz RW, Krainitzki H, Stoneking M, and Pääbo S (1997). Neandertal DNA sequences and the origin of modern humans. *Cell* **90,** 19–30.

Krueger A and Hennings IC (2006). Molecular phylogenetics of blackflies of the *Simulium damnosum* complex and cytophylogenetic implications. *Molecular Phylogenetics and Evolution* **39,** 83–90.

Krueger A, Kalinga AK, Kibweja AM, Mwaikonyole A, and Maegga BTA (2006a). Cytogenetic and PCR-based identification of *S. damnosum* 'Nkusi J' as the anthropophilic blackfly in the Uluguru onchocerciasis focus in Tanzania. *Tropical Medicine and International Health* **11,** 1066–74.

Krueger A, Mustapha M, Kalinga AK, Tambala PAJ, Post RJ, and Maegga BTA (2006b). Revision of the Ketaketa subcomplex of blackflies of the *Simulium damnosum* complex. *Medical and Veterinary Entomology* **20,** 76–92.

Kuiken T, Holmes EC, McCauley J, Rimmelzwaan GF, Williams CS, and Grenfell BT (2006). Host species barriers to influenza virus infections. *Science* **312,** 394–7.

Kulikova LA, McAlister MB, Ogden KL, Larkin MJ, and O'Hanlon JF (2002). Analysis of bacteria contaminating ultrapure water in industrial systems. *Applied and Environmental Microbiology* **68,** 1548–55.

Kulikova IV, Drovetski SV, Gibson DD, *et al.* (2005). Phylogeography of the mallard (*Anas platyrhynchus*): Hybridization, dispersal, and lineage sorting contribute to complex geographic structure. *Auk* **122,** 949–65.

Kulikova IV, Zhuravlev YN, and McCracken KG (2004). Asymmetric hybridization and sex-biased gene flow between eastern spot-billed ducks (*Anas zonorhyncha*) and mallards (*A. platyrhynchus*) in the Russian Far East. *Auk* **121,** 930–49.

Lack D (1947). *Darwin's finches.* Cambridge University Press, Cambridge.

Lahm SA, Kombila M, Swanepoel R, and Barnes RFW (2007). Morbidity and mortality of wild animals in relation to outbreaks of Ebola haemorrhagic fever in Gabon, 1994–2003. *Transactions of the Royal Society of Tropical Medicine and Hygiene* **101,** 64–78.

Lambert DM and Millar CD (2006). Ancient genomics is born. *Nature* **444,** 275–6.

Lanzaro GC, Touré YT, Carnahan J, *et al.* (1998). Complexities in the genetic structure of *Anopheles gambiae* populations in west Africa as revealed by microsatellite DNA analysis. *Proceedings of the National Academy of Sciences, USA* **95,** 14260–5.

Larbig KD, Christmann A, Johann A, *et al.* (2002). Gene islands integrated into tRNA[Gly] genes confer genome diversity on a *Pseudomonas aeruginosa* clone. *Journal of Bacteriology* **184,** 6665–80.

Larson G, Dobney K, Albarella U, *et al.* (2005). Worldwide phylogeography of wild boar reveals multiple centers of pig domestication. *Science* **307,** 1618–21.

Larson SG, Jungers WL, Morwood MJ, *et al.* (2007). *Homo floresiensis* and the evolution of the hominin shoulder. *Journal of Human Evolution* **53,** 718–31.

Lashermes P, Combes M-C, Robert J, *et al.* (1999). Molecular characterisation and origin of the *Coffea arabica* L. genome. *Molecular and General Genetics* **261,** 259–66.

Lau CH, Drinkwater RD, Yusoff K, Tan SG, Hetzel DJS, and Barker JSF (1998). Genetic diversity of Asian water buffalo (*Bubalus bubalis*): Mitochondrial DNA D-loop and cytochrome b sequence variation. *Animal Genetics* **29,** 253–64.

Lawrence JG and Ochman H (1998). Molecular archaeology of the *Escherichia coli* genome. *Proceedings of the National Academy of Sciences, USA,* **95,** 9413–17.

Lawson Handley L-J, Byrne K, Santucci F, *et al.* (2007). Genetic structure of European sheep breeds. *Heredity* **99,** 620–31.

Leach BJ, Foale MA, and Ashburner GR (2003). Some characteristics of wild and managed coconut palm populations and their environment in the Cocos (Keeling) Islands, Indian Ocean. *Genetic Resources and Crop Evolution* **50,** 627–38.

Lebot V (1999). Biomolecular evidence for plant domestication in Sahul. *Genetic Resources and Crop Evolution* **46,** 619–28.

Lebrun P, N'cho YP, Seguin M, Grivet L, and Baudouin L (1998). Genetic diversity in coconut (*Cocos nucifera* L.) revealed by restriction fragment length polymorphism (RFLP) markers. *Euphytica* **101,** 103–8.

Lecis R, Pierpaoli M, and Birò ZS (2006). Bayesian analyses of admixture in wild and domestic cats (*Felis silvestris*) using linked microsatellite loci. *Molecular Ecology* **15,** 119–31.

Lefèvre F and Charrier A (1993). Isozyme diversity within African *Manihot* germplasm. *Euphytica* **66,** 73–80.

Lehman N, Eisenhawer A, Hansen K, *et al.* (1991). Introgression of coyote mitochondrial DNA into sympatric North American gray wolf populations. *Evolution* **45,** 104–19.

Lehman T, Marcet PL, Graham DH, Dahl ER, and Dubey JP (2006). Globalization and the population structure of *Toxoplasma gondii*. *Proceedings of the National Academy of Sciences USA* **103,** 11423–8.

Lengeler KB, Cox GM, and Heitman J (2001). Serotype AD strains of *Cryptococcus neoformans* are diploid or aneuploid and are heterozygous at the mating-type locus. *Infection and Immunity* **69,** 115–22.

Leonard JA, Wayne RK, Wheeler J, Valadez R, Guillén S, and Vilà C (2002). Ancient DNA evidence for Old World origin of New World dogs. *Science* **298,** 1613–6.

Leroy EM, Rouquet P, Formenty P, *et al.* (2004). Multiple Ebola virus transmission events and rapid decline of central African wildlife. *Science* **303,** 387–90.

Levy DN, Aldrovandi GM, Kutsch O, and Shaw GM (2004). Dynamics of HIV–1 recombination in its natural target cells. *Proceedings of the National Academy of Sciences USA* **101,** 4204–9.

Lewis MD, Yousuf AA, Lerdthusnee K, Razee A, Chandranoi K, and Jones JW (2003). Scrub typhus reemergence in the Maldives. *Emerging Infectious Diseases* **9,** 1638–41.

Lewontin RC (1974). *The Genetic Basis of Evolutionary Change*. Columbia University Press, New York.

Lexer C, Kremer A, and Petit RJ (2006). Shared alleles in sympatric oaks: Recurrent gene flow is a more parsimonious explanation than ancestral polymorphism. *Molecular Ecology* **15,** 2007–12.

Li M-H, Tapio I, Vilkki J, *et al.* (2007). The genetic structure of cattle populations (*Bos taurus*) in northern Eurasia and the neighbouring Near Eastern regions: Implications for breeding strategies and conservation. *Molecular Ecology* **16,** 3839–53.

Liang X, Pham X-QT, Olson MV, and Lory S (2001). Identification of a genomic island present in the majority of pathogenic isolates of *Pseudomonas aeruginosa*. *Journal of Bacteriology* **183,** 843–53.

Lim KY, Kovarik A, Matyásek R, *et al.* (2007). Sequence of events leading to near-complete genome turnover in allopolyploid *Nicotiana* within five million years. *New Phytologist* **175,** 756–63.

Lim KY, Matyásek R, Lichtenstein CP, and Leitch AR (2000). Molecular cytogenetic analyses and phylogenetic studies in the *Nicotiana* section Tomentosae. *Chromosoma* **109,** 245–58.

Lim KY, Skalicka K, Koukalova B, *et al.* (2004). Dynamic changes in the distribution of a satellite homologous to intergenic 26–18S rDNA spacer in the evolution of *Nicotiana*. *Genetics* **166,** 1935–46.

Lin X, Litvintseva AP, Nielsen K, *et al.* (2007). αADα hybrids of *Cryptococcus neoformans*: Evidence of same-sex mating in nature and hybrid fitness. *PLoS Genetics* **3,** 1975–90.

Lindblad-Toh K, Wade CM, Mikkelsen TS, *et al.* (2005). Genome sequence, comparative analysis and haplotype structure of the domestic dog. *Science* **438,** 803–19.

Ling B, Veazey RS, Luckay A, *et al.* (2002). SIV_{mac} pathogenesis in Rhesus macaques of Chinese and Indian origin compared with primary HIV infections in humans. *AIDS* **16,** 1489–96.

Linnaeus C (1760). *Disquisitio de Sexu Plantarum*. St. Petersburg.

Lintas C, Hirano J, and Archer S (1998). Genetic variation of the European eel (*Anguilla anguilla*). *Molecular Marine Biology and Biotechnology* **7,** 263–9.

Linz B, Balloux F, Moodley Y, *et al.* (2007). An African origin for the intimate association between humans and *Helicobacter pylori*. *Nature* **445,** 915–8.

Lister AM and Sher AV (2001). The origin and evolution of the wooly mammoth. *Science* **294,** 1094–7.

Liston A, Parker-Defeniks M, Syring JV, Willyard A, and Cronn R (2007). Interspecific phylogenetic analysis enhances intraspecific phylogeographical inference: A case study in *Pinus lambertiana*. *Molecular Ecology* **16,** 3926–37.

Liston A, Robinson WA, Piñero D, and Alvarez-Buylla ER (1999). Phylogenetics of *Pinus* (Pinaceae) based on nuclear ribosomal DNA internal transcribed spacer region sequences. *Molecular Phylogenetics and Evolution* **11,** 95–109.

Liti G, Barton DBH, and Louis EJ (2006). Sequence diversity, reproductive isolation and species concepts in *Saccharomyces*. *Genetics* **174,** 839–50.

Litvintseva AP, Lin X, Templeton I, Heitman J, and Mitchell TG (2007). Many globally isolated AD hybrid strains of *Cryptococcus neoformans* originated in Africa. *PLoS Pathogens* **3,** 1109–17.

Litvintseva AP, Thakur R, Vilgalys R, and Mitchell TG (2006). Multilocus sequence typing reveals three genetic subpopulations of *Cryptococcus neoformans* var. *grubii* (Serotype A), including a unique population in Botswana. *Genetics* **172,** 2223–38.

Lochouarn L, Dia I, Boccolini D, Coluzzi M, and Fontenille D (1998). Bionomial and cytogenetic heterogeneities of *Anopheles funestus* in Senegal. *Transactions of the Royal Society of Tropical Medicine and Hygiene* **92,** 607–12.

Loftus B, Anderson I, Davies R, *et al.* (2005). The genome of the protist parasite *Entamoeba histolytica. Nature* **433,** 865–8.

Loison A, Jullien J-M, and Menaut P (1999). Subpopulation structure and dispersal in two populations of chamois. *Journal of Mammalogy* **80,** 620–32.

Londo JP, Chiang Y-C, Hung K-H, Chiang T-Y, and Schaal BA (2006). Phylogeography of Asian wild rice, *Oryza rufipogon*, reveals multiple independent domestications of cultivated rice, *Oryza sativa. Proceedings of the National Academy of Sciences USA* **103,** 9578–83.

Lordkipanidze D, Jashashvili T, Vekua A, *et al.* (2007). Postcranial evidence from early *Homo* from Dmanisi, Georgia. *Nature* **449,** 305–10.

Lorenzen ED, de Neergaard R, Arctander P, and Siegismund HR (2007). Phylogeography, hybridization and Pleistocene refugia of the kob antelope (*Kobus kob*). *Molecular Ecology* **16,** 3241–52.

Lorenzen ED, Simonsen BT, Kat PW, Arctander P, and Siegismund HR (2006). Hybridization between subspecies of waterbuck (*Kobus ellipsiprymnus*) in zones of overlap with limited introgression. *Molecular Ecology* **15,** 3787–99.

Lotsy JP (1931). On the species of the taxonomist in its relation to evolution. *Genetica* **13,** 1–16.

Lowe PR (1936). The finches of the Galápagos in relation to Darwin's conception of species. *Ibis* **6,** 310–21.

Lu J, Li W-H, and Wu C-I (2003). Comment on "Chromosomal speciation and molecular divergence—Accelerated evolution in rearranged chromosomes." *Science* **302,** 988b.

Lubinski PM (2003). Rabbit hunting and bone bead production at a late prehistoric camp in the Wyoming basin. *North American Archaeologist* **24,** 197–214.

Lucier G and Jerardo A (2007). Vegetables and melons situation and outlook yearbook. pp. 1–191. United States Department of Agriculture.

Lukes J, Mauricio IL, Schönian G, *et al.* (2007). Evolutionary and geographical history of the *Leishmania donovani* complex with a revision of current taxonomy. *Proceedings of the National Academy of Sciences USA* **104,** 9375–80.

Lumaret R, Tryphon-Dionnet M, Michaud H, *et al.* (2005). Phylogeographical variation of chloroplast DNA in cork oak (*Quercus suber*). *Annals of Botany* **96,** 853–61.

Luo M-C, Yang Z-L, You FM, Kawahara T, Waines JG, and Dvorak J (2007). The structure of wild and domesticated emmer wheat populations, gene flow between them, and the site of emmer domestication. *Theoretical and Applied Genetics* **114,** 947–59.

Lyons D and D'Andrea AC (2003). Griddles, ovens, and agricultural origins: An ethnoarchaeological study of bread baking in highland Ethiopia. *American Anthropologist* **105,** 515–30.

Ma W, Vincent AL, Gramer MR, *et al.* (2007). Identification of H2N3 influenza A viruses from swine in the United States. *Proceedings of the National Academy of Sciences USA* **104,** 20949–54.

Machado CA and Ayala FJ (2001). Nucleotide sequences provide evidence of genetic exchange among distantly related lineages of *Trypanosoma cruzi. Proceedings of the National Academy of Sciences USA* **98,** 7396–401.

MacHugh DE and Bradley DG (2001). Livestock genetic origins: Goats buck the trend. *Proceedings of the National Academy of Sciences USA* **98,** 5382–4.

Magri D, Fineschi S, Bellarosa R, *et al.* (2007). The distribution of *Quercus suber* chloroplast haplotypes matches the palaeogeographical history of the western Mediterranean. *Molecular Ecology* **16,** 5259–66.

Mahé L, Le Pierrès D, Combes M-C, and Lashermes P (2007). Introgressive hybridization between the allotetraploid *Coffea arabica* and one of its diploid ancestors, *Coffea canephora,* in an exceptional sympatric zone in New Caledonia. *Genome* **50,** 316–24.

Makarova K, Slesarev A, Wolf Y, *et al.* (2006). Comparative genomics of the lactic acid bacteria. *Proceedings of the National Academy of Sciences USA* **103,** 15611–6.

Malaty HM and Graham DY (1994). Importance of childhood socioeconomic status on the current prevalence of *Helicobacter pylori* infection. *Gut* **35,** 742–5.

Mallet J (2005). Hybridization as an invasion of the genome. *Trends in Ecology and Evolution* **20,** 229–37.

Manceau V, Després L, Bouvet J, and Taberlet P (1999). Systematics of the genus *Capra* inferred from mitochondrial DNA sequence data. *Molecular Phylogenetics and Evolution* **13,** 504–10.

Manguin S, Kengne P, Sonnier L, *et al.* (2002). SCAR markers and multiplex PCR-based identification of isomorphic species in the *Anopheles dirus* complex in Southeast Asia. *Medical and Veterinary Entomology* **16,** 46–54.

Manimekalai R and Nagarajan P (2006a). Assessing genetic relationships among coconut (*Cocos nucifera* L.) accessions using inter simple sequence repeat markers. *Scientia Horticulturae* **108,** 49–54.

Manimekalai R and Nagarajan P (2006b). Interrelationships among coconut (*Cocos nucifera* L.) accessions using RAPD technique. *Genetic Resources and Crop Evolution* **53,** 1137–44.

Mank JE and Avise JC (2003). Microsatellite variation and differentiation in North Atlantic eels. *Journal of Heredity* **94,** 310–4.

Mank JE, Carlson JE, and Brittingham MC (2004). A century of hybridization: Decreasing genetic distance between American black ducks and mallards. *Conservation Genetics* **5,** 395–403.

Mansiangi P, Kiyombo G, Mulumba P, Josens G, and Krueger A (2007). Molecular systematics of *Simulium squamosum*, the vector in the Kinsuka onchocerciasis focus (Kinshasa, Democratic Republic of Congo). *Annals of Tropical Medicine and Parasitology* **101,** 275–9.

Mariac C, Luong V, Kapran I, *et al.* (2006). Diversity of wild and cultivated pearl millet accessions (*Pennisetum glaucum* [L] R Br) in Niger assessed by microsatellite markers. *Theoretical and Applied Genetics* **114,** 49–58.

Marquès-Bonet T, Cáceres M, Bertranpetit J, Preuss TM, Thomas JW, and Navarro A (2004). Chromosomal rearrangements and the genomic distribution of gene-expression divergence in humans and chimpanzees. *Trends in Genetics* **20,** 524–9.

Marroig G, Cropp S and Cheverud JM (2004). Systematics and evolution of the jacchus group of marmosets (Platyrrhini). *American Journal of Physical Anthropology* **123,** 11–22.

Martin GS, Mannino DM, Eaton S, and Moss M (2003). The epidemiology of sepsis in the United States from 1979 through 2000. *New England Journal of Medicine* **348,** 1546–54.

Martin NH, Bouck AC, and Arnold ML (2005). Loci affecting long-term hybrid survivability in Louisiana Irises: Implications for reproductive isolation and introgression. *Evolution* **59,** 2116–21.

Martin NH, Bouck AC, and Arnold ML (2006). Detecting adaptive trait introgression in Louisiana Irises. *Genetics* **172,** 2481–9.

Martinón-Torres M, Bermúdez de Castro JM, Gómez-Robles A, *et al.* (2007). Dental evidence on the hominin dispersals during the Pleistocene. *Proceedings of the National Academy of Sciences USA* **104,** 13279–82.

Masembe C, Muwanika VB, Nyakaana S, Arctander P, and Siegismund HR (2006). Three genetically divergent lineages of the Oryx in eastern Africa: Evidence for an ancient introgressive hybridization. *Conservation Genetics* **7,** 551–62.

Masneuf I, Hansen J, Groth C, Piskur J, and Dubourdieu D (1998). New hybrids between *Saccharomyces* sensu stricto yeast species found among wine and cider production strains. *Applied and Environmental Microbiology* **64,** 3887–92.

Mason-Gamer RJ (2001). Origin of North American *Elymus* (Poaceae: Triticeae) allotetraploids based on granule-bound starch synthase gene sequences. *Systematic Botany* **26,** 757–68.

Masterson J (1994). Stomatal size in fossil plants: Evidence for polyploidy in majority of angiosperms. *Science* **264,** 421–4.

Mathews KH Jr and Vandeveer M (2007). Beef production, markets, and trade in Argentina and Uruguay—an overview. pp. 1–12. Economic Research Service, United States Department of Agriculture.

Matisoo-Smith E and Robins JH (2004). Origins and dispersals of Pacific peoples: Evidence from mtDNA phylogenies of the Pacific rat. *Proceedings of the National Academy of Sciences USA* **101,** 9167–72.

Matsuoka Y, Vigouroux Y, Goodman MM, G Sanchez J, Buckler E, and Doebley J (2002). A single domestication for maize shown by multilocus microsatellite genotyping. *Proceedings of the National Academy of Sciences USA* **99,** 6080–4.

Matthee CA and Robinson TJ (1999). Mitochondrial DNA population structure of roan and sable antelope: Implications for the translocation and conservation of the species. *Molecular Ecology* **8,** 227–38.

Matyásek R, Fulnecek J, Lim KY, Leitch AR, and Kovarík A (2002). Evolution of 5S rDNA unit arrays in the plant genus *Nicotiana* (Solanaceae). *Genome* **45,** 556–62.

Mayr E (1942). *Systematics and the Origin of Species.* Columbia University Press, New York.

Mayr E (1963). *Animal Species and Evolution.* Belknap Press, Cambridge, Massachusetts.

Mayr E (1992). A local flora and the biological species concept. *American Journal of Botany* **79,** 222–38.

McBrearty S and Brooks AS (2000). The revolution that wasn't: A new interpretation of the origin of modern human behavior. *Journal of Human Evolution* **39,** 453–563.

McGovern PE and Hartung U (1997). The beginnings of winemaking and viniculture in the ancient Near East and Egypt. *Expedition* **39,** 3–22.

McKay DL and Blumberg JB (2006). A review of the bioactivity and potential health benefits of peppermint tea (*Mentha piperita* L.). *Phytotherapy Research* **20,** 619–33.

Meadows JRS, Cemal I, Karaca O, Gootwine E, and Kijas JW (2007). Five ovine mitochondrial lineages identified from sheep breeds of the Near East. *Genetics* **175,** 1371–9.

Meadows JRS, Li K, Kantanen J, *et al.* (2005). Mitochondrial sequence reveals high levels of gene flow between breeds of domestic sheep from Asia and Europe. *Journal of Heredity* **96,** 494–501.

Meinke DW, Cherry JM, Dean C, Rounsley SD, and Koornneef M (1998). *Arabidopsis thaliana*: A model plant for genome analysis. *Science* **282,** 662–82.

Meireles CM, Czelusniak J, Schneider MPC, *et al.* (1999). Molecular phylogeny of ateline New World monkeys (Platyrrhini, Atelinae) based on γ-globin gene sequences: Evidence that *Brachyteles* is the sister group of *Lagothrix*. *Molecular Phylogenetics and Evolution* **12,** 10–30.

Mellars P (2004). Neanderthals and the modern human colonization of Europe. *Nature* **432,** 461–5.

Mellars P, Gravina B, and Ramsey CB (2007). Confirmation of Neanderthal/modern human interstratification at the Chatelperronian type-site. *Proceedings of the National Academy of Sciences USA* **104,** 3657–62.

Melo-Ferreira J, Boursot P, Randi E, *et al.* (2007). The rise and fall of the mountain hare (*Lepus timidus*) during Pleistocene glaciations: Expansion and retreat with hybridization in the Iberian peninsula. *Molecular Ecology* **16,** 605–18.

Melo-Ferreira J, Boursot P, Suchentrunk F, Ferrand N, and Alves PC (2005). Invasion from the cold past: Extensive introgression of mountain hare (*Lepus timidus*) mitochondrial DNA into three other hare species in northern Iberia. *Molecular Ecology* **14,** 2459–64.

Mercure A, Ralls K, Koepfli KP, and Wayne RK (1993). Genetic subdivisions among small canids: Mitochondrial DNA differentiation of swift, kit, and arctic foxes. *Evolution* **47,** 1313–28.

Messier W and Stewart C-B (1997). Episodic adaptive evolution of primate lysozymes. *Nature* **385,** 151–4.

Meyer A, Salzburger W, and Schartl M (2006). Hybrid origin of a swordtail species (Teleostei: *Xiphophorus clemenciae*) driven by sexual selection. *Molecular Ecology* **15,** 721–30.

Michel AP, Grushko O, Guelbeogo WM, *et al.* (2006). Divergence with gene flow in *Anopheles funestus* from the Sudan savanna of Burkina Faso, West Africa. *Genetics* **173,** 1389–95.

Michel AP, Ingrasci MJ, Schemerhorn BJ, *et al.* (2005). Range-wide population genetic structure of the African malaria vector *Anopheles funestus*. *Molecular Ecology* **14,** 4235–48.

Mignouna HD and Dansi A (2003). Yam (*Dioscorea* ssp.) domestication by the Nago and Fon ethnic groups in Benin. *Genetic Resources and Crop Evolution* **50,** 519–28.

Minin VN, Dorman KS, Fang F, and Suchard MA (2007). Phylogenetic mapping of recombination hotspots in human immunodeficiency virus via spatially smoothed change-point processes. *Genetics* **175,** 1773–85.

Mir C, Toumi L, Jarne P, Sarda V, Di Giusto F, and Lumaret R (2006). Endemic North African *Quercus afares* Pomel originates from hybridisation between two genetically very distant oak species (*Q. suber* L. and *Q. canariensis* Willd.): Evidence from nuclear and cytoplasmic markers. *Heredity* **96,** 175–84.

Mitchell TG and Perfect JR (1995). Cryptococcosis in the era of AIDS—100 years after the discovery of *Cryptococcus neoformans*. *Clinical Microbiology Reviews* **8,** 515–48.

Mona S, Randi E, and Tommaseo-Ponzetta M (2007). Evolutionary history of the genus *Sus* inferred from cytochrome b sequences. *Molecular Phylogenetics and Evolution* **45,** 757–62.

Moncada P and McCouch S (2004). Simple sequence repeat diversity in diploid and tetraploid *Coffea* species. *Genome* **47,** 501–9.

Moncayo A (1999). Progress towards interruption of transmission of Chagas disease. *Memórias do Instituto Oswaldo Cruz* **94** (Suppl. I), 401–4.

Monda K, Simmons RE, Kressirer P, Su B, and Woodruff DS (2007). Mitochondrial DNA hypervariable region-1 sequence variation and phylogeny of the concolor gibbons, *Nomascus*. *American Journal of Primatology* **69,** 1285–306.

Monteiro FA, Donnelly MJ, Beard CB, and Costa J (2004). Nested clade and phylogeographic analyses of the Chagas disease vector *Triatoma brasiliensis* in northeast Brazil. *Molecular Phylogenetics and Evolution* **32,** 46–56.

Moore CM, Janish C, Eddy CA, Hubbard GB, Leland MM, and Rogers J (1999). Cytogenetic and fertility studies of a Rheboon, Rhesus macaque (*Macaca mulatta*) X baboon (*Papio hamadryas*) cross: Further support for a single karyotype nomenclature. *American Journal of Physical Anthropology* **110,** 119–27.

Moore GA (2001). Oranges and lemons: Clues to the taxonomy of *Citrus* from molecular markers. *Trends in Genetics* **17,** 536–40.

Morales JC and Melnick DJ (1998). Phylogenetic relationships of the macaques (Cercopithecidae: *Macaca*), as revealed by high resolution restriction site mapping of mitochondrial ribosomal genes. *Journal of Human Evolution* **34,** 1–23.

Morales-Hojas R, Post RJ, Cheke RA, and Wilson MD (2002). Assessment of rDNA IGS as a molecular marker in the *Simulium damnosum* complex. *Medical and Veterinary Entomology* **16,** 395–403.

Mora-Urpí J, Weber JC, and Clement CR (1997). Peach palm. *Bactris gasipaes* Kunth. Promoting the conservation and use of underutilized and neglected crops. 20.

Institute of Plant Genetics and Crop Plant Research, Gatersleben/International Plant Genetic Resources Institute, Rome, Italy.

Morin PA, Moore JJ, Chakraborty R, Jin L, Goodall J, and Woodruff DS (1994). Kin selection, social structure, gene flow, and the evolution of chimpanzees. *Science* **265,** 1193–201.

Moritz C, Patton JL, Schneider CJ, and Smith TB (2000). Diversification of rainforest faunas: An integrated molecular approach. *Annual Review of Ecology and Systematics* **31,** 533–63.

Morrison HG, McArthur AG, Gillin FD, *et al.* (2007). Genomic minimalism in the early diverging intestinal parasite *Giardia lamblia*. *Science* **317,** 1921–6.

Morwood MJ, Soejono RP, Roberts RG, *et al.* (2004). Archaeology and age of a new hominin from Flores in eastern Indonesia. *Nature* **431,** 1087–91.

Motamayor JC, Risterucci AM, Heath M, and Lanaud C (2003). Cacao domestication II: Progenitor germplasm of the Trinitario cacao cultivar. *Heredity* **91,** 322–30.

Motamayor JC, Risterucci AM, Lopez PA, Ortiz CF, Moreno A, and Lanaud C (2002). Cacao domestication I: The origin of the cacao cultivated by the Mayas. *Heredity* **89,** 380–6.

Mueller S, Wimmer E, and Cello J (2005). Poliovirus and poliomyelitis: A tale of guts, brains, and an accidental event. *Virus Research* **111,** 175–93.

Muir CC, Galdikas BMF, and Beckenbach AT (2000). mtDNA sequence diversity of orangutans from the islands of Borneo and Sumatra. *Journal of Molecular Evolution* **51,** 471–80.

Myers RH and Shafer DA (1979). Hybrid ape offspring of a mating of gibbon and siamang. *Science* **205,** 308–10.

Nassar NMA (2003). Gene flow between cassava, *Manihot esculenta* Crantz, and wild relatives. *Genetics and Molecular Research* **2,** 334–47.

National Agricultural Statistics Service (2007a). Crop values—2006 summary. 48 pp. United States Department of Agriculture.

National Agricultural Statistics Service (2007b). Trout production. 17 pp. United States Department of Agriculture.

National Research Council (1989). *Lost Crops of the Incas. Little Known Plants of the Andes With Promise for Worldwide Cultivation*. National Academy Press, Washington, D.C.

Navarro A and Barton NH (2003a). Accumulating postzygotic isolation genes in parapatry: A new twist on chromosomal speciation. *Evolution* **57,** 447–59.

Navarro A and Barton NH (2003b). Chromosomal speciation and molecular divergence—accelerated evolution in rearranged chromosomes. *Science* **300,** 321–4.

Ndenga B, Githeko A, Omukunda E, *et al.* (2006). Population dynamics of malaria vectors in western Kenya highlands. *Journal of Medical Entomology* **43,** 200–6.

Negro JJ, Torres MJ, and Godoy JA (2001). RAPD analysis for detection and eradication of hybrid partridges (*Alectoris rufa* × *A. graeca*) in Spain. *Biological Conservation* **98,** 19–24.

Nesbø CL, Dlutek M, and Doolittle WF (2006). Recombination in *Thermotoga*: Implications for species concepts and biogeography. *Genetics* **172,** 759–769.

Newman TK, Jolly CJ, and Rogers J (2004). Mitochondrial phylogeny and systematics of baboons (*Papio*). *American Journal of Physical Anthropology* **124,** 17–27.

Nicolosi E, Deng ZN, Gentile A, La Malfa S, Continella G, and Tribulato E (2000). Citrus phylogeny and genetic origin of important species as investigated by molecular markers. *Theoretical and Applied Genetics* **100,** 1155–66.

Nielsen EE, Grønkjær P, Meldrup D, and Paulsen H (2005). Retention of juveniles within a hybrid zone between North Sea and Baltic Sea Atlantic cod (*Gadus morhua*). *Canadian Journal of Fisheries and Aquatic Sciences* **62,** 2219–25.

Nielsen EE, Hansen MM, Ruzzante DE, Meldrup D, and Grønkjær P (2003). Evidence of a hybrid-zone in Atlantic cod (*Gadus morhua*) in the Baltic and the Danish Belt Sea revealed by individual admixture analysis. *Molecular Ecology* **12,** 1497–508.

Nielsen EE, Hansen MM, Schmidt C, Meldrup D, and Grønkjær P (2001). Population of origin of Atlantic cod. *Nature* **413,** 272.

Nielsen EE, Nielsen PH, Meldrup D, and Hansen MM (2004). Genetic population structure of turbot (*Scophthalmus maximus* L.) supports the presence of multiple hybrid zones for marine fishes in the transition zone between the Baltic Sea and the North Sea. *Molecular Ecology* **13,** 585–95.

Nieves M, Ascunce MS, Rahn MI, and Mudry MD (2005). Phylogenetic relationships among some *Ateles* species: The use of chromosomic and molecular characters. *Primates* **46,** 155–64.

Nijman IJ, Otsen M, Verkaar ELC, *et al.* (2003). Hybridization of banteng (*Bos javanicus*) and zebu (*Bos indicus*) revealed by mitochondrial DNA, satellite DNA, AFLP and microsatellites. *Heredity* **90,** 10–16.

Noda R, Kim CG, Takenaka O, *et al.* (2001). Mitochondrial 16S rRNA sequence diversity of hominoids. *Journal of Heredity* **92,** 490–6.

Noonan JP, Coop G, Kudaravalli S, *et al.* (2006). Sequencing and analysis of Neanderthal genomic DNA. *Science* **314,** 1113–8.

Normile D and Enserink M (2004). Avian influenza makes a comeback, reviving pandemic worries. *Science* **305,** 321.

Nosenko T and Bhattacharya D (2007). Horizontal gene transfer in chromalveolates. *BMC Evolutionary Biology* **7,** 173.

Obenauer JC, Denson J, Mehta PK, *et al.* (2006). Large-scale sequence analysis of avian influenza isolates. *Science* **311,** 1576–80.

O'Brien SJ and Mayr E (1991). Bureaucratic mischief: Recognizing endangered species and subspecies. *Science* **251,** 1187–8.

Ochman H (2005). Genomes on the shrink. *Proceedings of the National Academy of Sciences USA* **102,** 11959–60.

Ochman H, Lawrence JG, and Groisman EA (2000). Lateral gene transfer and the nature of bacterial innovation. *Nature* **405,** 299–304.

Ochman H, Lerat E, and Daubin V (2005). Examining bacterial species under the specter of gene transfer and exchange. *Proceedings of the National Academy of Sciences, USA* **102,** 6595–9.

Ofuya ZM and Akhidue V (2005). The role of pulses in human nutrition: A review. *Journal of Applied Sciences and Environmental Management* **9,** 99–104.

Ogata H, Renesto P, Audic S, *et al.* (2005). The genome sequence of *Rickettsia felis* identifies the first putative conjugative plasmid in an obligate intracellular parasite. *PLoS Biology* **3,** 0001–12.

Oh JD, Kling-Bäckhed H, Giannakis M, *et al.* (2006). The complete genome sequence of a chronic atrophic gastritis *Helicobacter pylori* strain: Evolution during disease progression. *Proceedings of the National Academy of Sciences USA* **103,** 9999–10004.

O'hUigin C, Satta Y, Takahata N, and Klein J (2002). Contribution of homoplasy and of ancestral polymorphism to the evolution of genes in anthropoid primates. *Molecular Biology and Evolution* **19,** 1501–13.

O'Kane SL and Al-Shehbaz IA (1997). A synopsis of *Arabidopsis* (Brassicaceae). *Novon* **7,** 323–7.

O'Kane SL, Schaal BA, and Al-Shehbaz IA (1996). The origins of *Arabidopsis suecica* (Brassicaceae) as indicated by nuclear rDNA sequences. *Systematic Botany* **21,** 559–66.

Oliveira R, Godinho R, Randi E, Ferrand N, and Alves PC (2007). Molecular analysis of hybridisation between wild and domestic cats (*Felis silvestris*) in Portugal: Implications for conservation. *Conservation Genetics* **9,** 1–11.

Olmstead RG and Sweere JA (1994). Combining data in phylogenetic systematics: An empirical approach using three molecular data sets in the Solanaceae. *Systematic Biology* **43,** 467–81.

O'Lorcain P and Holland CV (2000). The public health importance of *Ascaris lumbricoides*. *Parasitology* **121,** S51–71.

Olsen KM (2004). SNPs, SSRs and inferences on cassava's origin. *Plant Molecular Biology* **56,** 517–26.

Olsen KM and Schaal BA (1999). Evidence on the origin of cassava: Phylogeography of *Manihot esculenta*. *Proceedings of the National Academy of Sciences USA* **96,** 5586–91.

ORCA-EU (2007). A report on IUU fishing of Baltic Sea cod. pp. 1–72. A Report Commissioned by FISH from ORCA-EU. The Fisheries Secretariat (FISH), Bromma, Sweden.

Osada N and Wu C-I (2005). Inferring the mode of speciation from genomic data: A study of the great apes. *Genetics* **169,** 259–64.

Ovchinnikov IV, Götherström A, Romanova GP, Kharitonov VM, Lidén K, and Goodwin W (2000). Molecular analysis of Neanderthal DNA from the northern Caucasus. *Nature* **404,** 490–3.

Pääbo S (1999). Human evolution. *Trends in Genetics* **15,** M13–M16.

Pääbo S (2003). The mosaic that is our genome. *Nature* **421,** 4–9–12.

Pääbo S, Poinar H, Serre D, *et al.* (2004). Genetic analyses from ancient DNA. *Annual Review of Genetics* **38,** 645–79.

Pan D, Chen J-H, Groves C, *et al.* (2007). Mitochondrial control region and population genetic patterns of *Nycticebus bengalensis* and *N. pygmaeus*. *International Journal of Primatology* **28,** 791–9.

Papa R and Gepts P (2003). Asymmetry of gene flow and differential geographical structure of molecular diversity in wild and domesticated common bean (*Phaseolus vulgaris* L.) from Mesoamerica. *Theoretical and Applied Genetics* **106,** 239–50.

Parker HG, Kim LV, Sutter NB, *et al.* (2004). Genetic structure of the purebred domestic dog. *Science* **304,** 1160–4.

Parsonnet J, Shmuely H, and Haggerty T (1999). Fecal and oral shedding of *Helicobacter pylori* from healthy infected adults. *Journal of the American Medical Association* **282,** 2240–5.

Pasquet RS (1999). Genetic relationships among subspecies of *Vigna unguiculata* (L.) Walp. based on a allozyme variation. *Theoretical and Applied Genetics* **98,** 1104–19.

Pastorini J, Forstner MRJ, and Martin RD (2001). Phylogenetic history of sifacas (Propithecus: Lemuriformes) derived from mtDNA sequences. *American Journal of Primatology* **53,** 1–17.

Pastorini J, Thalmann U, and Martin RD (2003). A molecular approach to comparative phylogeography of extant Malagasy lemurs. *Proceedings of the National Academy of Sciences USA* **100,** 5879–84.

Paterson AH, Freeling M, and Sasaki T (2005). Grains of knowledge: Genomics of model cereals. *Genome Research* **15,** 1643–50.

Patterson N, Richter DJ, Gnerre S, Lander ES, and Reich D (2006). Genetic evidence for complex speciation of humans and chimpanzees. *Nature* **441,** 1103–8.

Pedrosa S, Uzun M, Arranz J-J, Gutiérrez-Gil B, Primitivo FS, and Bayón Y (2005). Evidence of three maternal lineages in near eastern sheep supporting multiple domestication events. *Proceedings of the Royal Society of London B* **272,** 2211–7.

Peng W, Anderson TJC, Zhou X, and Kennedy MW (1998). Genetic variation in sympatric *Ascaris* populations from humans and pigs in China. *Parasitology* **117,** 355–61.

Peng W, Yuan K, Hu M, Zhou X, and Gasser RB (2005). Mutation scanning-coupled analysis of haplotypic variability in mitochondrial DNA regions reveals low gene flow between human and porcine *Ascaris* in endemic regions of China. *Electrophoresis* **26,** 4317–26.

Pereira F, Davis SJM, Pereira L, McEvoy B, Bradley DG, and Amorim A (2006). Genetic signatures of a Mediterranean influence in Iberian peninsula sheep husbandry. *Molecular Biology and Evolution* **23,** 1420–6.

Perera L, Russell JR, Provan J, and Powell W (2003). Studying genetic relationships among coconut varieties/populations using microsatellite markers. *Euphytica* **132,** 121–8.

Pérez T, Albornoz J, and Domínguez A (2002). Phylogeography of chamois (*Rupicapra* spp.) inferred from microsatellites. *Molecular Phylogenetics and Evolution* **25,** 524–34.

Perry RD and Fetherston JD (1997). *Yersinia pestis*—etiologic agent of plague. *Clinical Microbiology Reviews* **10,** 35–66.

Pesole G, Sbisá E, Preparata G, and Saccone C (1992). The evolution of the mitochondrial D-loop region and the origin of modern man. *Molecular Biology and Evolution* **9,** 587–98.

Peters JL, McCracken KG, Zhuravlev YN, *et al.* (2005). Phylogenetics of wigeons and allies (Anatidae: *Anas*): The importance of sampling multiple loci and multiple individuals. *Molecular Phylogenetics and Evolution* **35,** 209–24.

Peters JL and Omland KE (2007). Population structure and mitochondrial polyphyly in North American gadwalls (*Anas strepera*). *Auk* **124,** 444–62.

Peters JL, Zhuravlev Y, Fefelov I, Logie A, and Omland KE (2007). Nuclear loci and coalescent methods support ancient hybridization as cause of mitochondrial paraphyly between Gadwall and Falcated duck (*Anas* spp.). *Evolution* **61,** 1992–2006.

Petit M, Lim KY, Julio E, *et al.* (2007). Differential impact of retrotransposon populations on the genome of allotetraploid tobacco (*Nicotiana tabacum*). *Molecular Genetics and Genomics* **278,** 1–15.

Petren K, Grant PR, Grant BR, and Keller LF (2005). Comparative landscape genetics and the adaptive radiation of Darwin's finches: The role of peripheral isolation. *Molecular Ecology* **14,** 2943–57.

Philip CB (1948). Tsutsugamushi disease (scrub typhus) in World War II. *Journal of Parasitology* **34,** 169–91.

Pidancier N, Jordan S, Luikart G, and Taberlet P (2006). Evolutionary history of the genus *Capra* (Mammalia, Artiodactyla): Discordance between mitochondrial DNA and Y-chromosome phylogenies. *Molecular Phylogenetics and Evolution* **40,** 739–49.

Pierpaoli M, Biro ZS, Herrmann M, *et al.* (2003). Genetic distinction of wildcat (*Felis silvestris*) populations in Europe, and hybridization with domestic cats in Hungary. *Molecular Ecology* **12,** 2585–98.

Piganeau G, Gardner M, and Eyre-Walker A (2004). A broad survey of recombination in animal mitochondria. *Molecular Biology and Evolution* **21,** 2319–25.

Pinto MA, Rubink WL, Coulson RN, Patton JC, and Johnston JS (2004). Temporal pattern of africanization in a feral honeybee population from Texas inferred from mitochondrial DNA. *Evolution* **58,** 1047–55.

Pinto MA, Rubink WL, Patton JC, Coulson RN, and Johnston JS (2005). Africanization in the United States: Replacement of feral European honeybees (*Apis mellifera* L.) by an African hybrid swarm. *Genetics* **170,** 1653–65.

Piperno DR, Andres TC, and Stothert KE (2000). Phytoliths in *Cucurbita* and other Neotropical Cucurbitaceae and their occurrence in early archaeological sites from the lowland American tropics. *Journal of Archaeological Science* **27,** 193–208.

Pitra C, Hansen AJ, Lieckfeldt D, and Arctander P (2002). An exceptional case of historical outbreeding in African sable antelope populations. *Molecular Ecology* **11,** 1197–208.

Plagnol V and Wall JD (2006). Possible ancestral structure in human populations. *PLoS Genetics* **2,** 0972–9.

Pollack S and Perez A (2007). Fruit and Tree Nuts Situation and Outlook—yearbook 2007, pp. 1–200. Economic Research Service, United States Department of Agriculture.

Pourrut X, Kumulungui B, Wittmann T, *et al.* (2005). The natural history of Ebola virus in Africa. *Microbes and Infection* **7,** 1005–14.

Prager EM and Wilson AC (1975). Slow evolutionary loss of the potential for interspecific hybridization in birds: A manifestation of slow regulatory evolution. *Proceedings of the National Academy of Sciences USA* **72,** 200–4.

Prakash A, Walton C, Bhattacharyya DR, O'Loughlin S, Mohapatra PK, and Mahanta J (2006). Molecular characterization and species identification of the *Anopheles dirus* and *An. minimus* complexes in north-east India using r-DNA ITS-2. *Acta Tropica* **100,** 156–61.

Presa P, Pardo BG, Martínez P, and Bernatchez L (2002). Phylogeographic congruence between mtDNA and rDNA ITS markers in brown trout. *Molecular Biology and Evolution* **19,** 2161–75.

Prohens J, Anderson GJ, Blanca JM, Cañizares J, Zuriaga E, and Nuez F (2006). The implications of AFLP data for the systematics of the wild species of *Solanum* section *Basarthrum*. *Systematic Botany* **31,** 208–16.

Pryde D (1971). *Nunaga—Ten Years of Eskimo Life*. Walker and Company, New York.

Putnam AS, Scriber JM, and Andolfatto P (2007). Discordant divergence times among Z-chromosome regions between two ecologically distinct Swallowtail Butterfly species. *Evolution* **61,** 912–27.

Qiu X, Gurkar AU, and Lory S (2006). Interstrain transfer of the large pathogenicity island (PAPI-1) of *Pseudomonas aeruginosa*. *Proceedings of the National Academy of Sciences USA* **103,** 19830–5.

Rabarivola C, Meyers D, and Rumpler Y (1991). Distribution and morphological characters of intermediate forms between the black lemur (*Eulemur macaco macaco*) and the Sclater's lemur (*E. m. flavifrons*). *Primates* **32,** 269–73.

Radchenko OA (2004). Introgressive hybridization of chars of the genus *Salvelinus* as inferred from mitochondrial DNA variation. *Russian Journal of Genetics* **40,** 1392–8.

Ragone D (2001). Chromosome numbers and pollen stainability of three species of Pacific island breadfruit (*Artocarpus*, Moraceae). *American Journal of Botany* **88,** 693–6.

Raina SN and Mukai Y (1999). Genomic in situ hybridization in *Arachis* (Fabaceae) identifies the diploid wild progenitors of cultivated (*A. hypogaea*) and related wild (*A. monticola*) peanut species. *Plant Systematics and Evolution* **214,** 251–62.

Raina SN, Mukai Y, and Yamamoto M (1998). *In situ* hybridization identifies the diploid progenitor species of *Coffea arabica* (Rubiaceae). *Theoretical and Applied Genetics* **97,** 1204–9.

Randi E and Bernard-Laurent A (1999). Population genetics of a hybrid zone between the red-legged partridge and rock partridge. *Auk* **116,** 324–37.

Randi E and Lucchini V (2002). Detecting rare introgression of domestic dog genes into wild wolf (*Canis lupus*) populations by Bayesian admixture analyses of microsatellite variation. *Conservation Genetics* **3,** 31–45.

Randi E, Lucchini V, Fjeldsø M, *et al.* (2000). Mitochondrial DNA variability in Italian and East European wolves: Detecting the consequences of small population size and hybridization. *Conservation Biology* **14,** 464–73.

Randi E, Pierpaoli M, Beaumont M, Ragni B, and Sforzi A (2001). Genetic identification of wild and domestic cats (*Felis silvestris*) and their hybrids using Bayesian clustering methods. *Molecular Biology and Evolution* **18,** 1679–93.

Randi E, Tabarroni C, Rimondi S, Lucchini V, and Sfougaris A (2003). Phylogeography of the rock partridge (*Alectoris graeca*). *Molecular Ecology* **12,** 2201–14.

Randolph LF (1966). *Iris nelsonii*, a new species of Louisiana iris of hybrid origin. *Baileya* **14,** 143–69.

Randolph LF, Nelson IS, and Plaisted RL (1967). Negative evidence of introgression affecting the stability of Louisiana *Iris* species. *Cornell University Agricultural Experiment Station Memoir* **398,** 1–56.

Raoult D, Dutour O, Houhamdi L, *et al.* (2006). Evidence for louse-transmitted diseases in soldiers of Napoleon's Grand Army in Vilnius. *The Journal of Infectious Diseases* **193,** 112–20.

Raoult D and Roux V (1999). The body louse as a vector of reemerging human diseases. *Clinical Infectious Diseases* **29,** 888–911.

Ravel C, Cortes S, Pratlong F, Morio F, Dedet J-P, and Campino L (2006). First report of genetic hybrids between two very divergent *Leishmania* species: *Leishmania infantum* and *Leishmania major*. *International Journal for Parasitology* **36,** 1383–8.

Rawal KM (1975). Natural hybridization among wild, weedy and cultivated *Vigna unguiculata* (L.) Walp. *Euphytica* **24,** 699–707.

Rawson PD and Hilbish TJ (1998). Asymmetric introgression of mitochondrial DNA among European populations of blue mussels (*Mytilus* spp). *Evolution* **52,** 100–8.

Redenbach Z and Taylor EB (2002). Evidence for historical introgression along a contact zone between two species of char (Pisces: Salmonidae) in northwestern North America. *Evolution* **56,** 1021–35.

Reed DL, Smith VS, Hammond SL, Rogers AR, and Clayton DH (2004). Genetic analysis of lice supports direct contact between modern and archaic humans. *PLoS Biology* **2,** 1972–83.

Ren N and Timko MP (2001). AFLP analysis of genetic polymorphism and evolutionary relationships among cultivated and wild *Nicotiana* species. *Genome* **44,** 559–71.

Renno J-F, Hubert N, Torrico JP, *et al.* (2006). Phylogeography of *Cichla* (Cichlidae) in the upper Madera basin (Bolivian Amazon). *Molecular Phylogenetics and Evolution* **41,** 503–10.

Rhesus Macaque Genome Sequencing and Analysis Consortium (2007). Evolutionary and biomedical insights from the Rhesus macaque genome. *Science* **316,** 222–34.

Rieseberg LH (1997). Hybrid origins of plant species. *Annual Review of Ecology and Systematics* **28,** 359–89.

Rieseberg LH, Carter R, and Zona S (1990). Molecular tests of the hypothesized hybrid origin of two diploid *Helianthus* species (Asteraceae). *Evolution* **44,** 1498–511.

Rieseberg LH and Livingstone K (2003). Chromosomal speciation in primates. *Science* **300,** 267–8.

Rieseberg LH, Raymond O, Rosenthal DM, *et al.* (2003). Major ecological transitions in wild sunflowers facilitated by hybridization. *Science* **301,** 1211–6.

Rieseberg LH and Wendel JF (1993). Introgression and its consequences in plants. In RG Harrison, ed. *Hybrid Zones and the Evolutionary Process*,pp. 70–109. Oxford University Press, Oxford.

Riginos C and Cunningham CW (2005). Local adaptation and species segregation in two mussel (*Mytilus edulis* x *Mytilus trossulus*) hybrid zones. *Molecular Ecology* **14,** 381–400.

Riginos C, Hickerson MJ, Henzler CM, and Cunningham CW (2004). Differential patterns of male and female mtDNA exchange across the Atlantic Ocean in the blue mussel, *Mytilus edulis*. *Evolution* **58,** 2438–51.

Riginos C and McDonald JH (2003). Positive selection on an acrosomal sperm protein, M7 lysin, in three species of the mussel genus *Mytilus*. *Molecular Biology and Evolution* **20,** 200–7.

Riley HP (1938). A character analysis of colonies of *Iris fulva*, *Iris hexagona* var. *giganticaerulea* and natural hybrids. *American Journal of Botany* **25,** 727–38

Riley MA and Wertz JE (2002). Bacteriocin diversity: Ecological and evolutionary perspectives. *Biochimie* **84,** 357–64.

Ríos D, Ghislain M, Rodríguez F, and Spooner DM (2007). What is the origin of the European potato? Evidence from Canary Island landraces. *Crop Science* **47,** 1271–80.

Rivals F and Deniaux B (2005). Investigation of human hunting seasonality through dental microwear analysis of two Caprinae in late Pleistocene localities in southern France. *Journal of Archaeological Science* **32,** 1603–12.

Roberts-Nkrumah LB and Badrie N (2005). Breadfruit consumption, cooking methods and cultivar preference among consumers in Trinidad, West Indies. *Food Quality and Preference* **16,** 267–74.

Robins JH, Ross HA, Allen MS, and Matisoo-Smith E (2006). *Sus bucculentus* revisited. *Nature* **440,** E7.

Robinson JP, Harris SA, and Juniper BE (2001). Taxonomy of the genus *Malus* Mill. (Rosaceae) with emphasis on the cultivated apple, *Malus domestica* Borkh. *Plant Systematics and Evolution* **226,** 35–58.

Robinson TJ and Harley EH (1995). Absence of chromosomal variation in the roan and sable antelope and the cytogenetics of a naturally occurring hybrid. *Cytogenetics and Cell Genetics* **71,** 363–9.

Roca AL, Georgiadis N, and O'Brien SJ (2005). Cytonuclear genomic dissociation in African elephant species. *Nature Genetics* **37,** 96–100.

Roca AL, Georgiadis N, Pecon-Slattery J, and O'Brien SJ (2001). Genetic evidence for two species of elephant in Africa. *Science* **293,** 1473–7.

Rodiño AP, Santalla M, González AM, De Ron AM, and Singh SP (2006). Novel genetic variation in common bean from the Iberian Peninsula. *Crop Science* **46,** 2540–6.

Rodrigues DP, Filho SA, and Clement CR (2004). Molecular marker-mediated validation of morphologically defined landraces of pejibaye (*Bactris gasipaes*) and their phylogenetic relationships. *Genetic Resources and Crop Evolution* **51,** 871–82.

Rodriguez A and Spooner DM (1997). Chloroplast DNA analysis of *Solanum bulbocastanum* and *S. cardiophyllum*, and evidence for the distinctiveness of *S. cardiophyllum* subsp. *ehrenbergii* (sect. *Petota*). *Systematic Botany* **22,** 31–43.

Rodríguez F, Albornoz J, and Domínguez A (2007). Cytochrome *b* pseudogene originated from a highly divergent mitochondrial lineage in genus *Rupicapra*. *Journal of Heredity* **98,** 243–9.

Rodríguez-Burruezo A, Prohens J, and Nuez F (2003). Wild relatives can contribute to the improvement of fruit quality in pepino (*Solanum muricatum*). *Euphytica* **129,** 311–8.

Rogaev EI, Moliaka YK, Malyarchuk BA, *et al.* (2006). Complete mitochondrial genome and phylogeny of Pleistocene mammoth *Mammuthus primigenius*. *PLoS Biology* **4,** 0403–0410.

Rong J, Feltus FA, Waghmare VN, *et al.* (2007). Meta-analysis of polyploid cotton QTL shows unequal contributions of subgenomes to a complex network of genes and gene clusters implicated in lint fiber development. *Genetics* **176,** 2577–88.

Roos C and Geissmann T (2001). Molecular phylogeny of the major hylobatid divisions. *Molecular Phylogenetics and Evolution* **19,** 486–94.

Roos C, Schmitz J, and Zischler H (2004). Primate jumping genes elucidate strepsirrhine phylogeny.

Proceedings of the National Academy of Sciences USA 101, 10650–4.

Roosevelt T (1996). *Hunting Trips of a Ranchman*. Random House Inc, New York.

Ropiquet A and Hassanin A (2006). Hybrid origin of the Pliocene ancestor of wild goats. *Molecular Phylogenetics and Evolution* **41,** 395–404.

Rosenblum LL, Supriatna J, Hasan MN, and Melnick DJ (1997). High mitochondrial DNA diversity with little structure within and among leaf monkey populations (*Trachypithecus auratus* and *Trachypithecus cristatus*). *International Journal of Primatology* **18,** 1005–28.

Rossan RN and Baerg DC (1977). Laboratory and feral hybridization of *Ateles geoffroyi panamensis* Kellogg and Goldman 1944 and *A. fusciceps robustus* Allen 1914 in Panama. *Primates* **18,** 235–7.

Rosselló-Mora R and Amann R (2001). The species concept for prokaryotes. *FEMS Microbiology Reviews* **25**, 39–67.

Rouquet P, Froment J-M, Bermejo M, *et al.* (2005). Wild animal mortality monitoring and human Ebola outbreaks, Gabon and Republic of Congo, 2001–2003. *Emerging Infectious Diseases* **11**, 283–90.

Roux V, Rydkina E, Eremeeva M, and Raoult D (1997). Citrate synthase gene comparison, a new tool for phylogenetic analysis, and its application for the Rickettsiae. *International Journal of Systematic Bacteriology* **47,** 252–61.

Roy MS, Geffen E, Smith D, Ostrander EA, and Wayne RK (1994). Patterns of differentiation and hybridization in North American wolflike canids, revealed by analysis of microsatellite loci. *Molecular Biology and Evolution* **11,** 553–70.

Roy MS, Geffen E, Smith D, and Wayne RK (1996). Molecular genetics of pre-1940 red wolves. *Conservation Biology* **10,** 1413–24.

Rozen S, Skaletsky H, Marszalek JD, *et al.* (2003). Abundant gene conversion between arms of palindromes in human and ape Y chromosomes. *Nature* **423,** 873–6.

Ruas PM, Ruas CF, Rampim L, Carvalho VP, Ruas EA, and Sera T (2003). Genetic relationships in *Coffea* species and parentage determination of interspecific hybrids using ISSR (Inter Simple Sequence Repeat) markers. *Genetics and Molecular Biology* **26**, 319–27.

Rüber L, Meyer A, Sturmbauer C, and Verheyen E (2001). Population structure in two sympatric species of the Lake Tanganyika cichlid tribe Eretmodini: Evidence for introgression. *Molecular Ecology* **10**, 1207–25.

Rubin EJ, Lin W, Mekalanos JJ, and Waldor MK (1998). Replication and integration of a *Vibrio cholerae* cryptic plasmid linked to the CTX prophage. *Molecular Microbiology* **28,** 1247–54.

Saisho D and Purugganan MD (2007). Molecular phylogeography of domesticated barley traces expansion of agriculture in the Old World. *Genetics* **177,** 1765–76.

Salazar CA, Jiggins CD, Arias CF, Tobler A, Bermingham E, and Linares M (2005). Hybrid incompatibility is consistent with a hybrid origin of *Heliconius heurippa* Hewitson from its close relatives, *Heliconius cydno* Doubleday and *Heliconius melpomene* Linnaeus. *Journal of Evolutionary Biology* **18,** 247–56.

Salem A-H, Ray DA, Xing J, *et al.* (2003). Alu elements and hominid phylogenetics. *Proceedings of the National Academy of Sciences USA* **100,** 12787–91.

Säll T, Jakobsson M, Lind-Halldén C, and Halldén C (2003). Chloroplast DNA indicates a single origin of the allotetraploid *Arabidopsis suecica*. *Journal of Evolutionary Biology* **16,** 1019–29.

Sallum MAM, Foster PG, Li C, Sithiprasasna R, and Wilkerson RC (2007). Phylogeny of the Leucosphyrus group of *Anopheles* (*Cellia*) (Diptera: Culicidae) based on mitochondrial gene sequences. *Annals of the Entomological Society of America* **100,** 27–35.

Saltonstall K, Amato G, and Powell J (1998). Mitochondrial DNA variability in Grauer's gorillas of Kahuzi-Biega National Park. *Journal of Heredity* **89,** 129–35.

Salzburger W, Baric S, and Sturmbauer C (2002). Speciation via introgressive hybridization in East African cichlids? *Molecular Ecology* **11**, 619–25.

Samonte IE, Satta Y, Sato A, Tichy H, Takahata N, and Klein J (2007). Gene flow between species of Lake Victoria haplochromine fishes. *Molecular Biology and Evolution* **24,** 2069–80.

Santalla M, Rodiño AP, and De Ron AM (2002). Allozyme evidence supporting southwestern Europe as a secondary center of genetic diversity for the common bean. *Theoretical and Applied Genetics* **104,** 934–44.

Sarich VM and Wilson AC (1967). Immunological time scale for hominid evolution. *Science* **158,** 1200–3.

Sato A, O'hUigin C, Figueroa F, Grant PR, Grant BR, Tichy H, and Klein J (1999). Phylogeny of Darwin's finches as revealed by mtDNA sequences. *Proceedings of the National Academy of Sciences, USA* **96,** 5101–6.

Saunders JA, Mischke S, Leamy EA, and Hemeida AA (2004). Selection of international molecular standards for DNA fingerprinting of *Theobroma cacao*. *Theoretical and Applied Genetics* **110,** 41–7.

Savolainen P, Zhang Y-p, Luo J, Lundeberg J, and Leitner T (2002). Genetic evidence for an East Asian origin of domestic dogs. *Science* **298,** 1610–3.

Scannell DR, Frank AC, Conant GC, Byrne KP, Woolfit M, and Wolfe KH (2007). Independent sorting-out of thousands of duplicated gene pairs in two yeast species descended from a whole-genome duplication. *Proceedings of the National Academy of Sciences, USA* **104,** 8397–402.

Scarcelli N, Tostain S, Mariac C, *et al.* (2006a). Genetic nature of yams (*Dioscorea* sp.) domesticated by farmers in Benin (West Africa). *Genetic Resources and Crop Evolution* **53,** 121–30.

Scarcelli N, Tostain S, Vigouroux Y, Agbangla C, Daïnou O, and Pham J-L (2006b). Farmers' use of wild relative and sexual reproduction in a vegetatively propagated crop. The case of yam in Benin. *Molecular Ecology* **15,** 2421–31.

Schaschl H, Kaulfus D, Hammer S, and Suchentrunk F (2003). Spatial patterns of mitochondrial and nuclear gene pools in chamois (*Rupicapra r. rupicapra*) from the Eastern Alps. *Heredity* **91,** 125–35.

Schelly R, Salzburger W, Koblmüller S, Duftner N, and Sturmbauer C (2006). Phylogenetic relationships of the lamprologine cichlid genus *Lepidolamprologus* (Teleostei: Perciformes) based on mitochondrial and nuclear sequences, suggesting introgressive hybridization. *Molecular Phylogenetics and Evolution* **38,** 426–38.

Schillaci MA, Froehlich JW, Supriatna J, and Jones-Engel L (2005). The effects of hybridization on growth allometry and craniofacial form in Sulawesi macaques. *Journal of Human Evolution* **49,** 335–69.

Schluter D (1984). Morphological and phylogenetic relations among Darwin's finches. *Evolution* **38,** 921–30.

Schluter D (2000). *The Ecology of Adaptive Radiation*. Oxford University Press, Oxford.

Schwarz D, Matta BM, Shakir-Botteri NL, and McPheron BA (2005). Host shift to an invasive plant triggers rapid animal hybrid speciation. *Nature* **436**, 546–9.

Scientific Working Group on Chagas Disease (2005). pp 1–4. World Health Organization.

Seehausen O (2004). Hybridization and adaptive radiation. *Trends in Ecology and Evolution* **19**, 198–07.

Seehausen O (2006). African cichlid fish: A model system in adaptive radiation research. *Proceedings of the Royal Society of London B* **273**, 1987–98.

Seehausen O, Koetsier E, Schneider MV, *et al.* (2002). Nuclear markers reveal unexpected genetic variation and a Congolese-Nilotic origin of the Lake Victoria cichlid species flock. *Proceedings of the Royal Society of London B* **270**, 129–37.

Seelanan T, Schnabel A, and Wendel JF (1997). Congruence and consensus in the cotton tribe (Malvaceae). *Systematic Botany* **22,** 259–90.

Seijo G, Lavia GI, Fernández A, *et al.* (2007). Genomic relationships between the cultivated peanut (*Arachis hypogaea*, Leguminosae) and its close relatives revealed by double GISH. *American Journal of Botany* **94,** 1963–71.

Sena L, Schneider MPC, Brenig B, Honeycutt RL, Womack JE, and Skow LC (2003). Polymorphisms in MHC-*DRA* and -*DRB* alleles of water buffalo (*Bubalus bubalis*) reveal different features from cattle *DR* alleles. *Animal Genetics* **34,** 1–10.

Seong S-Y, Choi M-S, and Kim I-S (2001). *Orientia tsutsugamushi* infection: Overview and immune responses. *Microbes and Infection* **3,** 11–21.

Serre D, Langaney A, Chech M, *et al.* (2004). No evidence of Neandertal mtDNA contribution to early modern humans. *PLoS Biology* **2,** 0313–7.

Shao P-L, Huang L-M, and Hsueh P-R (2007). Recent advances and challenges in the treatment of invasive fungal infections. *International Journal of Antimicrobial Agents* **30,** 487–95.

Sharakhov IV, White BJ, Sharakhova MV, *et al.* (2006). Breakpoint structure reveals the unique origin of an interspecific chromosomal inversion (*2La*) in the *Anopheles gambiae* complex. *Proceedings of the National Academy of Sciences USA* **103,** 6258–62.

Shimada MK, Panchapakesan K, Tishkoff SA, Nato Jr AQ, and Hey J (2007). Divergent haplotypes and human history as revealed in a worldwide survey of X-linked DNA sequence variation. *Molecular Biology and Evolution* **24,** 687–98.

Shotake T (1981). Population genetical study of natural hybridization between *Papio anubis* and *P. hamadryas*. *Primates* **22,** 285–308.

Shotake T, Nozawa K, and Tanabe Y (1977). Blood protein variations in baboons. I. Gene exchange and genetic distance between *Papio anubis, Papio hamadryas* and their hybrid. *Japanese Journal of Genetics* **52,** 223–37.

Sihavong A, Phouthavane T, Lundborg CS, Sayabounthavong K, Syhakhang L, and Wahlström R (2007). Reproductive tract infections among women attending a gynecology outpatient department in Vientiane, Lao PDR. *Sexually Transmitted Diseases* **34,** 791–5.

Simon MV, Benko-Iseppon A-M, Resende LV, Winter P, and Kahl G (2007). Genetic diversity and phylogenetic relationships in *Vigna* Savi germplasm revealed by DNA amplification fingerprinting. *Genome* **50,** 538–47.

Simonsen L, Viboud C, Grenfell BT, *et al.* (2007). The genesis and spread of reassortment human influenza A/H3N2 viruses conferring adamantane resistance. *Molecular Biology and Evolution* **24,** 1811–20.

Singh PN, Shukla SK, and Bhatnagar VK (2007). Optimizing soil moisture regime to increase water use efficiency of sugarcane (*Saccharum* spp. hybrid complex) in subtropical India. *Agricultural Water Management* **90,** 95–100.

Skaletsky H, Kuroda-Kawaguchi T, Minx PJ, *et al.* (2003). The male-specific region of the human Y chromosome is a mosaic of discrete sequence classes. *Nature* **423,** 825–37.

Skalická K, Lim KY, Matyásek R, Koukalová B, Leitch AR, and Kovarík A (2003). Rapid evolution of parental rDNA in a synthetic tobacco allotetraploid line. *American Journal of Botany* **90,** 988–96.

Skalická K, Lim KY, Matyásek R, Matzke M, Leitch AR, and Kovarík A (2005). Preferential elimination of repeated DNA sequences from the paternal, *Nicotiana tomentosiformis* genome donor of a synthetic, allotetraploid tobacco. *New Phytologist* **166,** 291–303.

Skibinski DOF, Ahmad M, and Beardmore JA (1978). Genetic evidence for naturally occurring hybrids between *Mytilus edulis* and *Mytilus galloprovincialis*. *Evolution* **32,** 354–64.

Skorecki K, Selig S, Blazer S, *et al.* (1997). Y chromosomes of Jewish priests. *Nature* **385,** 32.

Slotman MA, Reimer LJ, Thiemann T, Dolo G, Fondjo E, and Lanzaro GC (2006). Reduced recombination rate and genetic differentiation between the M and S forms of *Anopheles gambiae* s.s. *Genetics* **174,** 2081–93.

Small E and Cronquist A (1976). A practical and natural taxonomy for *Cannabis*. *Taxon* **25,** 405–35.

Smith DG, George D, Kanthaswamy S, and McDonough J (2006). Identification of country of origin and admixture between Indian and Chinese Rhesus macaques. *International Journal of Primatology* **27,** 881–98.

Smith DG and McDonough J (2005). Mitochondrial DNA variation in Chinese and Indian rhesus macaques (*Macaca mulatta*). *American Journal of Primatology* **65,** 1–25.

Smith JL and Fonseca DM (2004). Rapid assays for identification of members of the *Culex (Culex) pipiens* complex, their hybrids, and other sibling species (Diptera: Culicidae). *American Journal of Tropical Medicine and Hygiene* **70,** 339–45.

Smith PF, Konings AD, and Kornfield I (2003). Hybrid origin of a cichlid population in Lake Malawi: Implications for genetic variation and species diversity. *Molecular Ecology* **12,** 2497–504.

Solomon AM (1983). Pollen morphology and plant taxonomy of white oaks in eastern North America. *American Journal of Botany* **70,** 481–94.

Soltis DE and Soltis PS (1993). Molecular data and the dynamic nature of polyploidy. *Critical Reviews in Plant Sciences* **12**, 243–73.

Soltis DE, Soltis PS, and Tate JA (2003). Advances in the study of polyploidy since *Plant Speciation*. *New Phytologist* **161,** 173–91.

Sorek R, Zhu Y, Creevey CJ, Francino MP, Bork P, and Rubin EM (2007). Genome-wide experimental determination of barriers to horizontal gene transfer. *Science* **318,** 1449–52.

Spielman A (2001). Structure and seasonality of Nearctic *Culex pipiens* populations. *Annals of the New York Academy of Sciences* **951**, 220–34.

Spooner DM, Núñez J, Trujillo G, del Rosario Herrera M, Guzmán F, and Ghislain M (2007). Extensive simple sequence repeat genotyping of potato landraces supports a major reevaluation of their gene pool structure and classification. *Proceedings of the National Academy of Sciences USA* **104,** 19398–403.

Spoor F, Leakey MG, Gathogo PN, *et al.* (2007). Implications of new early *Homo* fossils from Ileret, east of Lake Turkana, Kenya. *Nature* **448,** 688–91.

Springer SA and Crespi BJ (2007). Adaptive gamete-recognition divergence in a hybridizing *Mytilus* population. *Evolution* **61,** 772–83.

Stanley SL Jr (2003). Amoebiasis. *Lancet* **361**, 1025–34.

Stebbins GL Jr (1947). Types of polyploids: Their classification and significance *Advances in Genetics* **1,** 403–29.

Stebbins GL Jr (1950). *Variation and Evolution in Plants*. Columbia University Press, New York.

Stebbins GL Jr (1959). The role of hybridization in evolution. *Proceedings of the American Philosophical Society* **103**, 231–51.

Stebbins GL Jr, Matzke EB, and Epling C (1947). Hybridization in a population of *Quercus marilandica* and *Quercus ilicifolia*. *Evolution* **1,** 79–88.

Steiger DL, Nagai C, Moore PH, Morden CW, Osgood RV, and Ming R (2002). AFLP analysis of genetic diversity within and among *Coffea arabica* cultivars. *Theoretical and Applied Genetics* **105**, 209–15.

Stevens J, Blixt O, Tumpey TM, Taubenberger JK, Paulson JC, and Wilson IA (2006). Structure and receptor specificity of hemagglutinin from an H5N1 influenza virus. *Science* **312,** 404–10.

Stiefkens L, Bernardello G, and Anderson GJ (1999). Karyotypic studies in artificial hybrids of *Solanum* sections *Anarrhichomenum* and *Basarthrum* (Solanaceae). *Australian Journal of Botany* **47**, 147–55.

Stone AC, Griffiths RC, Zegura SL, and Hammer MF (2002). High levels of Y-chromosome nucleotide diversity in the genus *Pan*. *Proceedings of the National Academy of Sciences USA* **99,** 43–8.

Stringer CB and Andrews P (1988). Genetic and fossil evidence for the origin of modern humans. *Science* **239,** 1263–8.

Struik PC, Amaducci S, Bullard MJ, Stutterheim NC, Venturi G, and Cromack HTH (2000). Agronomy of fibre hemp (*Cannabis sativa* L.) in Europe. *Industrial Crops and Products* **11,** 107–18.

Stuart AJ, Sulerzhitsky LD, Orlova LA, Kuzmin YV, and Lister AM (2002). The latest wooly mammoths

(*Mammuthus primigenius* Blumenbach) in Europe and Asia: A review of the current evidence. *Quaternary Science Reviews* **21,** 1559–69.

Stump AD, Fitzpatrick MC, Lobo NF, *et al.* (2005). Centromere-proximal differentiation and speciation in *Anopheles gambiae*. *Proceedings of the National Academy of Sciences USA* **102,** 15930–5.

Suerbaum S and Achtman M (2004). *Helicobacter pylori*: Recombination, population structure and human migrations. *International Journal of Medical Microbiology* **294,** 133–9.

Suerbaum S, Maynard Smith J, Bapumia K, *et al.* (1998). Free recombination within *Helicobacter pylori*. *Proceedings of the National Academy of Sciences USA* **95,** 12619–24.

Suerbaum S and Michetti P (2002). *Helicobacter pylori* infection. *New England Journal of Medicine* **347,** 1175–86.

Sundqvist A-K, Björnerfeldt S, Leonard JA, *et al.* (2006). Unequal contribution of sexes in the origin of dog breeds. *Genetics* **172,** 1121–8.

Susnik S, Weiss S, Odak T, Delling B, Treer T, and Snoj A (2007). Reticulate evolution: Ancient introgression of the Adriatic brown trout mtDNA in softmouth trout *Salmo obtusirostris* (Teleostei: Salmonidae). *Biological Journal of the Linnean Society* **90,** 139–52.

Sutton M, Sternberg M, Koumans EH, McQuillian G, Berman S, and Markowitz L (2007). The prevalence of *Trichomonas vaginalis* infection among reproductive-age women in the United States, 2001–2004. *Clinical Infectious Diseases* **45,** 1319–26.

Swisher CC III, Rink WJ, Antón SC, *et al.* (1996). Latest *Homo erectus* of Java: Potential contemporaneity with *Homo sapiens* in Southeast Asia. *Science* **274,** 1870–4.

Swofford DL (1998). *PAUP**: Phylogenetic Analysis Using Parsimony (*and Other Methods).* Version 4. Sinauer Associates, Sunderland, Massachusetts.

Syring J, Farrell K, Businsky R, Cronn R, and Liston A (2007). Widespread genealogical nonmonophyly in species of *Pinus* subgenus *Strobus*. *Systematic Biology* **56,** 163–81.

Szmulewicz MN, Andino LM, Reategui EP, Wooley-Barker T, Jolly CJ, Disotell TR, and Herrera RJ (1999). An Alu insertion polymorphism in a baboon hybrid zone. *American Journal of Physical Anthropology* **109,** 1–8.

Tacconelli E, Tumbarello M, Bertagnolio S, *et al.* (2002). Multidrug-resistant *Pseudomonas aeruginosa* bloodstream infections: Analysis of trends in prevalence and epidemiology. *Emerging Infectious Diseases* **8,** 2.

Tagliaro CH, Schneider MPC, Schneider H, Sampaio IC, and Stanhope MJ (1997). Marmoset phylogenetics, conservation perspectives, and evolution of the mtDNA control region. *Molecular Biology and Evolution* **14,** 674–84.

Takahashi S, Furukawa T, Asano T, *et al.* (2005). Very close relationship of the chloroplast genomes among *Saccharum* species. *Theoretical and Applied Genetics* **110,** 1523–9.

Takahata N, Lee S-H, and Satta Y (2001). Testing multiregionality of modern human origins. *Molecular Biology and Evolution* **18,** 172–83.

Tamiru M, Becker HC, and Maass BL (2007). Genetic diversity in yam germplasm from Ethiopia and their relatedness to the main cultivated *Dioscorea* species assessed by AFLP markers. *Crop Science* **47,** 1744–53.

Tang T, Lu J, Huang J, *et al.* (2006). Genomic variation in rice: Genesis of highly polymorphic linkage blocks during domestication. *PLoS Genetics* **2,** 1824–33.

Tapio M, Marzanov N, Ozerov M, *et al.* (2006). Sheep mitochondrial DNA variation in European, Caucasian, and central Asian areas. *Molecular Biology and Evolution* **23,** 1776–83.

Tarlinton RE, Meers J, and Young PR (2006). Retroviral invasion of the koala genome. *Nature* **442,** 79–81.

Tattersall I and Schwartz JH (1999). Hominids and hybrids: The place of Neanderthals in human evolution. *Proceedings of the National Academy of Sciences USA* **96,** 7117–9.

Taylor EB, Redenbach Z, Costello AB, Pollard SM, and Pacas CJ (2001). Nested analysis of genetic diversity in northwestern North American char, Dolly Varden (*Salvelinus malma*) and bull trout (*Salvelinus confluentus*). *Canadian Journal of Fisheries and Aquatic Sciences* **58,** 406–20.

Teixeira ARL, Nascimento RJ, and Sturm NR (2006). Evolution and pathology in Chagas disease—a review. *Memórias do Instituto Oswaldo Cruz* **101,** 463–91.

Tejedor MT, Monteagudo LV, Hadjisterkotis E, and Arruga MV (2005). Genetic variability and population structure in Cypriot chukar partridges (*Alectoris chukar cypriotes*) as determined by microsatellite analysis. *European Journal of Wildlife Research* **51,** 232–6.

Tejedor MT, Monteagudo LV, Mautner S, Hadjisterkotis E, and Arruga MV (2007). Introgression of *Alectoris chukar* genes into a Spanish wild *Alectoris rufa* population. *Journal of Heredity* **98,** 179–82.

Teklu Y and Tefera H (2005). Genetic improvement in grain yield potential and associated agronomic traits of tef (*Eragrostis tef*). *Euphytica* **141,** 247–54.

Telci I, Sahbaz N (I), Yilmaz G, and Tugay ME (2004). Agronomical and chemical characterization of spearmint (*Mentha spicata* L.) originating in Turkey. *Economic Botany* **58,** 721–8.

Templeton AR (2002). Out of Africa again and again. *Nature* **416,** 45–51.

Templeton AR (2005). Haplotype trees and modern human origins. *Yearbook of Physical Anthropology* **48,** 33–59.

Templeton AR (2007a). Genetics and recent human evolution. *Evolution* **61,** 1507–19.

Templeton AR (2007b). Population biology and population genetics of Pleistocene hominins. In W Henke, H. Rothe and I Tattersall, eds. *Handbook of Palaeoanthropology*, pp. 1825–59. Springer-Verlag, Berlin.

Temu EA, Minjas JN, Tuno N, Kawada H, and Takagi M (2007). Identification of four members of the *Anopheles funestus* (Diptera: Culicidae) group and their role in *Plasmodium falciparum* transmission in Bagamoyo coastal Tanzania. *Acta Tropica* **102,** 119–25.

Teulat B, Aldam C, Trehin R, *et al.* (2000). An analysis of genetic diversity in coconut (*Cocos nucifera*) populations from across the geographic range using sequence-tagged microsatellites (SSRs) and AFLPs. *Theoretical and Applied Genetics* **100,** 764–71.

Thalmann O, Fischer A, Lankester F, Pääbo S, and Vigilant L (2007). The complex evolutionary history of gorillas: Insights from genomic data. *Molecular Biology and Evolution* **24,** 146–58.

Thalmann O, Hebler J, Poinar HN, Pääbo S, and Vigilant L (2004). Unreliable mtDNA data due to nuclear insertions: A cautionary tale from analysis of humans and other great apes. *Molecular Ecology* **13,** 321–35.

This P, Lacombe T, and Thomas MR (2006). Historical origins and genetic diversity of wine grapes. *Trends in Genetics* **22,** 511–9.

Thulin C-G, Jaarola M, and Tegelström H (1997). The occurrence of mountain hare mitochondrial DNA in wild brown hares. *Molecular Ecology* **6,** 463–7.

Thulin C-G and Tegelström H (2002). Biased geographical distribution of mitochondrial DNA that passed the species barrier from mountain hares to brown hares (genus *Lepus*): An effect of genetic incompatibility and mating behaviour? *Journal of the Zoological Society of London* **258,** 299–306.

Ting N, Tosi AJ, Li Y, Zhang Y-P, and Disotell TR (2008). Phylogenetic incongruence between nuclear and mitochondrial markers in the Asian colobines and the evolution of the langurs and leaf monkeys. *Molecular Phylogenetics and Evolution* **46,** 466–74.

Tishkoff SA, Kidd KK, and Risch N (1996). Interpretations of multiregional evolution: Response. *Science* **274,** 705–7.

Tishkoff SA and Verrelli BC (2003). Patterns of human genetic diversity: Implications for human evolutionary history and disease. *Annual Review of Genomics and Human Genetics* **4,** 293–304.

Tocheri MW, Orr CM, Larson SG, *et al.* (2007). The primitive wrist of *Homo floresiensis* and its implications for Hominin evolution. *Science* **317,** 1743–5.

Toro J, Innes DJ, and Thompson RJ (2004). Genetic variation among life-history stages of mussels in a *Mytilus edulis—M. trossulus* hybrid zone. *Marine Biology* **145,** 713–25.

Tosi AJ and Coke CS (2007). Comparative phylogenetics offer new insights into the Biogeographic history of *Macaca fascicularis* and the origin of the Mauritian macaques. *Molecular Phylogenetics and Evolution* **42,** 498–504.

Tosi AJ, Detwiler KM, and Disotell TR (2005). Y-chromosome markers suitable for noninvasive studies of guenon hybridization. *International Journal of Primatology* **26,** 685–96.

Tosi AJ, Morales JC, and Melnick DJ (2000). Comparison of Y chromosome and mtDNA phylogenies leads to unique inferences of macaque evolutionary history. *Molecular Phylogenetics and Evolution* **17,** 133–44.

Tosi AJ, Morales JC, and Melnick DJ (2002). Y-chromosome and mitochondrial markers in Macaca fascicularis indicate introgression with Indochinese M. mulatta and a biogeographic barrier in the Isthmus of Kra. *International Journal of Primatology* **23,** 161–78.

Tosi AJ, Morales JC, and Melnick DJ (2003). Paternal, maternal, and biparental molecular markers provide unique windows onto the evolutionary history of macaque monkeys. *Evolution* **57,** 1419–35.

Toumi L and Lumaret R (1998). Allozyme variation in cork oak (*Quercus suber* L.): The role of phylogeography and genetic introgression by other Mediterranean oak species and human activities. *Theoretical and Applied Genetics* **97,** 647–56.

Trinkaus E (2007). European early modern humans and the fate of the Neandertals. *Proceedings of the National Academy of Sciences USA* **104,** 7367–72.

Tucker AO and Fairbrothers DE (1990). The origin of *Mentha* × *gracilis* (Lamiaceae). I. Chromosome numbers, fertility, and three morphological characters. *Economic Botany* **44,** 183–213.

Tucker AO, Hendriks H, Bos R, and Fairbrothers DE (1991). The origin of *Mentha* x *gracilis* (Lamiaceae). II. Essential oils. *Economic Botany* **45,** 200–15.

Turner TL and Hahn MW (2007). Locus- and population-specific selection and differentiation between incipient species of *Anopheles gambiae. Molecular Biology and Evolution* **24,** 2132–8.

Turner TL, Hahn MW, and Nuzhdin SV (2005). Genomic islands of speciation in *Anopheles gambiae. PLoS Biology* **3,** e285.

UNAIDS (2007). 07 AIDS epidemic update. 60 pp.

UNCTAD (2005). Citrus fruit. 6 pp.

United States Department of Agriculture (2006). The economics of food, farming, natural resources, and rural America—Hogs. Economic Research Service.

United States Department of Agriculture/Foreign Agricultural Service (2006). World Apple Situation and Outlook. Horticultural & Tropical Products Division.

Usinger RL, Wygodzinsky P, and Ryckman RE (1966). The biosystematics of Triatominae. *Annual Review of Entomology* **11,** 309–30.

Valbuena-Carabaña M, González-Martínez SC, Hardy OJ, and Gil L (2007). Fine-scale spatial genetic structure in mixed oak stands with different levels of hybridization. *Molecular Ecology* **16,** 1207–19.

Valdez R, Nadler CF, and Bunch TD (1978). Evolution of wild sheep in Iran. *Evolution* **32,** 56–72.

Vallinoto M, Schneider MPC, Silva, Iannuzzi L, and Brenig B (2004). Molecular cloning and analysis of the swamp and river buffalo leptin gene. *Animal Genetics* **35,** 462–504.

van de Guchte M, Penaud S, Grimaldi C, *et al.* (2006). The complete genome sequence of *Lactobacillus bulgaricus* reveals extensive and ongoing reductive evolution. *Proceedings of the National Academy of Sciences USA* **103,** 9274–9.

Vega FE, Rosenquist E, and Collins W (2003). Global project needed to tackle coffee crisis. *Nature* **425,** 343.

Verardi A, Lucchini V, and Randi E (2006). Detecting introgressive hybridization between free-ranging domestic dogs and wild wolves (*Canis lupus*) by admixture linkage disequilibrium analysis. *Molecular Ecology* **15,** 2845–55.

Verkaar ELC, Nijman IJ, Beeke M, Hanekamp E, and Lenstra JA (2004). Maternal and paternal lineages in cross-breeding bovine species. Has wisent a hybrid origin? *Molecular Biology and Evolution* **21,** 1165–70.

Vershinin AV, Allnutt TR, Knox MR, Ambrose MJ, and Ellis THN (2003). Transposable elements reveal the impact of introgression, rather than transposition, in *Pisum* diversity, evolution, and domestication. *Molecular Biology and Evolution* **20,** 2067–75.

Vigilant L and Bradley BJ (2004). Genetic variation in gorillas. *American Journal of Primatology* **64,** 161–72.

Vigilant L, Stoneking M, Harpending H, Hawkes K, and Wilson AC (1991). African populations and the evolution of human mitochondrial DNA. *Science* **253,** 1503–7.

Vigne J-D, Guilaine J, Debue K, Haye L, and Gérard P (2004). Early taming of the cat in Cyprus. *Science* **304,** 259.

Vilà C, Maldonado JE, and Wayne RK (1999). Phylogenetic relationships, evolution, and genetic diversity of the domestic dog. *Journal of Heredity* **90,** 71–7.

Vilà C, Savolainen P, Maldonado JE, *et al.* (1997). Multiple and ancient origins of the domestic dog. *Science* **276,** 1687–9.

Vilà C, Seddon J, and Ellegren H (2005). Genes of domestic mammals augmented by backcrossing with wild ancestors. *Trends in Genetics* **21,** 214–8.

Vilà C, Walker C, Sundqvist A-K, *et al.* (2003). Combined use of maternal, paternal and bi-parental genetic markers for the identification of wolf–dog hybrids. *Heredity* **90,** 17–24.

Vilà C and Wayne RK (1999). Hybridization between wolves and dogs. *Conservation Biology* **13,** 195–8.

Viosca P, Jr (1935). The irises of southeastern Louisiana—a taxonomic and ecological interpretation. *Bulletin of the American Iris Society* **57,** 3–56.

Volkov RA, Borisjuk NV, Panchuk II, Schweizer D, and Hemleben V (1999). Elimination and rearrangement of parental rDNA in the allotetraploid *Nicotiana tabacum*. *Molecular Biology and Evolution* **16,** 311–20.

Volkov RA, Komarova NY, Panchuk II, and Hemleben V (2003). Molecular evolution of rDNA external transcribed spacer and phylogeny of sect. *Petota* (genus *Solanum*). *Molecular Phylogenetics and Evolution* **29,** 187–202.

vom Brocke K, Christinck A, Weltzien E, Presterl T, and Geiger HH (2003). Farmers' seed systems and management practices determine pearl millet genetic diversity patterns in semiarid regions of India. *Crop Science* **43,** 1680–9.

von Meurers R (1999). *Buffalo, Elephant, & Bongo*. Safari Press Inc, Long Beach, California.

Wagner PL and Waldor MK (2002). Bacteriophage control of bacterial virulence. *Infection and Immunity* **70,** 3985–93.

Wagner WH, Jr. (1970). Biosystematics and evolutionary noise. *Taxon* **19,** 146–51.

Wahlroos S (2001). *Mutiny and Romance in the South Seas: A Companion to the Bounty Adventure*. Backinprint.com

Wakeley J and Hey J (1997). Estimating ancestral population parameters. *Genetics* **145,** 847–55.

Waldor MK and Mekalanos JJ (1996). Lysogenic conversion by a filamentous phage encoding cholera toxin. *Science* **272,** 1910–4.

Walker EP, Warnick F, Hamlet SE, *et al.* (1975). Mammals of the World, Third Edition. Johns Hopkins University Press, Baltimore.

Wall JD and Kim SK (2007). Inconsistencies in Neanderthal genomic DNA sequences. *PLoS Genetics* **3,** 1862–6.

Walsh PD, Abernethy KA, Bermejo M, *et al.* (2003). Catastrophic ape decline in western equatorial Africa. *Nature* **422,** 611–4.

Walton C, Handley JM, Collins FH, *et al.* (2001). Genetic population structure and introgression in *Anopheles dirus* mosquitoes in south-east Asia. *Molecular Ecology* **10,** 569–80.

Walton C, Handley JM, Tun-Lin W, *et al.* (2000). Population structure and population history of *Anopheles dirus* mosquitoes in south-east Asia. *Molecular Biology and Evolution* **17,** 962–74.

Wang Y-Q and Su B (2004). Molecular evolution of *microcephalin,* a gene determining human brain size. *Human Molecular Genetics* **13,** 1131–7.

Ward DM, Cohan FM, Bhaya D, Heidelberg JF, Kühl M, and Grossman A (2008). Genomics, environmental genomics and the issue of microbial species. *Heredity* **100,** 207–19.

Ward TJ, Bielawski JP, Davis SK, Templeton JW, and Derr JN (1999). Identification of domestic cattle hybrids in wild cattle and bison species: A general approach using mtDNA markers and the parametric bootstrap. *Animal Conservation* **2,** 51–7.

Warren KS, Verschoor EJ, Langenhuijzen S *et al.* (2001). Speciation and intrasubspecific variation of Bornean orangutans, *Pongo pygmaeus pygmaeus. Molecular Biology and Evolution* **18,** 472–80.

Wayne RK and Gittleman JL (1995). The problematic red wolf. *Scientific American* **273,** 26–31.

Wayne RK and Jenks SM (1991). Mitochondrial DNA analysis implying extensive hybridization of the endangered red wolf *Canis rufus. Nature* **351,** 565–8.

Wayne RK, Lehman N, Allard MW, and Honeycutt RL (1992). Mitochondrial DNA variability of the gray wolf: Genetic consequences of population decline and habitat fragmentation. *Conservation Biology* **6,** 559–69.

Wayne RK and Ostrander EA (2007). Lessons learned from the dog genome. *Trends in Genetics* **23,** 557–67.

Weeden NF (2007). Genetic changes accompanying the domestication of *Pisum sativum*: Is there a common genetic basis to the "Domestication Syndrome" for legumes? *Annals of Botany* **100,** 1017–25.

Wei F, Coe E, Nelson W, *et al.* (2007). Physical and genetic structure of the maize genome reflects its complex evolutionary history. *PLoS Genetics* **3,** 1254–63.

Weir BS and Cockerham CC (1984). Estimating *F*-statistics for the analysis of population structure. *Evolution* **38,** 1358–70.

Wells S (2002). *The Journey of Man: A Genetic Odyssey.* Random House Inc, New York.

Wendel JF (1989). New World tetraploid cottons contain Old World cytoplasm. *Proceedings of the National Academy of Sciences USA* **86,** 4132–6.

Westenberger SJ, Barnabé C, Campbell DA, and Sturm NR (2005). Two hybridization events define the population structure of *Trypanosoma cruzi. Genetics* **171,** 527–43.

White GB (1974). *Anopheles gambiae* complex and disease transmission in Africa. *Transactions of the Royal Society of Tropical Medicine and Hygiene* **68,** 278–98.

White MJD (1978). *Modes of Speciation.* W.H. Freeman and Company, San Francisco.

Whitfield CW, Behura SK, Berlocher SH, *et al.* (2006). Thrice out of Africa: Ancient and recent expansions of the honey bee, *Apis mellifera. Science* **314,** 642–5.

Whitney KD, Randell RA, and Rieseberg LH (2006). Adaptive introgression of herbivore resistance traits in the weedy sunflower *Helianthus annuus. American Naturalist* **167,** 794–807.

Whittaker DJ, Morales JC, and Melnick DJ (2007). Resolution of the *Hylobates* phylogeny: Congruence of mitochondrial D-loop sequences with molecular, behavioral, and morphological data sets. *Molecular Phylogenetics and Evolution* **45,** 620–8.

Whittemore AT and Schaal BA (1991). Interspecific gene flow in sympatric oaks. *Proceedings of the National Academy of Sciences USA* **88,** 2540–4.

Wiehlmann L, Wagner G, Cramer N, *et al.* (2007). Population structure of *Pseudomonas aeruginosa. Proceedings of the National Academy of Sciences USA* **104,** 8101–6.

Wildman DE, Bergman TJ, al-Aghbari A, *et al.* (2004). Mitochondrial evidence for the origin of hamadryas baboons. *Molecular Phylogenetics and Evolution* **32,** 287–96.

Wildman DE, Uddin M, Liu G, Grossman LI, and Goodman M (2003). Implications of natural selection in shaping 99.4% nonsynonymous DNA identity between humans and chimpanzees: Enlarging genus *Homo. Proceedings of the National Academy of Sciences USA* **100,** 7181–8.

Williams GC and Koehn RK (1984). Icelandic eels: Evidence for a single species of *Anguilla* in the North Atlantic. *Copeia* **1984,** 221–3.

Williams I, Reeves GH, Graziano SL, and Nielsen JL (2007). Genetic investigation of natural hybridization between rainbow and coastal cutthroat trout in the Copper River Delta, Alaska. *Transactions of the American Fisheries Society* **136,** 926–42.

Williams JH, Boecklen WJ, and Howard DJ (2001). Reproductive processes in two oak (*Quercus*) contact zones with different levels of hybridization. *Heredity* **87,** 680–90.

Willis SC, Nunes MS, Montaña CG, Farias IP, and Lovejoy NR (2007). Systematics, biogeography, and evolution of the Neotropical peacock basses *Cichla* (Perciformes: Cichlidae). *Molecular Phylogenetics and Evolution* **44,** 291–307.

Winney BJ, Hammond RL, Macasero W, *et al.* (2004). Crossing the Red Sea: Phylogeography of the hamadryas baboon, *Papio hamadryas hamadryas. Molecular Ecology* **13,** 2819–27.

Wirth T and Bernatchez L (2001). Decline of North Atlantic eels: A fatal synergy? *Proceedings of the Royal Society of London* B **270,** 681–8.

Wirth T and Bernatchez L (2003). Genetic evidence against panmixia in the European eel. *Nature* **409,** 1037–40.

Wittmann TJ, Biek R, Hassanin A, *et al.* (2007). Isolates of Zaire ebolavirus from wild apes reveal genetic lineage and recombinants. *Proceedings of the National Academy of Sciences USA* **104,** 17123–7.

Wolpoff MH (1996). Interpretations of multiregional evolution. *Science* **274,** 704–5.

Wolpoff MH, Hawks J, Frayer DW, and Hunley K (2001). Modern human ancestry at the peripheries: A test of the replacement theory. *Science* **291,** 293–7.

Won Y-J and Hey J (2005). Divergence population genetics of chimpanzees. *Molecular Biology and Evolution* **22,** 297–307.

Won Y-J, Sivasundar A, Wang Y, and Hey J (2005). On the origin of Lake Malawi cichlid species: A population genetic analysis of divergence. *Proceedings of the National Academy of Sciences, USA* **102,** 6581–6.

Wood AR, Turner G, Skibinski DOF, and Beaumont AR (2003). Disruption of doubly uniparental inheritance of mitochondrial DNA in hybrid mussels (*Mytilus edulis* × *M. galloprovincialis*). *Heredity* **91,** 354–60.

Wood NJ and Phua SH (1996). Variation in the control region sequence of the sheep mitochondrial genome. *Animal Genetics* **27,** 25–33.

World Agricultural Outlook Board (2007). World Agricultural Supply and Demand Estimates. 40 pp. United States Department of Agriculture.

World Health Organization (2001). Regional strategic plan for elimination of lymphatic filariasis (2000–2004). 36 pp.

World Health Organization (2003). Influenza—report by the Secretariat. 4 pp.

World Health Organization (2004). Malaria—disease burden in SEA region. 2 pp.

World Health Organization (2005). HIV/AIDS epidemiological surveillance report for the WHO African Region—2005 update. 74 pp.

World Health Organization (2007). Cholera. 3 pp.

World Wildlife Fund (2006). Cork screwed? Environmental and economic impacts of the cork stoppers market. 34 pp. WWF/MEDPO.

Wrigley G (1995). Coffee—*Coffea* spp. (Rubiaceae). In J Smartt and NW Simmonds, eds. *Evolution of Crop Plants,* Second Edition, pp. 438–43. Longman Scientific & Technical, Essex, England.

Wyner YM, Johnson SE, Stumpf RM, and DeSalle R (2002). Genetic assessment of a White-Collared × Red-Fronted lemur hybrid zone at Andringitra, Madagascar. *American Journal of Primatology* **67,** 51–66.

Xia X and Palidwor G (2005). Genomic adaptation to acidic environment: Evidence from *Helicobacter pylori. American Naturalist* **166,** 776–84.

Xu J, Luo G, Vilgalys RJ, Brandt ME, and Mitchell TG (2002). Multiple origins of hybrid strains of *Cryptococcus neoformans* with serotype AD. *Microbiology* **148,** 203–12.

Xu J and Mitchell TG (2003). Comparative gene genealogical analyses of strains of serotype AD identify recombination in populations of serotypes A and D in the human pathogenic yeast *Cryptococcus neoformans. Microbiology* **149,** 2147–54.

Xu Q, Wen X, and Deng X (2007). Phylogenetic and evolutionary analysis of NBS-encoding genes in Rosaceae fruit crops. *Molecular Phylogenetics and Evolution* **44,** 315–24.

Xu X and Arnason U (1996). The mitochondrial DNA molecule of Sumatran orangutan and a molecular proposal for two (Bornean and Sumatran) species of orangutan. *Journal of Molecular Evolution* **43,** 431–7.

Yamamoto S, Kitano S, Maekawa K, Koizumi I, and Morita K (2006). Introgressive hybridization between Dolly Varden *Salvelinus malma* and white-spotted charr *Salvelinus leucomaenis* on Hokkaido Island, Japan. *Journal of Fish Biology* **68** (Supplement A), 68–85.

Yang Y-W, Tai P-Y, Chen Y, and Li W-H (2002). A study of the phylogeny of *Brassica rapa, B. nigra, Raphanus sativus,* and their related genera using noncoding regions of chloroplast DNA. *Molecular Phylogenetics and Evolution* **23,** 268–75.

Yawson AE, Weetman D, Wilson MD, and Donnelly MJ (2007). Ecological zones rather than molecular forms predict genetic differentiation in the malaria vector *Anopheles gambiae* s.s. in Ghana. *Genetics* **175,** 751–61.

Yoder AD and Yang Z (2004). Divergence dates for Malagasy lemurs estimated from multiple gene loci: Geological and evolutionary context. *Molecular Ecology* **13,** 757–73.

Yoder JS and Beach MJ (2007). Giardiasis surveillance—United States, 2003–2005. Centers for Disease Control and Prevention. *Morbidity and Mortality Weekly Report* **56,** 11–8.

Yu N, Jensen-Seaman MI, Chemnick L, *et al.* (2003). Low nucleotide diversity in chimpanzees and bonobos. *Genetics* **164,** 1511–8.

Yuan Y-w, Zhang Z-y, Chen Z-d, and Olmstead RG (2006). Tracking ancient polyploids: A retroposon insertion reveals an extinct diploid ancestor in the polyploid origin of belladonna. *Molecular Biology and Evolution* **23,** 2263–7.

Zamudio JR, Mittra B, Foldynová-Trantírková S, *et al.* (2007). The 2′-*O*-Ribose methyltransferase for Cap 1 of spliced leader RNA and U1 small nuclear RNA in *Trypanosoma brucei*. *Molecular and Cellular Biology* **27,** 6084–92.

Zerega NJC, Ragone D, and Motley TJ (2004). Complex origins of breadfruit (*Artocarpus altilis*, Moraceae): Implications for human migrations in Oceania. *American Journal of Botany* **91**, 760–6.

Zerega NJC, Ragone D, and Motley TJ (2005). Systematics and species limits of breadfruit (*Artocarpus*, Moraceae). *Systematic Botany* **30**, 603–15.

Zhang J, Wang X, and Podlaha O (2004). Testing the chromosomal speciation hypothesis for humans and chimpanzees. *Genome Research* **14,** 845–51.

Zhao X, Ji Y, Ding X, Stelly DM, and Paterson AH (1998). Macromolecular organization and genetic mapping of a rapidly evolving chromosome-specific tandem repeat family (B77) in cotton (*Gossypium*). *Plant Molecular Biology* **38,** 1031–42.

Zhaxybayeva O, Lapierre P, and Gogarten JP (2004). Genome mosaicism and organismal lineages. *Trends in Genetics* **20**, 254–260.

Zhi L, Karesh WB, Janczewski DN, *et al.* (1996). Genomic differentiation among natural populations of orangutan (*Pongo pygmaeus*). *Current Biology* **6,** 1326–36.

Ziegler T, Abegg C, Meijaard E, *et al.* (2007). Molecular phylogeny and evolutionary history of southeast Asian macaques forming the *M. silenus* group. *Molecular Phylogenetics and Evolution* **42,** 807–16.

Zietkiewicz E, Yotova V, Gehl D, *et al.* (2003). Haplotypes in the dystrophin DNA segment point to a mosaic origin of modern human diversity. *American Journal of Human Genetics* **73,** 994–1015.

Zilhão J, d'Errico F, Bordes J-G, Lenoble A, Texier J-P, and Rigaud J-P (2006). Analysis of Aurignacian interstratification at the Châtelperronian-ype site and implications for the behavioral modernity of Neandertals. *Proceedings of the National Academy of Sciences USA* **103,** 12643–8.

Zizumbo-Villarreal D, Cardeña-Lopez R, and Piñero D (2002). Diversity and phylogenetic analysis in *Cocos nucifera* L. in Mexico. *Genetic Resources and Crop Evolution* **49,** 237–45.

Zizumbo-Villarreal D, Fernández-Barrera M, Torres-Hernández N, and Colunga-GarcíaMarín P (2005). Morphological variation of fruit in Mexican populations of *Cocos nucifera* L. (Arecaceae) under *in situ* and *ex situ* conditions. *Genetic Resources and Crop Evolution* **52,** 421–34.

Zizumbo-Villarreal D and Piñero D (1998). Pattern of morphological variation and diversity of *Cocos nucifera* (Arecaceae) in Mexico. *American Journal of Botany* **85,** 855–65.

Zizumbo-Villarreal D, Ruiz-Rodriguez M, Harries H, and Colunga-GarcíaMarín P (2006). Population genetics, lethal yellowing disease, and relationships among Mexican and imported coconut ecotypes. *Crop Science* **46,** 2509–16.

Index

Note: page numbers in *italics* refer to Figures and Tables, whilst those in **bold** refer to Glossary entries.